Inner Theatres of
Good and Evil

Inner Theatres of Good and Evil

The Mind's Staging of Gods, Angels and Devils

MARK PIZZATO

McFarland & Company, Inc., Publishers
Jefferson, North Carolina, and London

Part of chapter 5 was previously published as "A Post-9/11 Passion Play," in *English Language Notes* 44.1: 247–52. Copyright 2006 The Regents of the University of Colorado. All rights reserved. Used by permission. An earlier version of that essay also appeared as a film review of *The Passion of the Christ*, in *Pastoral Psychology* 53.4 (March 2005): 371–76.

LIBRARY OF CONGRESS CATALOGUING-IN-PUBLICATION DATA

Pizzato, Mark, 1960–
   Inner theatres of good and evil : the mind's staging of gods, angels and devils / Mark Pizzato.
      p.   cm.
   Includes bibliographical references and index.

   ISBN: 978-0-7864-4260-7
   softcover : 50# alkaline paper ∞

   1. Psychology, Religious.   2. Brain — Religious aspects.
3. Theological anthropology.   4. Neurosciences.
I. Title.
BL53.P54   2011
200.1'9 — dc22   2010035605

British Library cataloguing data are available

© 2011 Mark Pizzato. All rights reserved

No part of this book may be reproduced or transmitted in any form or by any means, electronic or mechanical, including photocopying or recording, or by any information storage and retrieval system, without permission in writing from the publisher.

Cover images © 2011 Shutterstock

Manufactured in the United States of America

McFarland & , Inc., Publishers
   Box 611, Jefferson, North Carolina 28640
   www.mcfarlandpub.com

To all the believers
and questioners
in my life:
relatives, friends,
teachers, and authors
who shaped the sense of God
(through various cultures' gods)
in the pathways and stages of my brain,
especially those who taught me
to enjoy a moment of ecstasy
in each breath
right now
— thank you

Acknowledgments

I wish to thank John Emigh and Richard Schechner for their encouragement and critique of this project. Thanks also to Yann-Pierre Montelle for his advice on visiting prehistoric caves. And to Andrew Hartley who gave me insights on Marlowe and Shakespeare, through our conversations and his staging of *Doctor Faustus*; and to Jorge Huerta for answers about Chicano theatre.

A special thanks to my UNC-Charlotte colleague, biologist David Bashor, who met with me often over beers and peanuts to share ideas in neuroscience. Paul Youngman, like David, read and commented on parts of this book, from his view in culture studies. Thanks to others who joined us at times — Al Maisto, James Tabor, Jon Marks, Marvin Croy, Tony Jackson, Mirsad Hadzikadic, Celine Latulipe, and Ted Carmichael — for discussing related issues in psychology, religion, primatology, philosophy, humanities, and computer science.

Thanks are due as well to UNC-Charlotte (especially to the Dean of Arts and Architecture, Ken Lambla, and Chair of Theatre, James Vesce) for a Reassignment of Duties leave I received in spring 2009 and for travel funding to research Paleolithic cave art in France. The leave also gave me valuable time to complete this book.

And thanks to my teenage son, J. Luke Pizzato, whose intense interest in the subject of my research, and performances as Ariel and Caliban, gave me new energy to continue.

Table of Contents

Acknowledgments	vii
Introduction: The Brain's Evolving Theatre	1
1. Neural and Prehistoric Signs of the Divine	17
2. Ancient Animal and Human Deities	54
3. One Medieval God — with His Angels and Devils	84
4. From Renaissance Rebirths to Postmodern Experiments	125
5. Cosmic Forces on the Movie Screen	175
6. *Millennium* in the Home Theatre	227
Conclusion: More Morality with Humans as Gods?	278
Notes	289
Bibliography	323
Index	345

Introduction:
The Brain's Evolving Theatre

Is God watching and acting in our world? Why has such a theatre been imagined by the human brain and its various cultures? This book explores the primal sources of political and religious conflict today, through inherited neural structures and key theatrical developments in Western culture, which is now a global, mass-media enterprise. It looks back from our modern versus postmodern "culture wars," and the sacrificial clashes of monotheistic civilizations, to theatrical drives within the evolving human brain and specific societies of the past. Such drives include mimicry and mimetic community, play and representation, self-awareness, ego illusion and conflict, deception, lead-actor competition, territoriality, identification ideals, and projections of desires and character types — involving divine aspirations of good or threats of evil, regarding bio-cultural survival and reproduction.[1] These drives will be investigated through the melodramatic and tragicomic characterizations of divinity in the Egypto-Greco-Roman, Judeo-Christian, and secular humanist traditions of theatre, leading to recent screen media. Prior shamanic performances in prehistoric times, as evidenced by cave art dating to tens of thousands of years ago, will also be considered. Without arguing for or against the existence of certain gods, I will focus on representative figures of supernatural power at significant points in the history of Afro-Euro-American performances, in ritual and art, onstage and onscreen. Such figures express the prior and continuing evolution of higher-order consciousness, in particular cultural ways, through the human brain and its altering of the natural environment.

Humans are highly social animals, dependent upon one another and upon the earth.[2] Yet, we can also be competitive and otherworldly, performing extreme violence by demonizing others, for the sake of transcendent, sacrificial ideals.[3] Our species has been tremendously successful in surviving, reproducing on, and transforming this planet — while evolving bigger brains, ever larger collaborative groups, and new technologies. But we have also become highly destructive through our vast creativity, like the many gods idolized and feared en route to modern monotheism and today's scientific powers. By evolving self-awareness, imaginative tool use, symbolic cultures, and increased behavioral flexibility, our ancestors gained a vital edge in the conflicts of nature. Yet they also developed a tragic flaw. Through their uncertainty of right actions (with less instinctual guidance), awareness of death, and reflective metacognition, humans evolved a new need: not just for survival and reproduction, but also for a purpose to existence, with the hope for or fear of an afterlife, and demands of sacrifice.[4] Various gods and devils emerged, representing ideals of transcendence beyond mortal nature, plus the threat of natural and cultural forces around, within, and through human beings. Performances about good and evil gods reveal the gradual evolution of many interconnected brains, perceiving and projecting a growing awareness of self, community, and the Other(s) who may have caused existence or may be watching it — judging, directing, or playing in our "environmental theatre" of mortality.[5]

In our current millennial era, with various mass-media technologies, we sometimes have the illusion of being godlike ourselves: flying through space and time in the movie theater, watching the intimate lives of fictional and "reality television" characters, or interacting with such fantasies through videogames, where we can be reborn at the press of a button. The theatre of gods in prior cultures, extending our species' big-brain creativity and destructiveness toward the visionary spaces of a mythic cosmos, becomes immanent and immersive, polymorphous and web-like, individual and communal, liberating and sacrificial through current virtual-reality screens. Yet, Islamic fundamentalists also make the screen sacrifices all too real in their "holy war" against Judeo-Christian superpowers, sabotaging the divine illusions of Western humanism and its corporate materialism. Despite the claims of monotheists about knowing the one true God as Supreme Being, good and evil aspects of man's divine aspirations continue to evolve today — with collective energies driving them, whether or not personified as "gods" in the ancient sense. Religious and secular ideals become transcendent, yet performative in various theatres of reality, with democratic rights and terrorist wars,[6] miraculous technologies and ecological nightmares. Where do these angelic and demonic forces come from, in our common human nature and its ancestry? How do the past and present gods of Western culture extend the evolved structures of the brain's theatricality, creating new realities out of group identities and territories, in competition with others?

The Euro-American art form of "theatre" emerged from a specific tradition and history. But such theatre, like related performance traditions in other cultures, also reflects the makeup of the brain's inner world of perception, memory, dream,[7] and imagination — which has its own prehistory of evolving cultural stages in our hominid ancestors. In my initial attempt at charting the brain's inner performance elements and its evolutionary heritage, *Ghosts of Theatre and Cinema in the Brain*, I focused on spirit characters from ancient Egyptian, Greek, and Roman plays to Shakespeare's works, Japanese noh drama, and various films. I briefly considered certain gods in some of those dramas. In the present work, the characterization of good and evil gods becomes the primary concern, as traceable to the human brain's anatomy, its evolution, and Western theatre's specific history — involving theories, periods, plays, and screen performances beyond those explored in the earlier book.

I do not claim to give a comprehensive survey of all the godlike characters in Western theatre. Also, by not including most of Africa or Eastern Europe, the Middle East, Asia, Australia, Oceania, or South America, I do *not* mean to imply that certain African, European, or North American performances are the only way to view the cultural evolution of mankind. (*Nor* do I mean to suggest that technologically "advanced" cultures are better in all respects.) Given the limits of this book, I have chosen significant performance examples in the Western tradition and along its edges — developing toward the global media of film and television. I will focus on ritual and theatre in European prehistory, ancient Egypt and Greece, medieval and Renaissance England (plus a drama from colonial Mexico), modern to postmodern plays, and related screen media with recent American film and TV examples.[8] Below I will outline each of the chapters that follow. But first, I will summarize the findings of my previous book, which offer a foundation for this one.

A caveat is also in order here about the term "evolution." Species and specific mutations survive because they are better adaptations to their co-evolving environment, including the other species around them. However, humans have evolved the ability to radically transform their environment, as well as their own brains, through certain cultures and technologies — vastly increasing the rate of change in material and virtual realities, beyond nature's balancing acts and incremental adjustments.[9] Thus, the future of nature/culture co-evolution through the human species may or may not produce a better world for all humans or for other species. In the last half millennium, Euro-American culture has become a leading force in transforming the globe through its cultural ideas, new technologies, and imperial drives. Some people believe that this is due to "God's will" as a divine teleology. Others regard it as the work of the "Great Satan" of mass-media consumerism and Western militarism. But it may have been due initially (whether or not God or Satan was watching) to the luck of humans who had certain geographical advantages in advancing European cultures thousands of years ago. The given plants and animals, latitudinal trade routes across similar climates, and subsequent disease immunities with travel to new continents enabled Europeans to gain technological success and imperialistic dominance over others around the world (Diamond, *Guns*).[10] There is a tempting hubris to the primacy of Western cultural evolution and territorial expansion, masking its many sinister turns, from colonialism and the slave trade to modern genocide, ecological disasters, and current collateral damage in the "war on terror." Whether our species, with its multicultural mixing across various continents, develops further individual rights and ecological improvements, or brutal regimes of Social Darwinism and more environmental destruction, remains to be seen. Such a prospect of good or evil in the future of humanity, especially through Western culture and its expanding media, becomes a crucial issue as this book traces the gods and devils emerging within the theatre of the brain, through their past expressions onstage and onscreen.

Brain Anatomy, Evolutionary Stages, and Cultural Orders

In the current "cognitive turn" of the humanities, scholars of literature, theatre, and film are using brain science as a way to move beyond the psychoanalytic semiotics of postmodern theories that have pervaded those fields since the 1970s. The works of George Lakoff, Mark Johnson, Gilles Fauconnier, and Mark Turner relate cognitive science to philosophy, linguistics, and literature. Cognitivism has also been a leading approach in film studies for over a decade, as evidenced by David Bordwell, Noel Carroll, and their collaborators in *Post-Theory* (1996). Recently, Bruce McConachie and Elizabeth Hart have edited a volume of essays applying cognitive science to theatre studies, *Performance and Cognition: Theatre Studies After the Cognitive Turn* (2006). McConachie also applies cognitivism to specific plays and performances in his book *Engaging Audiences* (2008).[11] But the focus of such works on cognitive science, embodied metaphors, and linguistics can now be expanded to see the bigger picture of performances within the human brain and between brains—through further insights from related fields.

In the pages that follow, I use cognitive science along with neurology, psychoanalysis, evolutionary psychology, anthropology and primatology to consider the brain's anatomy, its structural development within individuals, and its collective staging of divine aspirations in relation to its animal heritage. I look for connections that ground the postmodern theories of psychoanalysis in recent research on the brain, especially regarding "god experiences."

I also extend such empirical interpretations of neuroanatomy and cognitive evolution — which claim certain universals of human nature[12] — to explore diverse issues in plays throughout Western history, via Lacanian psychoanalytic theory, which has greatly influenced cultural studies in the last quarter century.[13] Used with empirical science, psychoanalysis shows how individual family experiences and changing social structures redefine the brain's inherited mechanisms, especially regarding divine or devilish ideals, as they evolve historically. Like adults who influenced each of us in our childhood, shaping particular patterns in the drives and desires of our neural circuitry, which repeat throughout our lives, or become altered through cathartic awareness, the gods of the past continue to form foundations for cultural repetition or change, as symptoms of our interconnected brains. Thus, we can see how human beliefs and behaviors developed, with repeated, yet shifting patterns — from "baboon metaphysics" (reflecting our common primate ancestors)[14] to the supernatural realm of prehistoric cave art, to the stages of Western culture, involving gods that are layered in our brain theatres today.

My approach touches on cross-currents in the new field of "neurotheology." It also involves the research of many neurologists who connect their work with Freudian theory, as exemplified by the Centre for Neuro-Psychoanalysis in London and its *Journal of Neuro-Psychoanalysis*, published since 1999.[15] But I combine these strategies with "French Freud," thus using, while also further defining, certain terms from postmodern literary and cultural theory, such as the Other, the Symbolic, the Imaginary, and the Real.

Psychoanalysis helps us to understand the relations between hominid evolution over millions of years, neural evolution within each person's lifetime, and historical cultures that shape our evolutionary drives. From the natural and sexual selection of genetic codes, to the neural group selection of brain pathways (Edelman), to the cultural selection of social ideals and practices as "memes" (Dawkins),[16] Real bodies and environments, wherever humans have colonized the globe, are transformed by various Imaginary and Symbolic orders.[17] Consider, for example, the variety of dog breeds shaped by human desires — or how the different demands for human survival and our choices of sexual reproduction are fashioned by ego identities, social rules, and mass-media imagery. This sense of cultural evolution (in relation to biological legacies and neural patterns) is figured, too, by the drives and desires of "the Other," posited as a supernatural force or psychosocial network,[18] expressing human hopes and fears, within and beyond the natural world, through the gods of a particular society.

In his therapeutic research, drawing on Freud's discoveries and on continental philosophy (especially phenomenology), French psychiatrist Jacques Lacan posited three orders of human experience. Cultural "reality" involves the Symbolic order of language and law, as the social Other shaping each person's identity. It also involves the Imaginary order of self and other perceptions, memories, and fantasies — including the illusion of a whole ego, from the "mirror stage" onward, and its aggressive competitions.[19] Thus, human culture represses and transforms the animal passions and remnant drives of our prior existence in nature. With its tremendous powers to alter the environment and reconfigure each child's identity, culture radically changes human nature through language and self-awareness. A child's sense of self emerges through the parental Other's rules and mirroring desires, producing collective connections, yet also a fundamental alienation — and a tendency to seek the meaning of the mortal self in a further, transcendent Other.[20] This involves a third order that Lacan called the Real: the lack or want of being (*manque à être*) in being human, through the loss of primal instincts, natural experiences, and direct memories, with each person's gradual immersion in a certain culture's Imaginary and Symbolic constructions. Though mostly unconscious, the Real order, akin to Freud's notion of the id and its libido, may at times disrupt the Imaginary/Symbolic consensus and interpretations of cultural reality. This occurs through symptomatic actions, dreams, and artworks that involve emotions expressing something beyond the limits of conscious awareness.

I argue that these three Lacanian orders relate to the basic areas of neural anatomy: the left and right neocortex, plus the subcortical areas (from limbic system to brainstem).[21] Humans share with all pre-existing animals, at least as far back as reptiles, a core brainstem that regulates internal functions and processes instinctual responses to outside stimuli, such as the body's instant, unconscious reaction to danger. We share with mammals a limbic system (including the temporal lobes at the sides of the head) that evolved around the brainstem to process more complex emotions and learned behaviors.[22] Like other primates, we also have an expanded neocortex as the outermost layer of our brain (with occipital lobes in the back of the head, parietal lobes at the top rear, and frontal lobes).[23] However, humans evolved distinct functional areas on each side of the neocortex.[24] The left neocortex has audioverbal, linear, causal, executive, prosocial, routine functions, in contrast to the right hemisphere's visuospatial, holistic, intuitive, devil's advocate, anxiety-biased, novelty-detecting processes.[25] Distinctive language systems (syntax and semantics) are in the left hemisphere, in Broca's and Wernicke's areas,[26] in nearly all right-handed people and most left-handed.[27] The right brain has further ties to the emotional limbic system and instinctual brainstem, but the left tends to operate separately (especially in men[28]), expressing or inhibiting limbic emotions and right-cortical intuitions, through its rational language and executive controls. Specifically regarding theatrical mimesis, the left inferior parietal lobe (IPL) is used for recognizing "pantomimes executed by others" because it stores the "complex engrams" or schemas used in the "higher level intentional planning" of actions, while the right IPL is used for interpreting spatial orientation (Jacob and Jeannerod 253). Thus, certain left-cortical functions correlate with Lacan's Symbolic order of language, rules, and social codes, the right with the Imaginary, and the limbic system and brainstem areas with the Real. Yet these three orders are "inmixed" dimensions (Ragland-Sullivan 190), as are the corresponding areas of our brains. The Symbolic order resides primarily, but not solely within and between left brains, like the Imaginary in and between right hemispheres, and the Real in limbic systems and brainstems.[29]

I say "primarily" because there are also aspects of Symbolic language, involving imagery and emotions, in certain right-brain functions: making and interpreting metaphors, contextual meanings, puns, prosody, and nonverbal gestures (Ornstein 103–08; Cozolino, *Neuroscience of Psychotherapy* 109). Thus, the right brain is used more for language, along with the left, by "expert" readers (Wolf 162). While the right brain's Imaginary order is crucial for "self-image" (Ornstein 132, 175–76), the spatial sense of ego also depends upon the left brain's "orientation area," as I will consider in the first chapter.[30]

The general correspondence of Real, Imaginary, and Symbolic orders to the brainstem/limbic system, right hemisphere, and left hemisphere is confirmed by research on developmental growth spurts in the neocortex during childhood. As in Lacan's theory of the mirror stage, with the infant's Imaginary ego initially developing through preverbal communication with the (m)Other, neuroscience shows that right-brain to right-brain "attunement" between the mother and child, during its first two years of life, profoundly shapes its emotional and perceptual pathways, especially its sense of self in relation to others (Cozolino, *Neuroscience of Human* 38, 66–75, 84–85; *Neuroscience of Psychotherapy* 191–92). The "prosocial self" then shifts, through language development, into the left brain, with its growth in subsequent years (118; Wolf 185–88). This relates to the Lacanian Symbolic order of words and laws shaping the child more directly after the initial mirror stage, at 6–18 months. According to neuroscience, the self as a "distributed neural network that encompasses shared self-other representations" continues to be "right-hemisphere based" (Decety and Sommerville 527). Recognition of one's own face can be lost when the right hemisphere is anesthetized (529)—demonstrating that the Imaginary perception of ego (or the Freudian "imago"), and its possible fading or Lacanian "aphanisis," is based in the right cortex.[31]

Regarding our potential for therapeutic and theatrical catharsis,[32] there appears to be a crucial filter between Symbolic/Imaginary and Real orders (or superego/ego and id) in the prefrontal area of the neocortex, at the edge of the limbic system.[33] Neurologists locate a "stimulus barrier" between the Freudian superego and id in the "ventromesial [or ventromedial] regions of the prefrontal lobe [where it] merges into the limbic system" and protects the ego "from the incessant demands of instinctual life" (Kaplan-Solms and Solms 275–76).[34] Here, cathartic changes may occur in how remnant natural instincts are expressed (or transformed through greater awareness), from mostly unconscious, limbic, Real emotions, through right-brain, Imaginary perceptions and fundamental fantasies, to the Symbolic order of language, rules, and self identity in relation to the social Other.

Neurologists have also found four layers of the prefrontal cortex (PFC) with distinctive, nested, hierarchical functions (Koechlin et al.; Murphy and Brown 133–35). The premotor cortex, at the rear of the PFC, exerts sensory control, selecting specific motor (bodily action) responses to stimuli. The caudal lateral PFC, the next layer moving forward, adds contextual control regarding the current situation when stimuli are received. The rostral lateral PFC, a further anterior layer, then exerts episodic control over the other two, by tracking present and past information regarding general behavior, thus allowing for changing contingencies. (Murphy and Brown give the examples of answering the phone when it rings, not answering it at a friend's house, or answering it there because the friend is in the shower and asks you to, as illustrating these three levels of stimulus response.) A fourth area is posited in the frontopolar cortex, used for cognitive branching and controlling the shifts between different episodes of behavior, while exerting control over the other three layers. Likewise, the orbitofrontal cortex (OFC) determines "reward value" choices, including the selection of "stimuli on the basis of familiarity and [selection of] responses on the basis of a feeling of 'rightness'" (Elliott et al. 308). The lateral regions of the OFC are involved with "the suppression of previously rewarded responses." Brain imaging studies find that these areas are "fundamental" in behavioral choices, especially in "unpredictable situations."

One might argue that the Lacanian Symbolic and Imaginary orders of cultural rules and personal perceptions connect with the Real of stimuli and actions through these areas of the PFC (just behind and above the ventromedial). The brain responds to familiar or unpredictable stimuli with inner theatrical representations and outer performances, through shifting, time-bound, contextual, sensory controls. Such controls are shaped in each human brain through learned cultural experiences of the social Other, which create further top-down constraints utilized by the PFC's layered functions, in relation to bottom-up stimuli. And yet, theatrical performances are ways that the Other, as well as the individual, may change. A culture can explore extended possibilities of Symbolic and Imaginary shifts in situation, context, and sensation, using a collective dreamlike space. This may also involve divine and demonic characterizations of top-down or bottom-up forces, experienced in nature, in the body and brain, or in social networks.

Lacan's three orders relate not only to the brain's anatomy, but also to cognitive psychologist Merlin Donald's theory about the evolutionary stages of cultural development in our hominid ancestors. About two million years ago, early hominids evolved beyond the "episodic" experience of other animals (and prior australopithecines)—with the "mimetic" stage of human evolution.[35] Donald cites the evidence of increasing brain size in our hominid ancestors,[36] the first stone tools, big game hunting, a more group-oriented way of life, and thus "a cultural strategy for remembering and problem solving" (*Mind* 261).[37] Instead of being "immersed in a stream of raw episodic experience, from which they ... [could not] gain any distance," early hominids developed a new cognitive capacity, "mimetic skill, which was an extension of conscious control into the domain of action. It enabled playacting, body language, precise imitation, and gesture" (120, 261). This also included prosody, which is processed today in the brain's right hemisphere: "deliberately raising and lowering the voice, and producing imitations of emotional sounds."[38]

About a half million years ago, archaic *Homo sapiens* gradually evolved a "mythic" stage of culture and brain development, culminating with the emergence of our own subspecies, *Homo sapiens sapiens*, about 125,000 years ago (Donald, *Mind* 261). The mythic stage is evidenced by a much higher rate of innovation than in prior hominids: sophisticated tools, "beautifully crafted objects, improved shelters and hearths, and elaborate graves" (261–62). This stage included oral traditions of language and narrative thought—beyond the gesture, mime, and imitation of prior mimetic hominids, or the basic awareness and event sensitivity of episodic primates (260).[39] It thus involved a fundamental change in the human brain (and vocal tract): an "invasion" of the left parietal lobe by language, replacing spatial perception and movement, which then became a more distinctive function of the right parietal lobe (LeDoux, *Synaptic* 303, 318).[40]

Donald's mythic stage shows the evolution of the Symbolic order of mind and society,[41] as well as our current left hemisphere functions. The mimetic stage correlates to right brain processing and the Lacanian Imaginary. Today's human brains also bear the remnant animal emotions and drives of primal episodic awareness in the limbic system and brainstem, as a lost yet disruptive Real or *chora*.[42] Indeed, each child moves through similar developmental stages, recapitulating hominid phylogeny: from primal episodic awareness to the mimetic "interlinking of the infant's attentional system with those of other people" and then to narrative speech (Donald, *Mind* 255). Or, in Lacanian terms, a child moves from the Real of natural being to the Imaginary order of mirrored illusions of ego in the (m)Other's desires and then, through verbal language, to the Symbolic order of superego incorporation, with the Other's discourse and social rules, via the Name and No of the Father.

This basic outline of Lacanian orders, brain anatomy, and hominid evolution shows that "theatre" (and dance) in the most primal sense — as Imaginary, mimetic performance — began about two million years ago. At that time, our ancestors developed a new skill that eventually became specialized in the visuospatial, prosodic, Imaginary functions of the right hemisphere, with ties to the emotional/instinctual Real of the limbic system and brainstem. Later hominids developed oral language and myth-making, as further Symbolic orders, through distinct areas of the left brain about a half million years ago. As with the modern child's development from primary to higher-order consciousness, through the Real and Imaginary dimensions of the mirror stage and the later Symbolic acquisition of language and rules, these layers of the brain and of hominid culture continue to interact today — with each human being transformed by a particular family and society.

As Donald points out, primal mimesis in early hominids relates not only to the current playacting of children (*Mind* 266), but also to the "many institutionalized versions of pretend play in theater and film, and [to the] imaginative role playing [that] is integral to adult social life" (263). A crucial aspect of this evolutionary skill is emotional regulation, which involves the germ of self-consciousness, through a "mimetic controller" in the brain, "a whole-body mapping capacity ... under unified command" (269). Thus, early hominids developed larger frontal lobes, setting the stage for the later evolution of a distinctive left hemisphere (271).[43] Like children today (starting with the Imaginary dimension of the Lacanian mirror stage), our hominid ancestors developed a "kinematic imagination" with the physical "image of self" becoming an anchor to experience and awareness (273). This involved rhythmic body movements, expressing temporal relations, through the intersubjective medium of performance, as a "public theatre of convention" (272–74). However, the full emergence of theatre as narrative performance began with oral storytelling during the hominid "mythic" stage, starting about a half million years ago.

Then, about forty thousand years ago, humans evolved a further, "theoretic" stage, through the "externalization of memory ... [using] symbolic devices to store and retrieve cultural knowledge" (Donald, *Mind* 262). During this current stage of hominid evolution, the tradition of recorded theatre and drama developed, along with other artistic technologies,[44] a "Symptom" of being human that has vastly expanded in recent centuries.[45] Thus, theatre in the theoretic sense may have started with Paleolithic cave art (as considered in the first chapter). Eventually, the theoretic technologies of theatre, externalizing and interconnecting the performance elements of the human brain, developed in various ways through different cultures — culminating in the current globalism of virtual media screens, often dominated by Western paradigms. Our theoretic stage with its evolving technologies continues to reshape the skills of prior stages and "liberate consciousness from the limitations of the brain's biological memory systems" (305). However, such an external memory field can also be a "Trojan Horse," Donald warns, "a device that invades the innermost personal spaces of the mind. It can play our cognitive instrument, directing our minds toward predetermined end states along a set course" (316). Such a Trojan Horse potential, with good and evil effects, becomes even more significant through divine characters and godlike ideals, at various points in Western history, from stage to screen performances, as explored throughout this book.

Donald's stages of cognitive psychology match with Stephen Mithen's archeological theories and research.[46] According to Mithen, the early hominid social intelligence of *Homo erectus*, 1.6 million years ago, involved the communication of "contentment, anger or desire" through a "wide range of sounds" (*Prehistory* 144) — as with the mimetic prosody theorized by Donald. Human verbal language with "a vast lexicon and a set of grammatical rules" began 500,000 to 200,000 years ago, with Neanderthals and archaic *Homo sapiens*, as evidenced by brain and throat structure, indicated in fossils of their bones (140–42, 208). This corresponds to Donald's mythic stage of hominid evolution. Mithen also cites archeological evidence that a dramatic shift occurred 40,000 years ago. Early humans in the Upper Paleolithic period changed from having separate types of intelligence — natural history intelligence (such as interpreting animal hoofprints), social intelligence (with intentional communication), and technical intelligence (producing artifacts from mental templates) — to a new cognitive fluidity between them, creating artifacts with "symbolic meanings ... i.e. art" (163–65).[47] This shows the beginning of Donald's theoretic stage and relates to the possible shamanic visions and performances evidenced by Paleolithic cave art.[48]

The evolutionary stages, neurological layers, and psychoanalytic orders of self and Other awareness, developing through shared cultural performances, reflect what might be called an "inner theatre" of the brain.[49] By this, I do not mean a "Cartesian theatre" with the mind inside the brain as a single ghostly spectator watching the machinery of inner scenes, or as a playwright-homunculus inhabiting a central control area (the pineal gland, according to Descartes, 400 years ago). This theory has been fully critiqued by cognitive philosophers, from Gilbert Ryle to Daniel Dennett, as well as by current neurological evidence.⁵⁰ However, cognitive scientist Bernard Baars uses theatrical terms in other ways to explain the global workspace of human consciousness. Less than 10 percent of brain activity is conscious, like a "spotlight" on the visible actors and scenery (*Theater* 46–47).⁵¹ The rest involves unconscious agents, like a legislative "audience," competing and collaborating to focus attention on particular perceptions and ideas onstage. There are Deep Goal and Conceptual Contexts, like "backstage" workers, as well as immediate expectations and intentions, forming an unconscious sense of self as "director" of the brain's inner theatre (144–45).⁵² Thus, the development of mimetic, narrative, and theoretic skills in our hominid ancestors, beyond the basic episodic awareness of other animals, gradually involved an Imaginary and Symbolic performance space within the brain (in Lacanian terms) as a higher order of consciousness.⁵² Dennett himself calls this an "inner selective environment" with "*designed* tools" (*Darwin's* 374–81), representing the outer world, the self as a body in space, and potential interactions with others, in the mind's eye, before or instead of performing outside, as well as in memories and dreams.

Like Baars, neurophysicist Eric Harth describes a "theater" within the brain, where "all the sensory cues and cerebral fantasies conspire to paint a scene. There is also an observer: the rest of the brain looking down, as it were, on what it has wrought" (144). Through active mirroring in the brain, sensory input is "multiply reflected"; it is reiterated via "associative connections in the cortex, which trigger new concepts ... fed into the self-referent loops." Such mirroring, theatrical loops also extend outward to include other people and their inner brain theatres: a "progression of mirrors," forming a "wider loop that adds an essential ingredient to the creative act" (109).

In our ancestors, tremendous possibilities arose through this inner and intersubjective theatre of the brain, with the freedom to change self, culture, and world far beyond instinctual reactions to the environment or the limited patterns of learned behavior in other animals.⁵⁴ The rate of evolutionary change, of mutation and selection, greatly increased through mimetic and mythic cultures. Using their "mirror neurons," hominids mimicked each other's newly invented skills and tools.⁵⁵ Beyond simple "mimicry," they developed complex, multimodal, and symbolic "imitation," by understanding the goals and intentions of others' actions, thus using their mirror neurons differently from other primates (Jacob and Jeannerod 230–34). In Lacanian terms, their desires became "the desires of the Other," instead of just instinctual. Hence, human physical and verbal communication created larger and larger groups of collective minds, working together to solve survival problems. Autobiographical memories and semantic (as well as procedural) plans extended from the brain's inner theatre toward the external records and creative technologies of theoretic cultures.⁵⁶

Brain cells competed for collective agency (according to Edelman's theory of Neural Darwinism) and distinctive brain areas evolved in our hominid ancestors to present mimetic, mythic, and theoretic figures on collaborative stages of consciousness. Individuals with such brains also competed for dominance — for their genes and ideas to survive and reproduce. Yet the human brain's greater flexibility, beyond natural instincts, and its evolving power to change the environment, not just adapt within it, also produced more uncertainty about the means and meanings of survival. Early humans, like hunter-gatherers today and unlike our primate relatives, may have formed "egalitarian" communities, while developing toward the cooperative creativity of Paleolithic cave-art performances (Boehm, "Conflict" 93).[57] Social factors, such as the punishment of freeloaders (involving the risk of reciprocal violence within the group) and the cooperative demands of warfare and Ice Age climate changes, led to the earliest forms of religion, involving dance, music, ritual, trance, sacrifice, and the extension of moral ideals and judgments to supernatural figures (Wade 47–97). As communities became settled and much larger with the invention of agriculture about 10,000 years ago, alpha leaders gained control of increasingly hierarchical groups. Greater actors were projected into the mimetic, mythic cosmos as organizing and chaotic principles, which spread through military and cultural imperialism. Creative and destructive gods represented the social powers of control and rebellion, as well as the forces of nature that continued to nurture and threaten humans. They also depicted unconscious energies of the Real within human brains: the remnant animal instincts and primal emotions that might erupt, in creative or destructive ways, through higher orders and stages of consciousness, despite cultural controls.

Theatre as an art form enables actors to externalize the inner performance realms of memory, dream, and fantasy through mimetic and narrative skills. The theatre artist gains new perspectives with the audience, regarding otherwise unconscious instincts, emotions, images, and symbols. But such a sharing and distancing, like the evolving theatre of the brain itself, becomes a double-edged sword. The left-cortical functions of the brain's internal theatre, along with the Symbolic order of cultural performances, may create myths and gods that censor and repress the right brain's Imaginary novelties, mimetic anxieties, and devil's advocate functions, plus the limbic system's emotional Real — through executive, linear, causal, and binary controls. Or the left brain and its mythic order may enable new perceptions and ideas to filter through and reach the spotlight of consciousness, while its Symbolic framework protects the self or society from being overwhelmed by the right hemisphere's intuitive, holistic Imaginary and the limbic system's Real passions. The synaptic ties between and within these brain areas are not "hard wired." The human brain's high degree of plasticity means that specific neural connections are continually "pruned" and rerouted by experience (Solms and Turnbull 146–48; also Doidge). The brain's inner theatre changes through acting in or watching a play, through fictional distance and intimacy, especially if the performance presents issues and images that touch upon repressed desires and drives, percolating at the edges between Symbolic/Imaginary and Real orders.[58]

Evolving Gods, Angels, and Devils

Chapter 1 of this book considers the new interdisciplinary field of "neurotheology." It makes possible connections, through neuroscience, cognitive authropology, evolutionary psychology, and Lacanian psychoanalysis, to certain theatrical sites inside and between human brains — as "neurotheatre." (My use of this term also relates to the new fields of "neuroaesthetics," "neuroethics," and "neuroeconomics.")[59] The first chapter then looks at the cultural remains of humans from tens of thousands of years ago: the cave art that may have recorded and then again triggered shamanic trances, as a prehistoric form of theatre within and between the evolving brains of our species.

Chapter 2 investigates the written record of performance roles, dialogues, and mythic actions in an Egyptian coronation rite, an apparent form of theatre from four thousand years ago. It also considers the good and evil, nurturing and threatening interactions with gods in ancient Greek drama, focusing on Aeschylus's *Prometheus Bound* and *The Oresteia*, and Euripides's *The Bacchae*. These plays will be analyzed in relation to the basic anatomy of the human brain, which we share with the ancient Egyptians and Greeks, to consider how their malicious, mischievous, and yet at times benevolent gods reflect the theatre within our heads: its Real animal drives transformed by Imaginary mimesis and Symbolic myth-making toward mystical communion or causal, binary rationality.

Chapter 3 looks at the development of angel and devil figures in biblical, miracle, and morality dramas of the Middle Ages, including *The Play of the Sacrament*, the Wakefield/Towneley Cycle, and *Everyman*, vis-à-vis the kingly, judicial persona of a Christian God. Issues of guilt, fate, and sacrifice in ancient drama — involving the gods' rivalries and demands — are refocused through one God (plus his angelic servants) as alpha ruler, with mischievous devils tempting humans, and yet enabling their free will. This chapter will continue the book's overall concern about how such gods, angels, devils, and spiritual issues arise from material structures in the human brain and prior evolutionary stages, transformed by new developments in historical cultures.

The idea of "free will" becomes complex in the Renaissance period. Western civilization continues to be based in Christian beliefs about the soul's journey through life toward or away from God. But Christopher Marlowe's demonic characters and spiritual temptations also emphasize the choice of indulging in regressive animal passions (while lacking the guidance of natural instincts) or progressing toward a rational order with humans becoming more and more godlike in this world. Likewise, William Shakespeare's magical colonizer, Prospero, displays the primal drives of fear, rage, and longing in the limbic brain fueling both tragic errors and tragicomic wisdom, through holistic anxieties and causal binaries in the right and left neocortex. These heroes exemplify how new humanist ideals gained godlike power in the Renaissance, through theatrical magic in summoning ancient and medieval spirits.

As European empires grew across the Atlantic, they clashed with Amerindian traditions, strange rituals, and foreign gods. The fourth chapter considers not only the mixture of medieval and early modern ideals in Marlowe and Shakespeare, but also the meeting of Old and New Worlds in Sor Juana Inés de la Cruz's allegorical *loa* to *The Divine Narcissus*. It then forms a bridge to the final chapters of this book, surveying various centuries, from the divine ideologies of the early modern period to the transcendent ideals and existential despair of modernist drama, absurdist theatre, and postmodern performances. How do the ancient deities and medieval devils in the plays of Shakespeare and Marlowe correspond to later, progressive and yet reflective steps: from Renaissance to Romantic, High Modernist, absurdist, and postmodern relations of nature, culture, and the human brain — with its animal passions and divine aspirations?

Chapter 5 shifts the discussion of evolving gods in live theatre to the newer medium of cinema. It considers recent films involving God, angels, and devils, via the "metaphysics" of social rank knowledge in our primate relatives — showing an evolutionary basis for good versus evil ideals, displayed in human behaviors onscreen today. Then it explores the biblical film, *The Passion of the Christ*, as a postmodern return to medieval paradigms explored in chapter 3, through the increasing power and sacrificial temptation of screen performances. How do Christian interpretations of God and other supernatural figures appear in traditional, yet new ways through cinema? Do such cosmic forms also relate to ancient deities, or make the movie screen akin to the Paleolithic cave wall, as a membrane for shamanic visions?

Chapter 6 turns to the medium of television, through a case study of *Millennium*, a series from the 1990s that exemplifies the popular interest in angels and devils at the turn of the millennium, which continues in the mass media today. *Millennium* did not depict God directly, yet dwelt on His mysterious absence, or distant spectatorship, through the apocalyptic context of rival religious cults and supernatural forces vying for the fate of the world. It also became prophetic for the renewed devotion to human sacrifice in our current "war on terror," raising fundamental questions about the further evolution of our brain's performance spaces.

Finally, the conclusion to this book considers how the human species continues its development today, in our cultural remaking of the world, through our good and evil natures. How do we create heaven and hell on earth, as extensions of the theatres in our brains? Is more morality onstage and onscreen, with violent melodramatic victories of good over evil, of superheroes over arch-villains, the best way for us to sculpt the inner performance spaces of each new generation? Or do tragicomic antiheroes and other complex figures better enable us to reshape the neural connections between animal drives and higher-order ideals, through evolving performances of mortal awareness, immortal desires, and divine technologies?[60]

1

Neural and Prehistoric Signs of the Divine

In many cultures, there are myths about gods creating the world. Such myths are ways to codify a culture's values in relation to its power structures, knowledge of nature, and ritual behaviors for surviving in and transforming the environment. Modern science functions in a similar way, though with more of a claim to universal truth and the spread of Western neoclassical values, through repeatable "ritualized" experiments,[1] rational causes, and technological powers. According to the "myth" of science, how were the gods created—at least as human ideals?

Anthropologist Walter Burkert, famous for his *Homo necans* theory of human cultural development through hunting packs, explores religion through evolutionary psychology. Using the work of Niklas Luhmann, Burkert argues that religion reduces the anxiety-ridden complexity of human awareness into dualistic, causal, and hierarchical concepts as "containers" (*Creation* 26).[2] Mythic notions of the gods as "superfather or mother," along with mimetic rites, have long functioned as coping devices—focusing fear in order to contain a more alarming sense of panic, though also involving "aggression to overcome anxiety" through sacrifice (30). Burkert views myths about demons and the "evil eye" as drawing on primal anxieties about predators in our ancestor's evolution (42–43). He compares ancient Greek and Roman "rites of aversion," offering a scapegoat to demonic pursuers as a way to avert their evil, with a lizard losing its tail or a zebra herd escaping after one of its members is taken by a predator (41–55). Likewise, he explains the ancient rationale of sacrifices to the gods as a giving and getting arrangement (137). In Western and Eastern cultures, sacrifices are like "bribery" and the gods hasten to offerings that they have long missed, assembling "'like flies'" (143–44). Thus, in Lacanian terms, the Other order of the gods is lacking, even if immortal, like an ultra high order of consciousness that still desires what humans can offer. The gods prey upon or cultivate the offerings of men and women — like humans, with higher-order consciousness, do to wild and domesticated animals, which have "primary consciousness."[3] And yet, through sacrificial worship, humans gain from the gods symptomatic organizing principles for their own lack of being, simplifying the anxious complexity of human neural and cultural plasticity.

As Burkert points out, humans have lost a direct connection to nature's signs. "In more primitive organisms ... all the signals processed ... have simple and unequivocal 'natural' meanings" (*Creation* 161). But humans have lost "the immediacy of practical meanings." Instead, they seek cosmic meaning from the gods, often via animal (or human) sacrifices.[4] Burkert calls such divination "regressive"— as when superstition is evoked by "alarm or panic" (162). Yet, there is also the possibility for growth. "As the self-sufficiency of a normal, closed cultural system is shattered, it gives way to uncommon openness to signs hitherto disregarded, which may offer a chance for reorientation."

Through such openness to new signs, the art of theatre emerged in ancient Greece and elsewhere, beyond (yet still involving) ritual sacrifice and religious divination.[5] Today, as then, it continues to build upon the neural and cultural evolution of earlier hominid ancestors: from mimetic skills and mythic orders to the further technologies of our theoretic stage, with live performances, movies, television shows, and computerized virtual realities. Such theatrical displays may provoke political and spiritual questioning, for further complexity and reorientation. But they often serve to confirm social orthodoxy and its reassuring simplifications. Even secular spectators are tempted to simplify the uncertainty of their lives by believing in good or evil stereotypes and godlike celebrities.

The forces that evoked ancient gods can be seen today in the theatres of our brains and cultures.[6] Binary, causal, and executive agents in the left hemisphere (and in current myths) work to contain the right brain's holistic, intuitive "Devil's Advocate," with its limbic anxiety (Ramachandran and Blakeslee 135–47).[7] Yet, right-hemisphere anomaly detectors, interacting between brains, may also create new perspectives and possibilities for change. Though many of us understand our mortal fate differently now, the plays of earlier periods with devils, angels, and rebellious or authoritative gods still relate to the social interactions of right and left brains today (and their ties to limbic emotions and brainstem instincts). The plays of the past, when viewed through current neuroscience, offer valuable insights about the human remaking of natural signs as divine, a process that continues today in our evolving brains and cultures.

As biologist Terence Deacon puts it, we humans have a "savant-like compulsion to see symbols in everything ... [and] imagine ourselves as symbols, as tokens of a deeper discourse of the world" (437). Paleontologist and theologian Teilhard de Chardin argued, a half century ago, that human evolution moves in this discourse toward a teleological union with God.[8] Whether or not such a god exists and watches us, in the theatre of the cosmos, we inherit from our ancestors the neural apparatus for experiencing such mystical communion and emotional faith. That apparatus, with its cultural connections, is still developing through mimesis, myth, and theory, in our current mass-media theatres.

God(s) in the Brain Matter

The field of "neurotheology" involves many arguments about human belief in the existence of God (or an immortal human soul), using recent discoveries in brain research and theories of evolutionary psychology.[9] Robert Buckman acknowledges that the "facts" about God cannot be settled by science, but he argues (like Burkert) that "religious beliefs serve the function of a coping strategy" (208–09).[10] He explains the concept of God as "a transitional object ... [like] a childhood teddy bear or a security blanket ... giv[ing] you comfort during periods of stress and change" (209). Matthew Alper takes a harsher view against religion, finding that "spiritual consciousness represents 'nature's white lie,' an inherited misperception selected into our species for the purpose of alleviating the anxiety caused by our painful awareness of death" (175). He sees this white lie as producing the dangerous divisiveness of human religions, with each belief system regarding others as "an immortal threat" to the soul (181). "[O]ur species tends to engage in ... *religious tribalism* ... compel[ling] us to initiate the types of religious wars and atrocities that have marked our species' history." Offering more optimism about the neurology of spirituality, Michael Persinger has located the "God Experience" in the brain's temporal lobes, as an evolutionary adaptation with positive effects upon the human immune system. Yet he also theorizes that harnessing such a "powerful force within each human being," as a scientific and therapeutic mechanism, "would require the acceptance that God may not exist or that the experiences are illusionary products of the human brain" (*Neuropsychological* 156).

On the other hand, Carol Albright and James Ashbrook find evidence in neurotheology to support their belief in a Judeo-Christian God. They see certain aspects of God's existence in various areas of the human brain: an attending and ever-present God in the brainstem; a relating, nurturing, remembering, and meaning-making God in the limbic system; an organizing and versatile God in the left and right hemispheres of the neocortex; and a purposeful God in the frontal lobes.[11] With a narrower scientific focus, V. S. Ramachandran, like Persinger, considers the temporal lobes and other limbic areas as the brain's primary apparatus for mystical experiences of divinity.[12] Epileptics with focal seizures in these areas have spiritual experiences "ranging from intense ecstasy to profound despair, a sense of impending doom or even fits of extreme rage and terror" (Ramachandran and Blakeslee 179). Some epileptic patients develop a "temporal lobe personality," with repeated electrical bursts kindling neural pathways that produce the "feeling of divine presence and the sense that they are in direct communion with God" (179–80).

Ramachandran leaves open the question of whether such a communion with God, through material areas of the brain, is "genuine" or "pathological" (Ramachandran and Blakeslee 179). But he tested two patients with symptoms of temporal lobe personality (hypergraphia and obsessive talk about metaphysical feelings) and found a "selective amplification of [galvanic skin] response to religious words" and icons (186). He concluded that there is not a general kindling that strengthens all temporal lobe and limbic connections, but a specific enhancement of "circuits in the brain that involve religious experience and ... become hyperactive in some epileptics" (188). Ramachandran speculates that these neural circuits evolved in the human brain through the selective benefit of "conformist behavior," as shown in sacrificial rites, submission to priestly authority, and belief in providence, contributing "to the stability of one's own social group — or 'kin' — who share the same genes" (184). In a more specific vein, geneticist Dean Hamer argues that a certain "God gene" (a mutation of VMAT2) corresponds with the tendency toward self-transcendent spirituality and optimism in some individuals, which developed for evolutionary reasons, though religion arose beyond genetics through cultural memes (11–13, 73).

Persinger also sees an evolutionary "survival value" in the God Experience of the temporal lobes, even if it has caused great destruction between human groups through violent religious convictions (*Neuropsychological* 138–40; "Neuropsychiatry" 519). Furthermore, he argues that individual brains may be affected by fluctuations in the earth's electromagnetic field, producing a "sensed presence" that could be interpreted as supernatural or paranormal — as God, angels, muses, aliens, or ghosts (Horgan 91–92). This may explain why certain places, such as haunted houses, trigger the perception of temporal-lobe phantoms. Since the late 1980s, Persinger has experimented with applying "low frequency, very weak (similar to the intensity generated by a computer screen) complex magnetic fields" to the right and left temporal lobes of various individuals ("Neuropsychiatry" 516). When the complex magnetic fields were applied over the right hemisphere, "most normal people who were not aware of the purpose of the experiment experienced a 'sensed presence' or sentient being" (517).¹³ Many felt such a presence "interact with their thinking and 'move in space' as they 'focused their thoughts' on it." Persinger theorizes that this right temporal lobe experience of the Other is the "equivalent of the left hemispheric sense of self." He also relates such experiences to the rapid eye movement (REM) state of dreaming, "when right hemispheric functions are more prominent," and to religious rituals that evoke "right hemispheric activity such as singing or chanting within large groups." (One might relate such right-hemisphere states, as well as the right temporal lobe experience of the Other's presence, to the Lacanian Imaginary as the order of fluid fantasies and primal maternal attachments.) Sensed presences are also reported by "the majority of patients who have sustained mild brain injury" and feel they are "'not the same'" as before (519). When these presences occur on the left side of the body (involving the right brain), they are usually perceived as "negative or aversive," causing a "fear of the 'dissolution' of self" and corresponding suicidal ideation. "When the sensed presence is attributed to the right side [left brain], the affect is neutral or positive...."¹⁴

Such "positive" or "aversive" specters on each side of the body may correspond to Friedrich Nietzsche's sense of Apollonian and Dionysian gods in ancient Greek theatre — or to harmonious and dangerous divinities in other cultures as well (though Nietzsche would value the latter type more). These may be cultural extensions of the "sensed presences" produced by temporal lobe activity in the left and right hemispheres of individual brains. Left- and right-brain divinities are not independent, universal archetypes. They depend upon intersubjective, neural connections in specific cultural contexts. Yet, they exemplify distinctive relations of the neocortex to the limbic system in each hemisphere of the human brain: executive, linear, narrative (audioverbal) control in the "dominant" left brain or intuitive, holistic, theatrical (visuospatial, somatic) arousal in the right. This lends support to Nietzsche's dialectic, in *The Birth of Tragedy*, of a dreamlike Apollonian harmony and intoxicating Dionysian wildness in ancient Greek (or modern) culture.¹⁵ His theory points to the chthonic passion and satyr chorus of the theatre god, Dionysus, as necessary for producing the civilized transcendence of the Apollonian ideal and the mythic hero onstage — like the limbic system, with its remnant animal instincts and Real emotions erupting primarily through the right brain's Imaginary, mimetic intuitions, before being interpreted or repressed by the Symbolic left.¹⁶ The "birth" of theatre in ancient Greece, through Nietzsche's nineteenth-century retrospective view, corresponds to the emergence of theatrical spirits in each human brain through its primal pruning by parental personalities — and by further social contexts shaping the maternal, anxious, right and patriarchal, prosocial, left hemispheres (Cozolino, *Neuroscience of Psychotherapy* 12, 80, 107, 190–96).¹⁷

Persinger explains the persistent "origin" theme of God Experiences, and the often-used religious metaphor of being "born again," through the fact that the body image is stored in the temporal lobe during the first few months of life, before it shifts to the parietal lobe. "During the time when the new being is dependent on the behaviors of the mother and father and the *pattern* of those behaviors, the temporal lobe integrates and grows with that information" (*Neuropsychological* 39). God Experiences may involve "images and protosensations long locked within the old contexts of the temporal lobe." Eventually, a particular "God Concept" evolves in each human brain, through specific parental and cultural influences, in the temporal lobe traces and parietal lobe perceptions of self and Other, as body image and sensed presence (67).[18]

The distinction between right- or left-hemisphere presences, and their corresponding God Concepts, involves not only the maternal versus patriarchal characteristics within specific human cultures and in primal personal experiences, but also certain anatomical differences between female and male brains. Anatomical differences in gender, although physically structured, are a matter of degree with each person and are affected by the evolution of cultural ideals in the particular developmental pruning of each generation of new brains. Yet, neurologist Rhawn Joseph summarizes the evidence that, on average, women's amygdala neurons are "smaller and more densely packed" than men's, firing more easily and frequently (163). Joseph relates this to the "fact that females are more religious, more emotional and more easily frightened than males." He cites over a dozen studies demonstrating that women are generally superior to men in "the expression, perception, and comprehension of social emotional nuances ... from the earliest stages of childhood." Women also have "more intense religious experiences" (164). According to Joseph, this may be due to the anterior commissure, which connects the right and left amygdala (in the temporal lobes on each side of the brain), being 18 percent larger in females. Thus, the woman's amygdala is more densely packed, with neurons firing more frequently, and its lobes "can more easily communicate and excite one another."[19]

Dionysian rites, with ties to primordial mother goddess beliefs, appealed especially to women in ancient Greece. Western theatre's invention there as an art form, with its Dionysian dithyrambic chorus in the orchestra and Apollonian ideals of heroic characters onstage, all performed by men, shows a specific externalization and yet containment of the theatre of gods in the brain. Thus, the ancient patriarchal expression of a more excitable amygdala in Dionysian fear and spirituality, as an entertaining threat or tragicomic insight, gave "birth" to theatre in a very different context from our own. But the same anatomical structures and their potential for sensed presences haunt today's theatres, from the cranium to the stage and screen. In the ancient context, the potential for a female character's fear and vengeance to turn into a violent rage, through the maternal *chora* of a Dionysian chorus, appears in the textual evidence of *The Oresteia* and *Medea*. Yet, such plays also show the patriarchal fear of limbic emotions and right-brain spirituality/sexuality. This has produced violence against women in many historical periods, including when Christians have attacked the "witches" of prior mother-goddess religions (Joseph 146–51).[20]

Despite the cathartic potential of tragic violence onstage, theatres in the brains of various human cultures have perpetuated the demand for actual bloodshed. Limbic fear and rage has been projected upon the sexual or spiritual Other as the sensed presence of evil, thus producing the repeated evils of religious righteousness and "holy war." The human species may have evolved godlike technologies for survival in and the transformation of various environments, far beyond adaptive limits. But in changing the rules of fitness for survival, the human brain's theatre has also become the paragon of animal destructiveness, evolving material demons along with its transcendent, male and female spirits.

As a scientist, Joseph considers the Darwinian aspect of adaptive spirituality: belief in the soul's survival may influence the brain's interpretation of near-death, out-of-body experiences and other sensed presences — increasing hope, obedience, and physical survival through the ideal of a transcendent ego.[21] "Thus the limbic system and temporal lobes insure the survival of the self by dreaming and hallucinating ghosts, spirits, and avenging angels, and by promoting the illusion of perpetual and eternal survival or damnation if taboos, rituals, and laws of God are obeyed" (280). Some of Joseph's ideas (about God as a guiding force) go far beyond his claim of having "scientific evidence" (289).[22] Yet his argument represents a counter-voice that has been a part of evolutionary theory from its very beginnings (and continues with the Intelligent Design debate today). Alfred Russel Wallace presented the first paper on natural selection jointly with Charles Darwin in 1850. But Wallace disagreed with Darwin, theorizing that culture was a unique force in human evolution and that the human brain's quintessential abilities of math and music could not have arisen through the random process of natural selection (Ramachandran and Blakeslee 189). Ramachandran adds to this argument the evolutionary paradox of Cro-Magnon and Neanderthal brains being bigger than ours today, which might suggest to some that God directed human evolution (191). Yet, Ramachandran offers a different possibility, using evidence from the extraordinary talents of idiot savants. "Maybe when the brain reaches a critical mass, new and unforeseen traits, properties that were not specifically chosen by natural selection, emerge" (196). The leaps in human evolution may have been caused by the mutant expansion of specific brain areas (such as the angular gyrus) in some individuals, producing the "bonus" of random, savant abilities in math and art — rather than by steady brain growth or the directions of divine providence.[23]

Joseph and Wallace present an argument that still appeals to many brains in our postmodern, science-based culture. It is hard to believe that one's sense of having a single self (or soul) with free will is illusory, and generated by the random firing of neurons,[24] or that the human brain evolved its godlike apprehension through the chance mutation of esoteric traits. Neurologists agree today that there is no command center as Author of the brain's theatre. Yet, Nobel Laureate Gerald Edelman describes a shifting "dynamic core" of human higher-order consciousness that involves "unity and inherent subjectivity" through each person's singular point of view and private experiences, which "cannot be completely shared" (Edelman and Tononi 146). Edelman does not include "spirits" or gods in this formula (*Wider* 78). But physicist Gerald Schroeder goes farther. He argues that a "form of consciousness may remain" after the body dies and that the "hidden face of God" can be glimpsed through the ordered complexities of both the universe and the human brain, as well as through the Bible's revelations (152, 173–87).

On the other hand, atheistic biologist Richard Dawkins, author of *The God Delusion*, would say that God is merely one of many cultural "memes" (gene-like, self-spreading, ideological elements) that use human bodies and brains as their "vehicles."[25] He claims, however, that we still have the freedom to "rebel against the tyranny of the selfish replicators" (qtd. in Blackmore 241).[26] Yet, psychologist Susan Blackmore uses Dawkin's theory to argue that humans have no "free will," since the self is also an illusion, "just a story that forms part of a vast memeplex, and a false story at that," like the deceptive meme complexes of religion (202–03, 237).[27] She recommends the trick of "meme-weeding," through concentration on the present moment (as in Buddhist meditation): "letting go of any thoughts that come up" to realize that the memes are "fighting it out to grab the information-processing neurons of the brain they might use for their propagation" (243–43). In Blackmore's view, although our brains are used by the memes "God," "self," and "free will" as their reproductive vehicles, we may choose between true and false, or good and bad ideals — through greater awareness of such memes.[28] And yet, one might argue that meme theory is itself another way to define a certain God ideal, with memes as immortal forces of the Other, moving through nature, the evolving brain, and human culture.[29]

As a reader of this book, you may be weeding the memes of its theories and arguments, selecting which seem more true to replicate and grow in your brain. But even the rejected memes that you weed out might leave a root in your unconscious, through the neural patterns of prior ideas and ideologies, especially through the influences of significant people in your life and the emotions that their ways of thinking continue to arouse. The limbic system in your brain may be a "transmitter to God" and to the souls you have known (Joseph 74, 187). Or your brain may be a vehicle of Gaia's evolution (Lovelock), with the earth itself and all of nature evolving through your higher-order consciousness, as a paragon of animals and quintessence of dust (in Hamlet's terms) — along with billions of other human brains on this planet in the present moment.[30] Yet, with human consciousness as the zenith of nature's creative experiment, reflecting a desire in Gaia for godlike self-awareness, there comes the additional burden of valuing both birth and death, youth and age, growth and loss, individual and communal in our changing lives and cultural environment.

The human brain has evolved an internal theatre of cooperating, yet competing forces, producing many external theatres — in art, religion, war, mass media, and other arenas. These theatres extend the *agon* (contest) of selfish genes to survive and replicate, through memes of compassion and cruelty, in the self's awareness of mortality and its desire for the Other as a transcendent, providential ideal. Such antagonism of collective genes and memes, extending from plant and animal life to various human cultures, involves communal cooperation (as in "one nation under God") and yet also group conflict, sometimes perpetuating a "holy war" between good and evil gods, when the brain's potential for divine experience becomes polarized and politicized. Thus, competition or revenge against an external threat, or the sacrifice of an internal scapegoat, may be useful for increasing group cooperation and stability. It can also lead to thought control by ostensibly righteous leaders, enforcing conformity through the brain's theatre of god experiences, and to violence against others who are "needed" as evil enemies (Volkan). But mystical experiences of the divine may provide valuable mutations in a culture's evolving awareness, especially when such perceptions of divinity rewire the brain's binary, causal, holistic, and emotional circuitry.

Mystical Operators

Neurologists Andrew Newberg and Eugene d'Aquili have researched various brain areas involved in the mystical experience of "Absolute Universal Being" (AUB).[31] They define the "causal operator" in the brain's left hemisphere, the "ability to think in terms of abstract causes," as the basis for myth-making and science, both of which may approach a mystical view of the Creator or Prime Mover (Newberg, d'Aquili, and Rause 63, 170–71, 196).[32] They also describe the "abstractive operator" in the left parietal lobe as having a similar function, forming "general concepts" and finding "links between two separate facts" (49). The "holistic operator" in the right parietal lobe uses images and emotions (through the "emotional value operator") to give "depth and authority" to intuitive ideas that agree with the logical causes and effects discovered by the left parietal lobe (65, 69). These structures produced the power of myth in various hominid cultures, from the cosmic aspirations of Neanderthals and early humans hundreds of thousands of years ago[33] to modern science and religion as ways of explaining specific, material causes and larger, metaphysical meanings (54–59). The left and right hemispheres, with their causal or holistic operators, may solve problems independently, in distinct ways: "the connectors between right and left are not conveying detailed input.... However, when the imaged solution on the right side of the brain matches the analyses on the left, both hemispheres declare the problem solved through electrical impulses to the emotional center of the brain" (21).

According to Newberg and d'Aquili, our brains also have a "binary operator" in the left inferior (lower) parietal lobe (Newberg, d'Aquili, and Rause 196). This relates to abstractions of good and evil: the personified gods, angels, or devils that human brains and various cultures produce. For, unlike animals, we "think of danger in the abstract," anticipating its possibility even when the threat is not at hand (59). The left brain's causal, abstractive, and binary operators make Symbolic sense of right-brain, Imaginary anxiety (in Lacan's terms) — by searching for sources and creating binary opposites to define them, especially regarding the endless potential dangers of the immediate environment and the long-term metaphysical concerns of mortality (Newberg, d'Aquili, and Rause 59–64). The binary operator helps us to "negotiate the environment confidently" by dividing space and time: above and below, inside and outside, before and after, etc. (64).[34] When the brain's "cognitive imperative, driven by some existential fear, directs the binary function to make sense of the metaphysical landscape, it obliges ... [with] pairs of irreconcilable opposites that become the key elements of myth: heaven and hell; good and evil; celebration and tragedy; birth, death, and rebirth; isolation and unity."

However, as Alper indicates in his critique of religion, this binary view of metaphysical forces has caused tremendous destruction between human groups, projecting evils upon an outside Other that must be fought and destroyed, through righteous crusades, pre-emptive strikes, and holy wars. The dominant melodramatic mode of violent drama in today's theatre — including film, television, and videogames — engages the binary operator in spectators' and players' brains, arousing limbic passions to identify with the threatened victim and the courageous hero in their conflict with clear-cut, evil villains. But tragedy, onstage or onscreen, is not just a binary opposite to celebration (contrary to the quotation above). Tragic or tragicomic theatre offers a complex involvement of personal associations and the cathartic integration of many brain areas in each spectator, unlike the simplistic, binary entertainment of good versus evil in melodrama or in superficial comedies and farces.[35] Likewise, religious traditions and mythic dramas, as well as modern extensions of godlike ideals, sometimes involve complex figures, not purely good and evil forces.

According to Newberg and d'Aquili, myths are created by the brain's causal and binary operators, while also involving existential and holistic neural areas. Myths frame environmental anxieties "as a pair of apparently irreconcilable opposites" (Newberg, d'Aquili, and Rause 62). Yet they also reconcile these opposites, "through the actions of gods or other spiritual powers, in a way that relieves our existential concerns." I would add to this an important distinction: myths may resolve binary opposites through simple or complex causes and effects. A melodramatic resolution prevails in most myths, from ancient to modern, with heroes beating the villains in the end, through divine intervention or providential confidence, plus self-asserting courage. This maintains a comforting separation between good and evil, while promoting moral heroism as more powerful than mortal threats and evil temptations. (As discussed above, religions often work to console believers' brains by simplifying higher-order awareness and anxiety toward causal and binary divinities.) But some dramas have tragic or tragicomic resolutions, even when transcendent ideals are involved, with the hero realizing a troubling paradox of evil errors in his own godlike goodness, not just in the villain's threats — and with suffering as a consequence of the hero's righteous violence, for victims on both sides.

"The causal operator is designed to promote survival, not necessarily to find the truth..." (Newberg, d'Aquili, and Rause 68). Likewise, I would argue that the left brain's binary operator may promote survival by oversimplifying right-brain intuitions and limbic-system anxieties: positioning them as good or evil, for or against the body, the self, and its group identifications. Belief in the benevolence of certain communal spirits, personal gods, or ancestral influences (against malevolent ones) can reassure the self that the transcendent Other will keep it whole in this life and beyond. This may combine left-brain, binary, causal and right-brain, holistic resolutions to answer the limbic's existential concerns. But this binary morality polarizes the numerous perceptions, memory associations, and emotional drives in various areas of the brain, as it abstracts them into Symbolic paradigms of good and evil — especially when those are shared by the group mind and propagated as its memes.

The development of imperialist cultures in the history of Christianity and Islam, inspired by an all-powerful God who will ultimately triumph over and judge the forces of evil, shows the confident yet destructive power of binary abstraction reforming holistic existential anxiety.[36] Primal panic, fear, rage, and seeking systems (Panksepp) are evoked in many limbic brains, collectively uniting with God or Allah (or with totem deities in polytheistic cultures) against the Other's devils. The cultural evolution from family ancestor worship to multiple gods and monotheism offers larger and larger groups an illusory answer to the fundamental alienation and mortal anxiety in being a big-brained human. Yet, collective union with God may also mean more control of individual brains by the group mind and increased defensiveness or pre-emptive strikes against others figured as evil.

Newberg and d'Aquili have focused their research on the brain's apparatus for mystical communion with the divine, as Absolute Universal Being, beyond such contending binary figures. In studies of Tibetan Buddhist meditators and Franciscan nuns at prayer, SPECT (single photon emission computed tomography) scans reveal changes in brain activity indicating that "mystical experience is biologically, observably, and scientifically real" (Newberg, d'Aquili, and Rause 7). Meditation and prayer increase activity in the "attention association area" of the frontal cortex, "the neurological seat of the will," which "allows the brain to screen out superfluous sensory input and concentrate upon a goal," as well as process and control emotions through ties to the limbic system (29–30). Meditation also decreases activity in the left side of the posterior superior (back upper) parietal lobe, or left "orientation association area," which creates "the brain's spatial sense of self, while the right side creates the physical space in which the self can exist" (4, 28). The left parietal lobe normally makes a binary "distinction between self and other," allowing the self to operate in space—both outside the body (with distinct objects or the emptiness around the body as other) and in the theatre of the brain, through the right parietal lobe's Imaginary representations of that physical space outside.[37] But in meditation, the intense focus of the frontal lobes on nothingness (by Buddhists) or on a religious icon (by Catholic nuns) triggers the left parietal lobe's decrease in activity, its "deafferentation," creating a sense of no self or of merging with the Other (117–23). Whether it is interpreted by the Buddhist subject as "melting into nothingness" or by the Catholic nun as "melting into Christ," the communion with AUB in each case involves the same brain mechanism in the left parietal lobe, after different approaches in the attention area of the frontal cortex (123). The left orientation area, crucial to the subjective sense of self, is not able to locate "the boundaries of the body" (119). The self becomes limitless and dissolves into "a timeless and spaceless void" or into "the transcendent reality of Jesus"—through the intense focus of the frontal cortex on the Other, as non-being or Savior (119, 122).

This mystical dissolution of self also involves two, counter-balancing branches of the autonomic nervous system and its regulation of the body through the brainstem, hypothalamus, and limbic system (Newberg, d'Aquili, and Rause 38–39; Solms and Turnbull 29, 124). The "sympathetic" (arousal) system enacts the body's fight-or-flight response through an adrenaline boost, faster heart rate, higher blood pressure, and increased muscle tone. The "parasympathetic" (quiescent) system conserves energy and keeps the body's basic functions in harmonic balance, by regulating sleep, inducing relaxation, promoting digestion, distributing vital nutrients, and governing the growth of cells.[38] Unitary states of communion with the Other can arise through either system, in a hyperquiescent or hyperaroused state (d'Aquili and Newberg, *Mystical* 25). But they become especially intense when hyperquiescence triggered by slow ritual (such as meditative chanting and prayer), or hyperarousal activated by rapid ritual (like Sufi dancing), reaches its "maximum capacity," producing a "spillover" effect that evokes the opposite system at the same time (25–26). This may even produce a combined hyperquiescent and hyperaroused state: "a complete breakdown of any discrete boundaries between objects, a sense of the absence of time, and the elimination of the self-other dichotomy ... related to the [Catholic] *unio mystica*, the perfect experience of the void or [Buddhist] Nirvana, or other absolute unitary states" (26).[39]

Newberg has also researched Pentecostal glossolalia or "speaking in tongues" (Newberg and Waldman 191–209). As in Catholic prayer and Buddhist meditation, there is increased activity in the thalamus, "involved in the transfer of sensory information from the world to different parts of the brain and the body ... [thus] making spiritual experiences feel real" (206). There is also increased activity in the temporal lobes, involving emotions, and in the midbrain, which regulates the autonomic nervous system, thus producing both a frenzied appearance externally and tranquil feelings internally, through arousal and quiescent "drives" (206–08). But unlike prayer and meditation, speaking in tongues shows a decrease in frontal lobe activity (with SPECT brain scans), perhaps because glossolalia is "incomplete speech," involving less accuracy (1). It also involves a slight increase in the brain's orientation area, unlike prayer and meditation, suggesting that "speakers in tongues do not lose their personal sense of self" (205). And yet, the deafferentation of the frontal lobe's attention area, through the believer's submission to God, along with the activation of the thalamus, temporal lobe, and midbrain, creates "a realistic sense that the Holy Spirit is communicating through the self" (209).

According to Newberg and d'Aquili, religious ecstasy is not completely removed from ordinary brain activity. Like prayer and meditation, relaxing in a warm bath, while listening to the slow, steady rhythms of soft music, may activate the quiescent system (and hippocampus)[40] to exert "a mild inhibition on neural flow, which would cause a slight deafferentation of the orientation area, and a mild unitary state," with the self dissolving into a pleasant serenity in the Other of water and music (Newberg, d'Aquili, and Rause 114). Both secular and religious activities with slow rhythms, like a warm bath, "reading a poem, rocking a baby, or praying," and faster repetitive actions, "such as distance running, having sex, or cheering along with a crowd of thousands at a football game," can drive the brain toward unitary states, through the different mechanisms of the quiescent and arousal systems (115). Thus, a "large spectrum of increasingly unitary states is possible," from secular to religious, physical to spiritual. "The arc of this continuum links the most profound experiences of the mystics with the smaller transcendent moments most of us experience every day...." I would add that theatre and cinema offer varying degrees of smaller transcendent or more profound mystical experiences, with the audience seated quiescently in a darkened chamber, yet aroused by the action onstage or onscreen — and its potential for unitary states in collective minds.

Such transcendent moments may produce a sense of oneness with all humanity and all of nature. But when reinterpreted by religious rituals and binary myths, they may involve warring cultural ideologies, competing for replication in the theatrical spaces of human brains. The God Experience of monotheism has been especially valuable in supporting the individual rights of many souls. Yet it has also destroyed lives through witch-hunts, inquisitions, and holy wars (Newberg, d'Aquili, and Rause 162–63). Newberg and d'Aquili theorize that religious intolerance is rooted "in the same transcendent experiences that foster belief in the absolute supremacy of personalized, partisan gods" (163). But such partisan God Experiences "rise out of incomplete states of neurobiological transcendence"— especially in monotheistic traditions of mystical union, in which the orientation area is not fully deafferented as in the Buddhist contemplation of nothingness (164). The image of Jesus or the name of Allah becomes all powerful in the communion of self and Other. Instead of realizing the transcendent unity of all religions in AUB (165), the ego may re-inflate through an incomplete mystical experience, via ritual belief in a personal, partisan God — with destructive results for those who believe otherwise. How did this mechanism for myth-making evolve, as both asset and liability, creating an inner theatre of mystical communion, yet producing partisan holy wars through binary morality and angry personal gods?

Newberg and d'Aquili speculate that myth-making may have started with *Australopithecus*, our apelike biped ancestor of several million years ago. Their skulls show a parietal area that could perform "rudimentary conceptualizations, which would enable thinking in terms of opposites, and understanding the concept of cause," but not enough to enable language (Newberg, d'Aquili, and Rause 65). Therefore, *Australopithecus* may have felt existential dread and groped toward binary causes and resolutions, constructing "a personal mythology based entirely on abstract, nonverbal symbols"— although they say this is only a remote possibility (66). They find more evidence for myth-making with the advent of neural structures for language and speech, as well as causal and binary thinking, in *Homo erectus* (and then archaic *Homo sapiens*) several hundred thousand years ago.[41] This revised timeline corresponds to the mythic stage in Merlin Donald's theory, mentioned above.

Newberg and d'Aquili present hypothetical scenarios for the development of mythic beliefs in the hunter-gatherer societies of these early hominids and their evolving brains. The neurologists describe a scene where the chief of a tribe, brooding over the dead body of one of its members, seeks a cause through his (or her) logical left hemisphere (Newberg, d'Aquili, and Rause 70–71). The amygdala interprets this distress by triggering the limbic fear response of existential dread, activating the arousal system, which intensifies as the chief ruminates beside the corpse. Perhaps he sees the flames of a fire also die out, with wisps of smoke rising upward, and his right brain associates the image of the rising smoke with the spirit of the fallen clan member.[42] This right-brain solution resonates with the left-brain binary, cause-and-effect notion of life after death — in the image and idea of an ascending spirit — thus sending "positive neural discharges racing through the limbic system, to stimulate pleasure centers in the hypothalamus," the master controller of the autonomic nervous system, at the upper end of the brainstem (43, 72). The hypothalamus then triggers the quiescent system, which the chief experiences as "a powerful surge of calmness and peace" (72). This occurs through the spillover of the arousal system and a correspondence between limbic Real, right-brain Imaginary, and left-brain Symbolic orders (adding Lacan's terms). With both the arousal and quiescent systems simultaneously active, the chief feels fear and rapture, ecstasy and awe — as if freed from his own bonds of mortality, through a mythic resolution of the life and death binary, believing that something is eternal in the dead clan member and himself, like the rising wisp of smoke.

Eventually, this personal myth may have become a communal belief, through the theatres in other clan members' brains and their perceptions of ghosts as wisps of smoke near dead bodies, plus further images and ideas. Such collective beliefs and experiences of ghosts might have been triggered by the savant-like esoteric talents of an individual brain in the tribe, with mutant expansions of its temporal lobe revealing sensed presences (as Ramachandran and Persinger suggest). But Newberg and d'Aquili argue that the basic "neurological machinery of transcendence" could have evolved for mating and sexual experience, prior to myth-making and shamanic visions (Newberg, d'Aquili, and Rause 125).[43] The same neural "structures and pathways involved in transcendent experiences — including the arousal, quiescent, and limbic systems — evolved primarily to link sexual climax to the powerful sensations of orgasm ... activated by repetitive, rhythmic stimulation." However, sexual bliss (or Lacanian *jouissance*) is generated by the hypothalamus, a primitive structure connecting the limbic system and brainstem, whereas transcendent experiences leading to intense unitary states "likely depend upon the involvement of higher cognitive structures." Even if mystical experience is "an accidental by-product" of hominid evolution, Newberg and d'Aquili argue that it may have connected humans with the ultimate truth of God or AUB, because "evolution stumbled in the right direction," like the earthbound development of birds' feathers for warmth, which then became useful for flying (126).[44] They insist that their scientific research on meditation and prayer "supports the possibility that a mind can exist without ego, that awareness can exist without self." Thus, they find evidence not just for temporal-lobe transmitters of God experiences, but also for the higher-order consciousness of self-less and non-divisive divinity in "the neurological substance of Absolute Unitary Being."

Newberg and d'Aquili assert that human brains do not merely "invent a powerful God ... to gain the feeling of control" (Newberg, d'Aquili, and Rause 133). They posit instead that God exists as AUB because such Being is "*experienced* in mystical spirituality." Yet, they recognize that the human brain evolved its transcendent mystical machinery (and apprehension of a divine theatre) due to evolutionary survival values. Unlike animals, early humans were burdened by the awareness of death, which "could have sent them spiraling into depression and apathy, leading to an early and unhappy end to the human saga" (132). The "promises of religion," including benevolent gods and transcendent souls, "protected early humans from such self-defeating fatalism" (133). Along with such benevolent gods, I would add, religion gave early humans a binary, causal, and holistic focus to anxieties about illness and death with the metaphysical threat of evil and vengeful gods who might be appeased or avoided. Yet, more complex, tragicomic gods also evolved in various cultures and time periods, as will be considered later in this book. It remains to be seen how the current clash of cultures, religions, alternative spiritualities, and materialist yet virtual media will destroy, restrict, or creatively improve our world. Indeed, one's view of that depends on the success of certain ideologies — defining the "good life" in relation to God or the gods, or various secular ideals, as the cosmic or collective Other watching and directing it all (or not).

Unlike Albright and Ashbrook, Newberg and d'Aquili do not argue that the brain's anatomy shows distinctive aspects of one God. But they view the evolution of the human brain toward its higher-order consciousness of divinity and its potential mystical union with AUB as a natural teleology, not as nature's "white lie" (Alper).[45] They hypothesize that religion — and thus, I would add, an early form of ritual theatre — originated with "spontaneous mystical occurrences" in prehistoric hunter-gatherer societies (Newberg, d'Aquili, and Rause 133–34).[46] A deer hunter, for example, might have focused on the image of his prey, especially during long periods of no success and clan famine, with long hours in the wilderness and little sleep, scanning the horizon for signs of game and picturing in the theatre of his mind "a magnificent stag, an animal large enough to feed the entire clan, and save all his family and friends from starvation" (134). That meme of the Great Stag, growing more powerful in the hunter's brain, through his genetic survival needs, might become a spectral figure that he sees in his environment, through the Real of his limbic drives and the Imaginary, holistic intuitions of his right brain, even when it is not physically there. The vision of the Stag could become a Symbolic "mantra," a repetitive thought focusing the hunter's frontal cortex, like "the contemplative techniques of religious mystics" in later human history, and triggering "the same biological chain of events" in his brain. Even more like the "active" ritual practices of whirling Islamic Sufis, shamanic dancers, or Christian evangelicals (in contrast to the "passive" meditation of quiet prayer), the prehistoric hunter's arousal system might spill over to create simultaneous hyperquiescence and hyperarousal, through his limbic seeking, right-brain spectral vision, and left-brain survival ideal of the Stag mantra.

This could have produced "a similar unitary state" to what mystics experience today, with the sensed presence of "a powerful, primal deity — one of the great animal spirits that was among humanity's first gods" (Newberg, d'Aquili, and Rause 134–35). The hope and joy in that god experience might sustain the clan longer than otherwise. If the hunter then found his prey, bringing home food to save the clan, as "a gift from the Great Stag," the mythical beliefs and totem rites of that cult would be born. This prehistoric form of theatre would engage the community's right and left brains to figure out the desires of the Great Stag as causes for the nurturing and threatening forces in their environment (136).

Cranial and Cave Theatres

Evidence for such mystical experiences may be found in prehistoric cave art, showing both an evolving theatre within the brain and the development of mimetic and mythic aspects of theoretic culture through shamanic ritual performances. 32,000 to 10,000 years ago, our Ice Age ancestors left paintings and engravings on the walls of many caves, mostly in southern France and northern Spain, but also at a few sites in Portugal, Italy, and England (Bahn).[47] A common notion today about this cave art comes from the theories of Henri Breuil in the 1950s. He argued that art began with animal masks, used by prehistoric humans as disguises while hunting and then, with such power, in ceremonies to gain further spiritual power over the desired prey (Breuil 21; Curtis 76–77). He viewed the cave art as made for a similar reason: big game hunters gaining confidence in their pursuit of rhinoceroses, mammoths, wild horses, bears, lions, and great stags. Later researchers found, however, that this was not the prey our ancestors were normally hunting—based on bones they left at communal sites (Lewis-Williams, *Mind* 258). Also, the animals painted on cave walls are always depicted as floating in space, sometimes layered over one another, not in natural settings with background scenery (120, 194–95). Only 5 to 15 percent of the cave figures have wounds, spears, or arrows; more would be expected with Breuil's hunting-magic thesis (Curtis 146; Lewis-Williams, *Mind* 47).

Another theory, proposed by Max Raphael in the 1940s, and developed later by Annette Laming-Emperaire, was that each cave showed the dominant totem animal of a certain clan—since a certain type of animal seems to be prevalent in each cave (Lewis-Williams, *Mind* 54, 57; Curtis 146–47).[48] Animals painted over others might show a new clan and its totem taking power over another. Ritual objects were also found in certain caves, left by people in the same time period as the art work on the walls. For example, bone fragments, bear teeth, and seashells (transported many miles from the coast) were left in rock crevices and stuck into the floor, apparently as sacrificial offerings (Lewis-Williams, *Mind* 246; Curtis 175–78). In one cave, Chauvet, a bear skull was left upon a burned spot on a rock, perhaps as a kind of altar, along with 45 other bear skulls set around it, in a large chamber with a low floor that gradually rises toward the walls, "almost in tiers ... like an amphitheater with the rock and skull at center stage" (Curtis 210–11).

In the 1990s, Jean Clottes and David Lewis-Williams compared Paleolithic images with more recent cave paintings made by San bushmen in southern Africa, using native testimony (transcribed by Europeans in the late 1800s) of how the latter were used for shamanic rites.[49] Clottes and Lewis-Williams relate three types of cave art—geometric, animal, and animal-human hybrid images—to current neurological research on the three stages of hallucinatory experience: shifting geometric forms, emotional objects, and (via a vortex experience) complex surreal visions, when an "altered state of consciousness" is induced in the human brain by drugs, isolation, sensory deprivation, or rhythmic sounds and dancing (12–17). They find representations of all three visionary stages, plus the transitional vortex, in both the San and prehistoric cave art. They theorize that Paleolithic humans, tens of thousands of years ago, like the San bushmen more recently, used cave art to record and perhaps re-trigger the visionary quests of shamans in trance, who traveled into a spiritual world and met animal-spirit guides, for knowledge and healing powers (92).[50] The cave walls and the images made on them where thus a performative membrane to another world. "In a sense, the complex San painted panels were like a stage set awaiting the shamanic actors whose altered states of consciousness would take up images of past visions, vivify them, and, by the addition of new insights, enhance the richness of the shamanic cosmos" (35).[51]

Recently, Yann-Pierre Montelle has emphasized the performance aspects of initiation rites that probably occurred in Paleolithic caves. Such rites, as "pure theater," involved potential observers, framed and restored behavioral patterns, a physically strenuous "ordeal" for and psychological effects upon the initiate, and thus the communication of esoteric knowledge ("Paleoperformance" 134–36). This prehistoric form of environmental theatre used specific ritual spaces within the caves,[52] for certain sights, sounds, and staging effects, involving music, singing, and dancing, plus spectacular art on the walls—for both communal and alienating experiences. Drawing from Van Gennep's historical anthropology of initiation rites, Montelle argues that the prehistoric cave artists selected and created theatrical spaces within the caves in order for initiates to experience "separation" from everyday life, entry into a secretive, mythic, liminal realm for special cultic knowledge, and then a reintegration into the larger social structure (136). Montelle also defines certain emotional states and consequent effects for actors in this ritual experience: anxiety through their seclusion, subordination to authority through their physical and psychological ordeals, revelations and deception tactics in the secrecy of the cult membership, and special skills and status via the esoteric knowledge (137). He recalls the prior research of John Pfeiffer in describing the experience of cave art as a performance space around the actor or explorer—involving an obstacle course, shocking displays, and tricks to imprint information.[53] Such a rite, according to current studies of shamanic initiation, becomes a "symbolic death," through the seclusion, humiliating ordeals, and "fear of abandonment by the mother" that leads to "psychotic manifestations" (Montelle, "Paleoperformance" 137). Thus, the shamanic "initiators" act out "terrorizing behavior and brutal attacks" upon the initiate, with others in the cult observing and authorizing the rite, like spectators and directors, through the "mise-en-scène of the initiation," which exploits the psychology of separation anxiety and liminal presence/absence.

Montelle finds evidence for such performances in the architectonic chambers chosen for cave art and the musical instruments created by Ice Age

(Cro-Magnon) humans of the same period, as well as analogous rites of hunter-gatherer cultures today. In arguing that the "use of the caves was primarily initiatory," he points to the "scopic and sonic values" in cave chamber selection and thus the development of "pedagogical routes" for initiates to experience "specific psychological stages" and "levels of information" ("Paleoperformance" 138–39). He also specifies the Paleolithic instruments that might have been used: percussive flint blades (struck together), bone-made whistles, grooved bones as scrapers that would "reconfigure the [cave] soundscape into an anxiety-producing environment," and swinging bullroarers whose "low-frequency sounds," like those of bison and aurochs (Ice Age cattle), probably made the "zoomorphic" cave art "persist in the initiate's memory" (141–47). For Australian aborigines today, the bullroarer with its eerie sound "embodies a demonic figure," which carries boys away from their mothers during initiation rites—although the sound's production by the bullroarer is eventually revealed to the boys, while being kept secret from the women and children (146). It is also associated with rumbling thunder and roaring winds, as well as the male genitals.

According to Montelle, the Paleolithic caves were "cultural reservoirs" for containing and communicating such elite knowledge through ritual performances ("Paleoperformance" 139). I would add that they show, in Donald's terms,[54] an early culmination of the various levels of human cultural evolution: episodic, mimetic, mythic, and theoretic—involving the Real, Imaginary, Symbolic, and Symptomatic orders of Lacanian psychoanalytic theory and corresponding areas of the brain. Prehistoric cave experiences manipulated the brainstem reflex instincts and limbic emotions of our ancestors, plus their right-hemisphere intuitive, holistic, visuospatial, Imaginary perceptions, and their left-hemisphere, rational, linear, audioverbal, Symbolic concepts. It also involved a sacrificial dimension for ritual actors and spectators through the collective, Symptomatic shaping of such inner theatres, with transformed signs of cave walls and musical instruments made of bone or rock, as well as other materials that did not survive. As initiates progressed into deeper and more secretive areas of the cave, they passed through liminal stages of Imaginary and Symbolic death, engaging the Real of otherworldly visions and altered states of consciousness—in isolation, sensory deprivation, rhythmic song and dance, and perhaps drug-induced trances (according to Clottes and Lewis-Williams).

If Montelle is correct that the caves induced "fear of abandonment" and other terrors in initiates, then their neural pathways for panic and fear, centered in the anterior cingulate gyrus and amygdala (Panksepp), were activated by the experience of selective natural chambers as sacrificial theatres. These involved painted rock formations as cultural signs, along with sound effects and music as limbic system conductors (like in cinema today). The panic or "separation-distress" system might have also been triggered. This system in the mammalian brain extends from the anterior cingulate gyrus (an organ near the center of the brain with two horn-like curls) to various areas involving sexual and maternal behavior. It is "intimately connected with *social bonding* and with the process of *parenting*" (Solms and Turnbull 129–30). Experiments with infant animals show that stimulation of this system produces distress vocalizations and separation calls, plus seeking behaviors (involving the brain's seeking system as well). However, after a period of aroused distress and seeking, the animal shifts to quiescent withdrawal — exhibiting symptoms akin to human depression. This may increase its evolutionary chances of survival, with the initial possibility of finding its mother and then the greater priority of hiding from potential predators (Kandel 514; Solms and Turnbull 131). Thus, while Montelle emphasizes male initiation rites, with men and boys as directors and actors in Paleolithic cave performances ("Paleoperformance" 138), women were also involved, at least as the maternal and communal origins from which initiates were separated, while being put through a ritual ordeal involving deep recesses in the womb-like earth.[55] The panic of such isolation, with the seeking of new identity and meaning in rock art and cave sounds, through submission to the terrifying rite, probably involved a reshaping of crucial left, right, and limbic brain areas — in the Paleolithic performance of Symbolic death, through an Imaginary realm of animal-human creations, producing a Real sense of rebirth.

In my own experience of prehistoric art in spring 2009, visiting 10 caves in southern France (plus the reproduction of Lascaux), I was impressed by the variety of spaces, images, and techniques, suggesting various theatrical practices. Font-de-Gaume has a narrow (vulva-shaped) opening and passages only wide enough for one or two people with paintings along the walls going up five meters. Scaffolding must have been used to paint and perhaps observe such art. The images, from 15,000 years ago, are in several colors: black, red, and yellow (from charcoal or manganese dioxide and ochre). Tubes made of bone were found in the cave, used for blowing paint, after the ochre was ground and mixed with water. The artists' hands created definite lines and degrees of shading for animal figures, or negative handprints, with paint blown around them. Remnants of rock sticks were also found, used for sketching images. Brushes, made from animal hair or vegetation, were probably used to paint the figures as well. Most depict bison, but also many mammoths, reindeer, and a few predators. Some images are multicolored with detailed shading for three-dimensionality and hair-like texture. Some also use the natural shapes of the rock wall. Animals are painted in a floating line (with no surrounding scenery), as if in parade, like humans walking through the cave passage. One bison, overlapping with others in painted perspective, turns back toward the viewer, at a carefully-crafted three-quarter angle. In another scene, a female reindeer (with shorter antlers) kneels before a standing male, who stands over her and licks her forehead, between her antlers. Such artful details, created before the Agricultural Revolution, display a dawning awareness of human skills in observing, emulating, and transcending the powers of animals and their natural environment. A key step toward domesticating certain species may have been the capturing of animals through art, along with rituals and myths using such images to create a theatre in the caves (and in prehistoric minds). This changed humans' relations with animals and with their own animal natures.

When the tour guide made a flickering light across the cave images, with her flashlight, mimicking a prehistoric torch or oil lamp, the animal images came even more to life, as if in a movie. Such dynamic animal figures are even more impressive in Lascaux II (the reproduction of Lascaux, now closed to the public) with larger chambers and overlapping images on the ceiling as well as the side walls. In one chamber, known now as the Hall of the Bulls, aurochs parade together overhead and, from within another bull with a slash on its leg, a bear appears. Elsewhere, there is a strange human figure, outlined in black with arms and legs, an erect penis, and a bird-like head. It lies between a bird outlined on a staff and a fully depicted bison, hair raised and bowels drooping outside its belly, with its head and horns pointed down toward the human, as if charging it, plus spear shafts nearby.[56] These and other dramatic figures were probably drawn from both natural experiences and supernatural visions — thus depicting abstract scenes and characters from the inner brain theatre, made collective through the painters' skills and other performances, perhaps with song and dance, in the echoing chambers and their womblike spaces.

In the Rouffignac cave, visitors today travel on an electric train because the charcoal sketches and etchings of mammoths, rhinoceroses, ibexes, horses, and other animals were made so deep inside the earth, ranging over several kilometers (Clottes, *Cave* 224). The cave floor was different in prehistoric times, too, just a meter or so from the ceiling, so that our ancestors must have crouched and crawled for a long distance to reach areas where they made the art and then showed it to others, perhaps involving initiates in painful shamanic ordeals and rites of passage. Here, as in other caves, the bulky herbivores are valued. Mammoths have large tusks, foreheads, and torsos, yet narrow legs. Aurochs, rhinos, and ibexes are also outlined with long horns, large bodies, and just a trace of limbs. While such images are few along the walls, sometimes in parade, they lead to a wider chamber even deeper inside the cave, where outlines of 65 animals overlap, with a large horse figure centered on the ceiling (229). Such "horsepower" along with other impressive figures, sketched and then viewed by prehistoric humans while on their backs, after a long and arduous passage into the earth, probably embodied good and evil, nurturing and threatening forces in nature. These spirits may have also been evoked by natural rock formations in the dim light and low oxygen level of the cave. Rouffignac has many strange, orange-colored flint nodules, as well as flat surfaces where the art is drawn or cut into the rock.

The Combarelles cave has a low ceiling forcing visitors to crouch in a narrow passageway and delicate, difficult to discern figures on the walls, involving very subtle, overlapping details, sometimes using the rock surfaces. One horse figure curves around a turn in the passageway. The modern tour guide becomes like a shamanic teacher, introducing initiates to the mythic figures (with a red laser pointer). Even then, they are hard to see. But once an image becomes visible, it is burned into memory, with certain neurons in the brain's temporal lobes devoted, for the viewer's entire life, to that holistic picture, traced out of many lines and surface edges.[57] Perhaps the theatrical rites in this cave were, in prehistoric times, a test for both body and brain: to endure the darkness, minimal air, and cramped passages, while glimpsing multiple layers of mythic figures on and through the walls. Combarelles also has human figures, a rarity in prehistoric cave art: two females outlined with hanging breasts, as they bend over, headless, also a handprint and near it a phallic sign, combining a rectangle and triangle.

A similar phallic sign appears in Bara-Bahau, with etchings in the rock from around 14,000 years ago (Bahn 74) and shapes that are very difficult to discern. After a tight crevice as the original opening, this cave has a wide domed cavern, with low chalk ceiling and outlines of horses, aurochs, and bears, plus a hand involving parallel lines to indicate fingers, perhaps using actual bear scratches on the rock. Such scratches are from bears that hibernated in the cave, in prehistoric times, and then woke and sharpened their claws. Bear teeth and bones have also been found. At the Saint-Cirq cave, there are just a few images, etched and drawn, but they include a so-called "sorcerer," who has narrow legs and a curved belly like the horses depicted nearby, perhaps as a hybrid figure or pregnant male, for it has a round head and erect penis.

Possible human figures can also be found at Cougnac, which has fantastic natural formations as well (numerous needle-like stalactites and ornate stalagmite pillars). Its paintings include mysterious "aviform" symbols, shaped like an abstract bird with square head and rectangular wings. These are also known today as "Placard-type signs," found in other caves such as Pech Merle and associated with "killed or wounded human figures" (Clottes, *Cave* 132). In Cougnac, two human-like figures appear in a wide chamber with animal images, sketched in red or black. One of the humanoid forms is inside the belly of a megaloceros deer; another within the outline of a mammoth's head. They could be animals, too, but appear to be two-legged creatures in bare outlines: torso, legs, perhaps arms, no head (or one may have an animal head), and yet with spears attached or power lines radiating from inside the image. There is also evidence of prehistoric hands, dipped in ochre and charcoal, having rubbed the rock surface around some of the painted figures (Lorblanchet and Jach 21; Clottes, *Cave* 82). Carbon 14 tests show that some of the art dates from about 14,000 years ago, but other figures, including the red megaloceros, were made as long as 25,000 years ago (Bahn 92). Thus, the animal and hybrid forms, including possible shamanic figures, were made and probably used, along with the natural scenery of the cave, for various types of performances, with changing rituals, myths, and paintings over a 10,000 year period — longer than recorded human history.

Pech Merle's art also dates from that earlier period, 25,000 years ago. There are again many, impressive, natural formations in this cave, with big caverns and flat floor areas where groups may have gathered for performances, involving art on the walls as visionary scenery or characters. Paleolithic footprints, left in the clay floor, offer evidence that children were taken into this cave (Clottes, "Sticking" 196–98). The mysterious aviform symbol appears here, too, as in Cougnac. Just below one of these signs is the red outline of a human-like figure with a large, round, yet beaked head, slender body with long legs, short arms or perhaps outstretched wings, and eight spear lines (Clottes, *Cave* 132). A bear head is etched clearly on another wall, with bear scratches also found elsewhere. In other places, distinctive collages were made with outlines of many horses, mammoths, and bison, etched first and then drawn with charcoal. There is the subtle outline of a human female on one part of the ceiling, with large buttocks, hanging breasts, and balloon-like head, also involving many other interwoven figures. They were produced by "finger fluting," with the artist's fingertips drawing in the soft cave surface, while standing on a nearby boulder (84). The most unique depiction in this cave is of two spotted horses in black, using a natural edge of rock for one horse's head, yet with dots at that edge, surrounding the outline of the horse, along with dots inside the body, as if to indicate its supernatural aura. Six human handprints (negatives made by placing a hand on the wall, while blowing paint over it and the rock) are situated around the horse images. A long fish figure in red also lies within the black outline of one of the horses, the one using the rock shape for its head and dots.

Such large chambers, as possible prehistoric theatre spaces with distinctive painted icons, can also be found in the Niaux cave, farther south in the Pyrenees, near the Spanish border. Halfway up a mountain, with views of

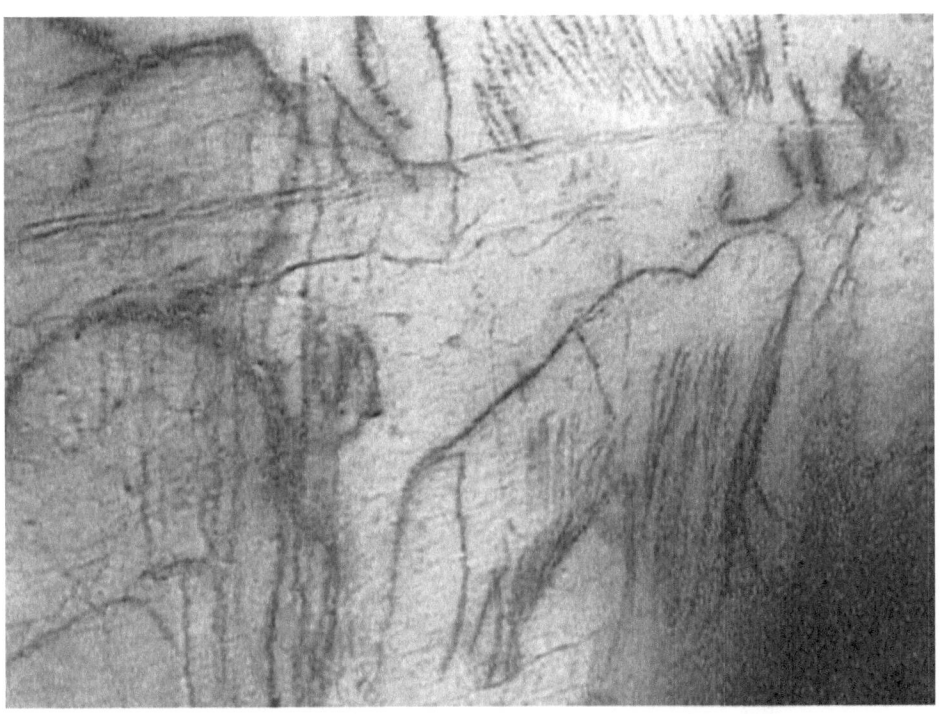

Top: Horse and woolly mammoth figures (with many hair lines) overlap on the cave wall in Pech Merle. *Bottom:* Overlapping horse and bison figures also appear in the Pech Merle cave.

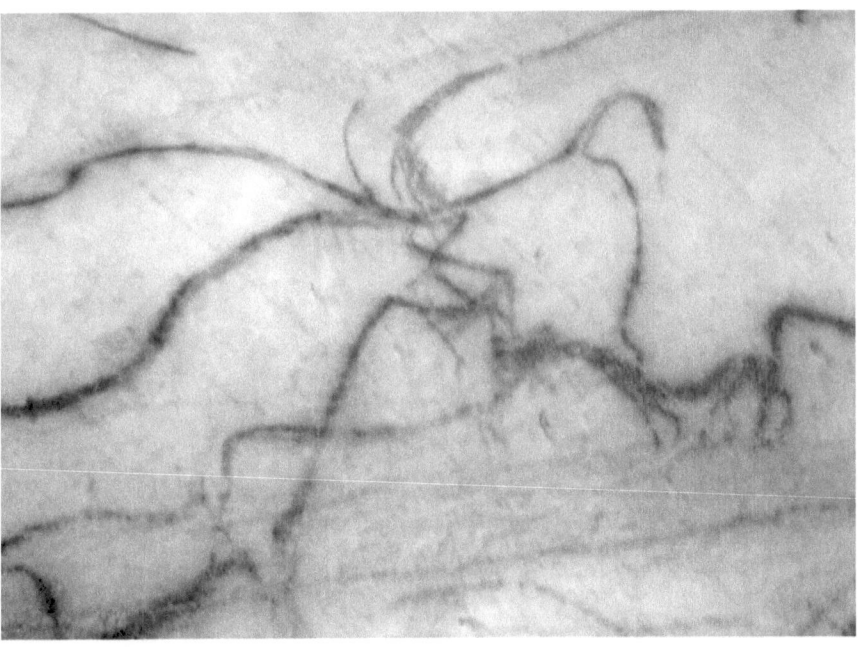

other peaks across a broad valley (and of the La Vache cave, a prehistoric habitation site), the mouth of this cave leads to a large initial cavern, with pools of water and drips from above, along with stalactite needles and stalagmite columns. Processions by torch and lamp light (using animal fat as fuel for the lamps) would have been beautiful and yet frightening in such a space, with shadows on the walls, reflections in the water, and slippery steps.[58] There are also several big columns of fountain-like rock. Even today, visitors must go up and down slopes, through two tight crevices (where a single body has to turn sideways), and then up a steep, sandy hill to a series of black and red, parallel bars and fingertip spots, painted on a boulder, like a map for the rest of the site. In one direction from this point is a cavern with mostly red images, including a rare weasel. In another direction lies the famous "Salon Noir" with black figures. The way to the latter leads up another hill to a huge dome, 20 meters wide, with a mostly flat floor, high ceiling, and acoustics that amplify voices and sounds (Clottes, *Cave* 184) — ideal for large group performances.

There are several groups of animals outlined on the far wall, including a bison apparently leaping over an ibex.[59] Most are bison, but some are

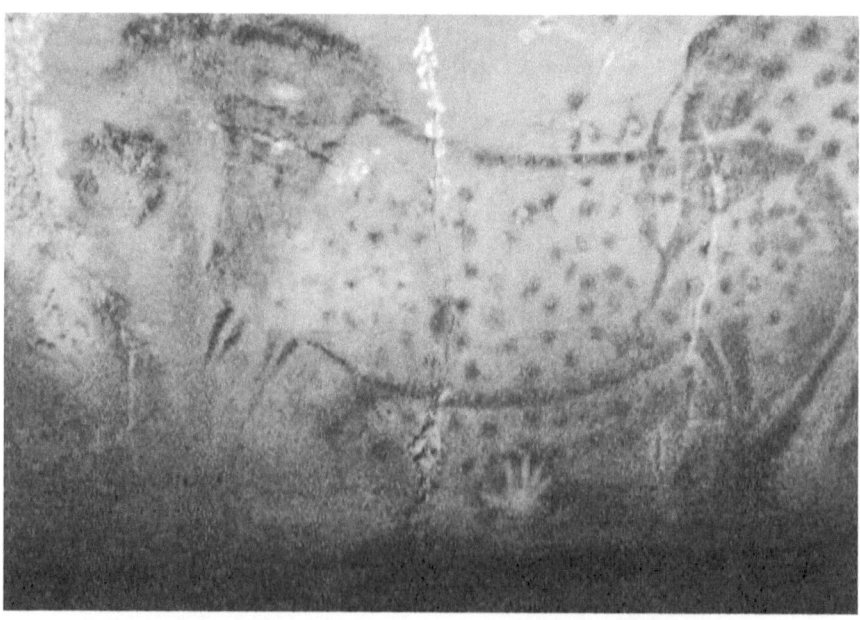

Handprints surround a spotted horse in Pech Merle, showing how prehistoric humans communed with the cave wall and its visionary figures, painting around or over their hands while touching it, perhaps making their hands disappear into the wall.

In the "Salon Noir" of the Niaux cave in the Pyrenees Mountains, the full outline of a bison appears with spear or power lines attached to it and other animals floating around it. There is ample room in this cave chamber for human performers and audience to have interacted with such images.

horses and mammoths. One bison head looks forward and back, like a modern cubist painting. Another looks toward the viewer — perhaps as a supernatural creature watching the humans perform. Some of the bison and horses have hair shading or full legs, even hooves. Yet, in prehistoric times, few horses, bison, or mammoths were seen in this mountainous area, for they lived in lower valleys. Ibexes and reindeer were hunted instead, but none of those are painted in the Niaux cave, where art was made 13,000 years ago. Thus, prehistoric artists were shifting their rarely observed subjects from natural contexts to others, using their brain's inner theatre, possibly through an altered state of consciousness, to create new worlds of representations and performances in the cave. This was a distinctive step from humans as an animal species interacting with others, through nature's survival and reproduction priorities, to humans as godlike creators, perhaps emulating iconic powers that they glimpsed in supernatural animal forms, as spirit helpers or challengers.

Through this evolving cultural power, transcending natural instincts and

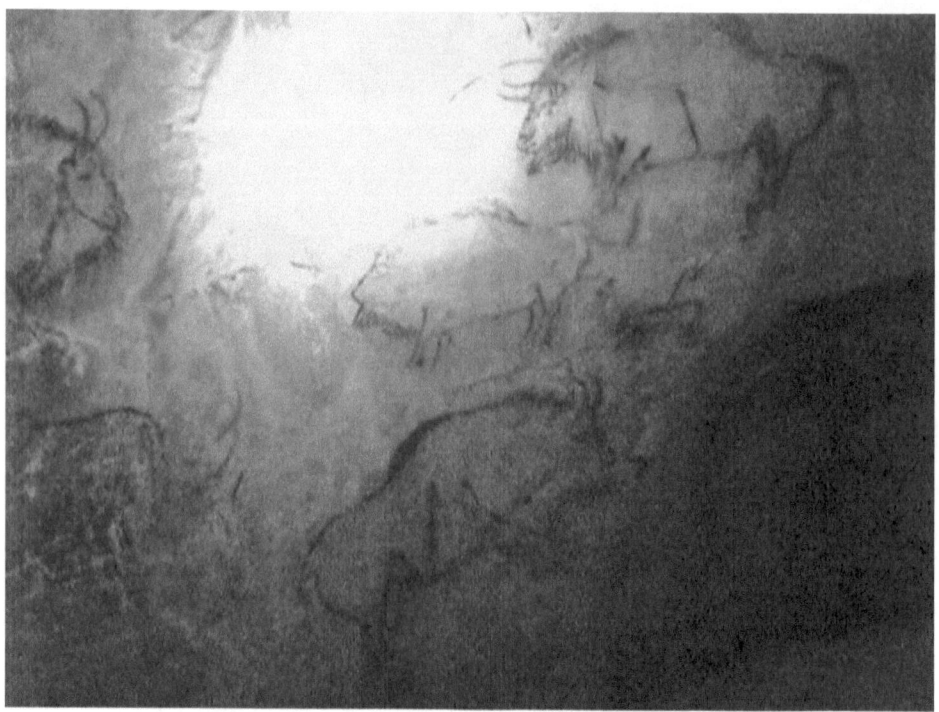

With the brightness of a modern flashlight, various animals are seen in the Niaux cave. The bison at the top right is turning its head toward the viewer, suggesting a prehistoric sense of three-dimensionality and interactivity, through the cave wall.

behavior patterns, came a distinctive tragic flaw: the human alienation of being, with fragile egos needing a purpose to existence, a greater meaning to life and death. Much larger human groups developed in later "civilizations," through further arts and technologies, building cities and monuments, for the living and the dead. But the beginning of our alienation from nature and the transforming of the animal within us, through the cultural remaking of our inner theatre and outer environment, can be seen in caves such as Niaux.

The nearby Grotte de la Bedeilhac also provides evidence of prehistoric humans selecting a large chamber for paintings and performances. In this cave, the artwork is reached only after a long trek (or procession) deep into the mountainside, past huge natural columns.[60] There are surfaces suitable for painting near to the mouth of the cave, yet none appears until an area that must have been specially selected, perhaps to create an arduous journey with theatrical effects along the way. In the final chamber, bison are both drawn and sculpted. There is also a positive handprint (made by putting paint on the hand and then applying it to the rock surface) with the little finger missing. And there are small clay sculptures of bison, and of phallic and vulvar symbols, some in the low ceiling edges of the large chamber.

Abri du Cap Blanc (near Font de Gaume and Combarelles) shows a different kind of artistry, in a cliff-side shelter used for habitation, unlike the deep caves.[61] A frieze of several almost life-size horses, overlapping with a sense of perspective, was sculpted there into the rock surface 15,000 years ago. Modern artists using the same tools have taken 10 hours to reproduce just one of the horse's heads. A skeleton of a woman from 13,800 years ago was also discovered there, plus evidence of further habitation 10,000 years ago. This art shows a skill and devotion quite distinct from the more quickly made cave paintings, etchings, and clay sculptures. Yet all these techniques, in various performative spaces, indicate a collective concern in prehistoric brains with the Otherness of animals in nature, and of humans evolving beyond animals. Early humans, as evidenced by the cave art of Europe, feared, but also aspired to the godlike power of a transcendent Other (or Others), as supernatural creator, actor, and spectator, who might give a greater meaning to life, beyond the human awareness of mortality. But even today, a visitor to such artwork, enclosed by cave walls and the earth's darkness, dependent on a shaman-like guide for safety and meaningful designs, may experience a ritual ordeal with a glimpse of transcendent power—and a sense of resurrection when emerging again into the light of day, with prehistoric patterns fixed in the brain's temporal lobes.

Rock Wall as Projection Screen

Lewis-Williams's most recent book, while focusing on the cave art, also considers the various types of performance spaces and possible musical instruments used, to theorize the entire theatrical experience. Along with flutes and bullroarers, as possible Paleolithic instruments, he adds drums, rattles, "stalactites [that] emit a deep booming sound when struck," and the human brain itself—with shamans in trance hearing "animal spirits ... speak to them ... [through] neurologically induced aural sensations" (*Mind* 223–24). He also invokes Plato's parable of the cave (in *The Republic*), with the prisoners chained in position to see shadows on the wall as their only reality. And yet, Lewis-Williams finds that the light creating such shadows, in the Paleolithic cave, streams through the inner theatres of visionary brains—mingling with "the shadows of the external objects" to create an alternate consciousness and spiritual communion, in the cave spaces and rock membranes (204). He recalls Steven Mithen's notion of cognitive fluidity between different types of hominid intelligence, as a radically new way that prehistoric humans began to recon-

A frieze of almost life-size, overlapping horses, sculpted into the rock shelter at Abri du Cap Blanc 15,000 years ago.

ceive the world (107–11). Thus, Lewis-Williams draws on archeological research, ethnographic analogies, philosophical parallels, and evolutionary psychology. Yet he also uses current neurology to explore the various performance spaces of Paleolithic caves, as diverse ritual theatres, while considering the brain's potential for god experiences in such environments.

As in his earlier collaboration with Clottes, Lewis-Williams argues in his new book that we share with Paleolithic peoples a "spectrum of consciousness," which is "wired" into the brain, but whose "content is mostly cultural" (*Mind* 126). The "normal trajectory" of consciousness may move from outward-directed attentiveness to inward-directed daydreaming or to dreaming and unconscious states while asleep (123–25). Yet, there is also an "intensified trajectory" that moves from waking states to "altered states"—involving entopic phenomena (internal imagery) and hallucinations. This may occur, as current research shows, through the quiescent, meditative means of sensory deprivation, which produces internal fantasies and "stimulus hunger," or through arousal with drumming, flashing lights, and rhythmic dancing (124). Fatigue, pain, and fasting may also contribute, as in shamanic rites, along with psychotropic drugs or "pathological states, such as schizophrenia and temporal lobe epilepsy."

Typically, the entopic phenomena come in three stages. First, there are abstract, geometrical dots, lines, and shapes, which "flicker, scintillate, expand, contract, and combine with one another" (Lewis-Williams, *Mind* 126). These brain-generated images or black holes in vision may be experienced with the eyes open, as well as closed, thus "projected onto ... perceptions of the environment." The eventual hallucinations may involve other senses, too, experienced "in the body" and through hearing, smell, and taste (127). In the second stage, the subject's brain tries to make sense of such hallucinatory perceptions by construing them as iconic forms from everyday life (127–28). With Paleolithic cave artists, the hallucinations involved geometric shapes from Stage 1 and various animal figures in Stage 2. Subjects today might experience something more familiar in the second stage, depending on personal disposition (or in Lacanian terms, the subject's unconscious signifiers producing symptomatic imagery). As subjects move to Stage 3, they often experience "a swirling vortex or rotating tunnel that seems to surround them and draw them into its depths" (128–29). Shamans in various cultures today speak of traveling to the spirit world through such a hole (129). Lewis-Williams sees tunnels in exemplary Paleolithic caves, such as Lascaux, as providing a physical experience that parallels the psychic vortex, though probably not simultaneously, since the passage would have been too difficult to negotiate while in trance (252). In the third stage, subjects experience "iconic images" derived from memory, often associated with powerful emotions (129). Through "fragmentation and integration, compound images are formed" (130). Not only that, but "subjects enter into and participate in their own imagery," becoming part of the hallucinatory realm, sometimes, even today, "turning into animals." I would add that Newberg and d'Aquili's research on Catholic prayer and Buddhist meditation suggests that a shift of brain activity from the left to right parietal lobes (and orientation areas) is probably involved—in the hallucinatory or mystical experience of self merging with the Other—as well as changes in the frontal-lobe attention areas, through the "intensified trajectory" of altered consciousness.

Such visionary experiences of communing with or becoming an animal-human hybrid help to make sense of the strange figures found in prehistoric cave art, such as those I saw and mentioned above. Lewis-Williams gives similar examples: the deer with human feet, hands, face, and beard in Les Trois Frères (looking down from near the ceiling), the bison-headed men in that cave and in Gabillou, and the bird-headed man in the Shaft of Lascaux. He also offers an explanation for the mysterious hand prints left in many caves. With paint sprayed over a hand held against the cave wall, the purpose was not just to leave a negative imprint of the hand, but to perform a merging of the hand with the wall as spiritual membrane (*Mind* 217–18). The paint seemed to dissolve the hand, sealing it into the wall and the world beyond. Lewis-Williams gives a similar reason for the finger fluting (traces of fingers touching the walls) and for the many lines drawn across certain animal images in cave art. They were marks to enact human connections with spiritual creatures. Paleolithic performers "*cut* or *scored* the 'membrane,' perhaps to allow power to seep through to them" (256–57).[62] Yet, such marks in the visionary scene might also relate to the lines attached to human figures on cave walls, which Lewis-Williams interprets as showing the prickling sensation under the skin, commonly felt with hallucinogenic drugs and associated with animal spirits or ghosts in many shamanic cultures (273–79). He argues that the image of a "Roaring Stag" in Lascaux demonstrates, along with parallel ethnographic evidence, that "the painted and engraved animals of Upper Palaeolithic art probably also 'spoke' to people in auditory hallucinations" (251).

Thus, visual, auditory, and tactile experiences of animal spirits, as guides or gods, may have been involved in prehistoric performances—as with historical shamanic rites and the Stage 3 hallucinations that our brains are capable of today. Lewis-Williams summarizes the architectural evidence of cave art's environmental spaces by showing that the larger chambers near the entrances of many caves were probably used for larger communal rituals and the narrow passages to deeper rooms involved select performers (*Mind* 267; "Of People" 150–51), like the initiates in Montelle's theory. This symbolized a passage through the vortex of the inner theatre, which was also involved in the (Stage 3) hallucinatory communion with animal figures on the walls. Naturally occurring carbon dioxide gas, as in the Shaft of Lascaux, might have induced hallucinations as well (*Mind* 266). As I can confirm, after visiting various caves, the images sometimes appear as active, three-dimensional forms, due to the painters' use of natural shapes in rock formations. Lewis-Williams gives the example of a pair of bison in Lascaux, "placed in a hollow bay in the rock wall ... [so that they] seem to be issuing forth, tails raised, from inside the rock" (259). The raised tails may show that the bison were males in the rutting season, trying to intimidate others, and the figures probably invoked fear in the human actor entering the chamber (260). At the "climax" of their vision quest, initiates or shamanic leaders, facing these spirit animals, had to demonstrate "a stoic disposition in the face of such danger if they wished to acquire power." More generally, such examples show that Paleolithic cave artists and performers "believed that a spirit realm filled with powerful animals lay just behind the rock wall" and could emerge through its membrane (259).

The many animal artifacts (bones, teeth, and shells) left in decorated chambers suggest that prehistoric performers were involved in a two-way ritual theatre, perhaps using the logic of restitution, after killing animals in the outside world. "They were drawing (in two senses) spirit animals through the 'membrane' and fixing them on the surface; they were also sending fragments of animals back through the 'membrane' into the spirit world" (Lewis-Williams, *Mind* 253). This relates to Mithen's theory of a new cognitive fluidity emerging 40,000 years ago — between natural, social, and technical types of intelligence in brain circuitry (as considered in the introduction here). Such communing with animal spirits, through performers' sacrificial ordeals as initiates, or with substitute offerings, continued to evolve, thousands of years later, into historical rites and technologies of theatre, outside of caves, yet still in natural settings in Egypt and Greece. As Western society shifted toward a greater technological confidence, in its "dominion" over nature and animals (as stated in the Bible), images of spiritual forces shifted, too. They changed from mostly animals in the prehistoric caves to more of the hybrid, animal-human gods in ancient Egypt, to mostly human-looking divinities in Greece, to the Judeo-Christian God, his Son, and their angel-warriors against devils on the medieval stage. Yet, sacrificial offerings were still demanded of human participants in order to make the fictional gods appear, through actors' bodies and characters' crises as extensions of spectators' inner theatres.

Even today, in our more entertainment-oriented media, characters often embody supernatural ideals. They emerge, not through cave wall membranes, but through the stage's mise-en-scène, the cinema's moving-camera diegesis, the home theatre's multi-channeled vortex, or the computer's interactive videogames and websites. The technology of transcendent performance membranes has changed greatly, but it draws on the same inner theatrical mechanism for god experiences within our brains, evolved over millions of years — now through the mass media's flexibility to create new divinities, yet also more Trojan Horses (in Donald's sense). The cave art surrounds us now, reshaping our brains through the daily television and computer communion rites, as well as periodic cinema and theatre ordeals, where the threats and pleasures of death, fragmentation, and rebirth may be experienced vicariously. It is now much easier for us to explore the virtual worlds of the Other. But our animal drives may emerge in newly creative and destructive ways: through sublimated, mass-media and cyberspace violence or in real-life, totemic figures and territorial, predator/prey ideologies — while influenced by the binary, causal, and holistic stereotypes of "good versus evil" melodramas onscreen.

We will never know the precise meanings of the mimetic displays, kinematic imaginings, and mythical performances of prehistoric cave art. But it offers us intriguing traces of the possible pruning of initiates' brains, through shamanic leaders' collective control (Hayden 147, 185) and through various performers' creativity. Such directors, scenic rock artists, and ritual actors, along with their spectators, developed hallucinatory performances inside the caves, in a crucial, transitional stage of neural and cultural evolution, across tens of thousands of years — with old and new powers idealized in totem animals and hybrid forms. This shows our human ancestors becoming aware of how they were transcending their animal nature, as they envisioned and identified with animal or animal-human gods in the rock-figure signs.[63] Thus, Paleolithic cave theatres show the early greatness of human culture, yet also an increasing lack of natural being that would lead, in later millennia, to vast destructiveness as well as creativity, with animal drives acquiring godlike technological powers, sometimes without the appropriate wisdom.

All four theories — Breuil's of hunting-magic, Raphael's of totem clans, Montelle's of initiation performances, and Lewis-Williams's of shamanic visions through the rock membrane — may correlate to various actual practices during the tens of thousands of years that the cave art was produced.[64] They are not mutually exclusive. Indeed, each may show a significant element in the prehistoric development of the human brain and its gods. With new hunting technologies, such as the spear thrower, Paleolithic humans were gaining power as predators, yet were still vulnerable, as a minority species amongst many other large animals in their Ice Age environment. Only about 50,000 humans lived in the area of what is now France during the period when the cave art was created (Curtis 25). But their feed or flee drives are not the only ones shown in the floating animals of the cave wall scenery. Their struggle for territorial identity as increasingly powerful animals is also shown in the layered figures on the cave walls, which might symbolize specific clan conflicts. Their creating of generational, socially stratified identities, controlled through initiation ordeals and esoteric knowledge in cave performances, would have involved limbic and brainstem systems of panic and separation-distress instincts, even while shaping higher cortical pathways of particular mimetic and mythic meanings through the theoretic devices of cave wall imagery and acoustics. Moreover, the discovery of further realms of dreamlike visions and sounds connected their inner brain theatres to the "parietal" membranes of natural cave formations — onto which they painted and etched geometric, animal, human-hand, and human-animal shapes.[65] The altered consciousness produced by darkness, flickering firelight, sensory deprivation, sound effects, and perhaps drug-induced trances in the cave theatres may have evoked individual and collective, mystical and religious experiences of divine power, and thus icons of animal gods, through the brain mechanisms (especially in the temporal lobes) that we share today with our prehistoric ancestors.

I would also suggest a connection between this prehistoric theatre of shamanic trance visions, involving cave-art icons, and the hyperaroused/ hyperquiescent overflow mechanism of mystical experience studied by Newberg and d'Aquili, along with their imagined primal scene of a Great Stag mantra.[66] The reading of "natural signs," in Burkert's sense, shifted from our animal ancestors' direct interaction with a stable environment to the distinctively human re-creation of a cultural world, through the brain's evolving, higher-order (and altered states of) consciousness. This apparently involved communing with godlike figures in the visionary cave theatre of shamanic rites.[67] Today, such mystical spectatorship and neural interactivity with godlike figures continues, in mild or intense forms, with or without the stimulation of rhythmic music, dance, trance, or drugs (and the sensory deprivation of caves), even in secular Western culture, through various types of theatrical performances, from stage to screen.[68]

Yet, the powers of higher-order awareness and divine communion in the altered states of evolving human consciousness involved a crucial theatrical trick: self and Other deception. Montelle suggests this regarding Paleolithic cave art as evidence of initiation rites, like those of hunter-gatherer cultures today. "The semantic of initiation is a playful and deceptive discourse in which men's frustrations are shrouded in layers of secrecy and their factual lies are forbidden to women and children. The deceptive tactics used by the male cults bring initiation one step closer to the premeditated and fictitious world of performance" ("Paleoperformance" 138). Researchers have also found such deception tricks in non-human primates, as they compete for food, sex, and dominance in complex social groups. According to theorists of "Machiavellian Intelligence," human consciousness and language evolved from the complex hierarchies, alliances, communications, mind-readings, and deceptions of our primate and hominid ancestors (Whiten and Byrne). In humans and other primates, the tactic of deception developed even to the point of deceiving oneself, as well as others, so as not to give away the truth with unconscious cues (Trivers). Through this "protean" ability, driven by sexual, not just natural selection (Miller), we continue today to play various roles, on and off-stage, transforming ourselves and our environment — with fictional characters and scenery that create new realities.

The archeological traces in cave art of shamanic ritual theatre show early expressions of evolving powers within the human brain and its many cultures. The brain's inner visionary environment and mystical apparatus may have been extended through rhythmic music and drug-induced trances, as well as icons in the caves. Shamanic performers may have changed their appearance for others watching as believers in a ritual context, along with changing their identities in another, dreamlike world. They interacted with and became human-animal gods, at least according to the beliefs and experiences of such actors in more recent times.⁶⁹ They then returned to the material, social world with healing power for others and political power for themselves. Such protean, Machiavellian transformations, through envisioning and communing with gods in the Other world, may have been merely a theatrical trick, deceiving spectators and the actors themselves. Yet, the evolution of an apparatus for such mystical, altered consciousness in our brains, along with the traces of its theatrical expression on cave walls, in communal areas and in deeper crevices of the earth's underground world, shows that humans make their gods have real effects, through individual idiosyncrasies and collective orders.

In our modern world, many people place more trust in objective science than in subjective states of religious belief and altered consciousness. But science is also showing us that mystical experiences past and present — performances involving gods in the brain's theatre and in communal rites — have material sources and effects.⁷⁰ The gods of specific cultures reveal collective identities, emotional projections, and ways of world ordering. Whether or not the gods (or our souls) have an independent existence in the cosmos, watching or acting with us, what we do with our inheritance of god-experience mechanisms, still evolving in our brains, makes a difference today and in the future.

Researchers have discerned an evolving "neophilia," a love of the new, in primate and human brains. "Both sensation-seeking and openness are moderately heritable.... Human neophilia is also the foundation of the art, music, television, film, publishing, drug, travel, pornography and fashion industries..." (Miller 331). As delineated above, the left side of the neocortex, with its executive, routine functions, may conflict with the right hemisphere's attraction to and anxiety about novelty (Geschwind and Iacoboni 62). According to Ramachandran, the right brain acts like a devil's advocate, scouting alternatives to the left brain's convictions as a war-room general (Ramachandran and Blakeslee 135–47). Thus, the neophilia of the right hemisphere, extending to the "sensed presence" of a mystical Other in Persinger's findings about temporal-lobe God experiences, biased as negative on that side of the brain (and the left side of the body), may relate to a devilish or Dionysian aspect in each person's neural theatre and in our collective, intersubjective spaces. Yet such sensation-seeking and openness to meme mutations — even if reconfigured by controlling, Apollonian, left-brain forces as the work of the devil — may be a crucial factor for human change, creatively or destructively.⁷¹

The theatrical extension of the brain's inner mystical performances results in conflicts not just of Machiavellian Intelligence, involving deception and neophilia, but also of Manichean (good versus evil) polarities in particular cultures, through the binary/causal and holistic operators of the left and right hemispheres.⁷² This continues to be the challenge for our evolving brains and conflicting societies: how to reach an ethical integration of various versions of reality, as cultural perceptions, ideologies, and actions compete in remaking our environment. In a neuro-psychoanalytic sense, we are tempted to "project" the negative within ourselves upon others as evil — through executive left-brain, devilish right-brain, and passionate limbic systems.[73] We may even be tempted to act out against the evil Other with pre-emptive violence, while sacrificing ourselves to the divine Spectator. The next chapter will begin an exploration of this tragic flaw in the theatre of the human brain, starting with the animal-human gods of ancient Egypt and the superhuman gods of ancient Greece, as depicted in written records and architectural remains of dramatic performances.

2

Ancient Animal and Human Deities

In the tens of thousands of years of human evolution from prehistoric to historical societies, the otherworldly creatures of cave paintings and etchings, neural vortex hallucinations, and shamanic performances became consolidated into collective gods — leaving mythic records and mimetic traces in certain cultures. The "sensed presences" of shamans' internal theatres, shared with initiates through ritual ordeals involving cave figures, or more collectively as performed visions of the world beyond the rock membrane, were eventually extrapolated as a cosmic theatre of sky and earth gods watching and sometimes performing with humans.[1] This assuaged the right brain's sensitivity to danger in the natural world and tempered the limbic system's panic, fear, and rage drives, by focusing existential, holistic anxieties through the left brain's binary, causal abstractions. It also provided an intersubjective binding of many brains, in larger and larger communities, to sense-making stories and performances, as well as temporal, political authorities. Yet, the devilish, neophilic right-brain continued to evolve mutant memes of personal, political, and cosmic awareness, rebelling, along with limbic passions, against left-brain controls. Thus, internal and external, fictional and actual theatres of change in the human brain, through many cultural permutations, produced new god and animal hybrids, as spectator, actor, and director/author/designer ideals — often embodying the binaries of good and evil, but sometimes with a more insightful, tragicomic complexity.

Such a creative struggle, within and between evolving brains in a particular cultural context, may be glimpsed through the written record of ancient Egyptian performances in the Ramesseum papyrus.[2] Like shamanic performances in various cultures today that use theatrical elements to illustrate the spirits being invoked or exorcised, the Egyptian "coronation drama" involved ritual gestures, props, and masks[3] that made the gods present — as natural and social forces, as well as objects of belief. Today's shamanic performances use magical tricks and ventriloquism, along with torture ordeals and animal masks, indicating how the art of theatre may have emerged in human prehistory (Kirby 9–22), as evidenced also by Paleolithic cave art from 10,000 to 30,000 years ago (considered in the last chapter). But specific performance choices in ancient Egyptian drama show a further stage in the evolving brain's god-making and world-ordering, four to five thousand years ago, with human rulers and natural powers combining as Promethean figures of good and evil, wholeness and dismemberment, control and rebellion, while also evoking more complex, non-binary identifications.

The Pharaoh's Other

Archeologist Steven Mithen argues that the human brain evolved an extraordinary "cognitive fluidity"— as shown in the human/animal hybrids of cave art — to combine different experiences in the environment, creating new, imaginative creatures ("Evolution" 48). Yet, this fluid imagination also found a "cognitive anchor" in religious art and ritual performance (50). The fluidity of the ancient Egyptian imagination developed numerous gods, with animal, human, and hybrid manifestations, but also anchored them in dominant figures, myths, and rites — lasting thousands of years. The remnant artworks and performance traces of ancient Egypt show not only an early articulation of evolving gods in the Western (or Afro-European, Mediterranean) imagination. They also point to the fluidity of animal instincts, passions, and perceptions that we bear in our human brains today: from the primal "reptilian" brainstem, basal ganglia, and cerebellum to the paleo-mammalian limbic system to the neo-mammalian cortex with its distinctive hemispheres — as considered earlier in this book. The fear or seeking of natural forces, as threatening or nurturing, the fight for territory, the struggle of alpha leaders for power, the self and other deceptions of Machiavellian Intelligence, the sacrifices demanded for collective identity and survival — all these animal/human drives are displayed, with their cognitive anchors, in a particular coronation drama performed by the pharaoh, his court, and his people on the banks of the Nile.

Lasting for thousands of years, ancient Egyptian civilization was organized in a pyramidal power structure. As with other monarchies, it reached a point of vulnerability at the turnover of power from one alpha leader to the next, traditionally from father to son, as ruling pharaoh. By the time of the

Middle Kingdom, 12th Dynasty reign of Sesostris I, when the coronation drama was first performed in approximately 1970 BCE (if not earlier),[4] Egyptian pharaohs had been ruling for a thousand years. No one knew then that their empire would continue another millennium and longer, until Alexander the Great conquered Egypt in 332 BCE. Yet, the Egyptians did believe that their civilization would end someday, that even their gods were mortal and would perish with it.[5] Chaotic forces in society and nature threatened the transfer of power from the dead pharaoh to his son, as well as the daily survival of a culture that depended on the fertile banks of the Nile and its life-giving water. Such threats were embodied in the god Set (or Seth), as antagonist to the father and son gods, Osiris and Horus.

According to the Ramesseum papyrus, the new pharaoh performed the role of the falcon-headed solar god Horus, while his father's mummy (and other props) represented the god of the dead, Osiris. A Chief Official played the ibis-headed scribe-god Thoth, who was also depicted in Egyptian iconography as a baboon. Thoth, or Djehuty, was the god of the moon as well as of writing, a servant to Osiris and healer of the injured Eye of Horus, which was also associated with the moon (Wilkinson 215). Other officials performed the part of the earth god Geb, an "umpire" in the battle between Horus and Set, and the falcon-headed god Sokar. Wailing women performed Isis and Nephthys (sisters and wives to Osiris and Set). Temple personnel played Set and his henchmen.

Set's image, as cognitive cultural anchor, usually took the hybrid form of a human body with a mysterious animal head, of uncertain identity today. It appears to be that of an aardvark, dog, or camel with a curved head and tall square-topped ears, or of a dangerous desert animal that is now extinct (Wilkinson 197–99; Budge 243). Set also appeared at times as a pig, hippopotamus, or other animals (Meeks and Favard-Meeks 72; Traunecker 86; Budge 247–48). His symbolic color was red, representing the dangerous desert sand, especially the threat of sandstorms that could cover and destroy the fertile lands along the Nile (Lesko 160). He was associated with the raging sea, as well as other storms: civil unrest, crimes, foreign invasion, rape, and disease (Wilkinson 198). As god of chaos, he was the mortal enemy of his brother Osiris, god of the dead and of fertility—of life born from decay—whose symbolic colors were black and green. According to a prevalent myth, Set killed and dismembered Osiris, dispersing his body around Egypt. But Osiris was reunified and brought back to life by his sister, Isis, who gave birth to their son, Horus. Horus battled his evil uncle Set, taking Set's testicles but losing his eye. (The taking of a rival's testicles parallels recent observations of violence between chimpanzees, our nearest primate relatives [de Waal, *Our* 44–48].) Various elements of this myth were performed in the coronation drama — to enact the passage of power from one pharaoh to the next, from father to son, and how the threat of chaos could be handled.

Horus was often depicted in hybrid form, with a human body and falcon head. He was the god of Lower Egypt (in the north around the fertile Nile Delta), while Set represented Upper Egypt (Wilkinson 197). Horus's victory over Set, re-enacted in the coronation drama, showed territorial conflict and the power to unify. Horus was a sky and sun god. His name meant "the one on high" and "the distant one," referring to the soaring flight of the hunting falcon that was his image — in animal or hybrid form (Wilkinson 200–01). In a neuro-evolutionary view, the battle of Set and Osiris, as mythical background to the coronation drama, shows these gods symbolizing the continued threat of nature, as desert sandstorm, and of rebellious social forces, to destroy Egyptian civilization and its fertile land, as the human brain made sense of its environment and fought to maintain collective order through a particular culture. Yet, the battle of Horus and Set shows a further dimension of neurotheatre as well. It not only represents the father's power as pharaoh (Osiris) continuing in his son, the new ruler (Horus), against the threat of potential usurpers (Set). It also shows the brain's executive, routine, prosocial controls in the left hemisphere, especially the frontal lobe (Horus as sky/sun/falcon god), rising above the "devil's advocate" neophilia of the right hemisphere and emotional drives of the limbic system (Set as chaotic sandstorm god). Thus, the Egyptian gods can be seen as figures for the animal/human hybridity within and between our brain theatres today, as well as in ancient times. For we share with the ancient Egyptians the same basic brain anatomy — although their cultural expression of animal, human, and divine forces in the brain was very different from ours.

The ancient Egyptian coronation drama used the banks of the Nile River as its theatre, celebrating its significance for trade and the gradual flooding of cropland that sustained the culture, against the threat of the desert beyond. Repeatedly, at various points along the Nile, this processional performance showed the new pharaoh in the role of Horus receiving certain offerings — each symbolizing his restored "Eye of Power." It also showed Set being mastered by symbolic props, which represented the continuing power of the deceased pharaoh, even as a mummy. The remnant fragments of the drama begin with the seasonal opening of the Nile for navigation and the equipping of the pharaoh's royal barge — showing his power, as top dog (or falcon) of Egyptian society, over and yet also through the forces of nature. As Horus, the pharaoh tells his followers: "Convey to me the Eye that by its power / This waterway may now be opened up" (Gaster 378). Jars are loaded onto the barge, representing the corpse of Osiris, still powerful in death, being put onto the back of Set. The scribe-god Thoth then tells Set: "thou canst endure and stand no more / Against this god who mightier is than thou" (379). This demonstrates the Promethean aspirations of the evolving human brain in a particular cultural and performance context. The theatricality of our self-consciousness, watching potential actions within the mind as it transforms animal instincts and the given environment, is represented as an Eye of Power and divine corpse, which control the forces of nature and survive beyond the threat of mortal dismemberment.

As the royal barge continued along the Nile, the new pharaoh (being crowned or confirmed in power) acted various other scenes, at different sites in his civilization, showing his divine role as Horus — in this ancient form of mass media. A ram was sacrificed and a portion of its body given to the pharaoh, again symbolizing his restored Eye of Power (Gaster 379–80). Grain, loaves, scepters, a crown, stave, and plumes were also offered to the pharaoh as his restored Eye or Eyes (380–83, 391). A goat and goose were beheaded, representing the decapitation of Set, with their heads then given to the powerful new pharaoh playing Horus (385). A punching match was staged to show the battle between Horus and Set, with the earth-god Geb intervening to end the fight (387). Both the nurturing and protective aspects of the pharaoh were displayed — as milk was poured, representing "the sweet influence of this mine Eye / poured forth," and sacrificial meat again offered, with "my protection shed o'er you" (388). At another point, while being given blood-red carnelian beads as a tribute, the pharaoh challenged Set, or any other potential rebel, with his Eyes of Power as a "fearsome gaze upon thy face" (390). The pharaoh as Horus even held two maces representing the testicles of Set, taken from him in combat (391). He was also given the thighbone of a sacrificed animal, as the thighbone of Set taken in battle.

The chaotic, yet fecund threat of Set was mastered by Horus onstage, as the rebellious forces in nature and society bowed to the new pharaoh's power. This mythical, theatrical rite, with its "Trojan Horse" memes, seems to have been an infectious mass-media gift, especially if it was performed repeatedly over hundreds of years, like the Abydos Passion Play (Brockett and Hildy 8). Thus, it extended the left hemisphere's control of, yet coordination with the right's alternative desires and the limbic system's animal drives — in ancient Egyptian brains — toward a divine form of political rule. For Egyptians also experienced the gods in their dreams, in the inner theatre of sleep, and expected to meet them in the next life, while submitting to godlike mortals in this one (Dunand and Zivie-Coche 22).

In the coronation drama's performance, a "fragrant bough" was put on the barge, symbolizing (like the earlier jars) the corpse of Osiris put onto the back of Set — with workmen staggering under the weight of the bough and Horus gazing on the corpse as "noble," "fair," and "beautiful" (Gaster 381–82).

The barge was painted red at one point, symbolizing Set as the devilish desert (384). The *djed* pillar, an image for Osiris (Wilkinson 121), was lowered in another scene, displaying "Set ... bow'd beneath Osiris' weight," as the pharaoh/Horus proclaimed (386). Eye salves in green and black, the colors of Osiris, were given to the new pharaoh (393). He also received a mummy cloth, binding him symbolically to his dead father as Osiris: "In close attachment ... [to] reunite his limbs once more" (395). Thus, the people and local authorities of Egypt gave offerings to their new pharaoh, as the falcon-headed god Horus, at various points along the banks of the Nile. Through their popular support, they restored his Eye of Power and the force of his dead father, as Osiris, over the dismembering storm-god Set. Yet they also confirmed the new ruler's binding and makeup in his role, through his mummy wrap and eye salve. Reuniting the land, after the death of his father, he embodied the Eye of the collective mind — as its projected, left-hemisphere, executive control — in this ancient intersubjective theatre, which fetishized the new alpha male, as object of power.

Gift of the Gods

The shamanic theatre of our prehistoric ancestors left traces of their hybrid spirits, visions, and rites on cave walls from tens of thousands of years ago. The ancient Egyptians left fragmentary scripts of a ritual theatre from four thousand years ago, associated with myths and images of their gods in animal, human, or animal-headed human forms. But the ancient Greeks left more complete scripts from 2,400 years ago and the ruins of open-air performance spaces built into the hillsides, near temples to their gods. Our word "theatre" comes from the Greek word *theatron*, the "seeing place" of the audience. Unlike the Egyptian coronation drama, staged in a ceremonial procession along the Nile, an ancient Greek play was performed in a single collective space,[6] with a stage and background *skene* (scene house) for the lead actors, an orchestra (dancing area) for the chorus, a *theatron* with benches and seats for thousands, and the trees, hills, and sky as cosmic backdrop. Yet, like Egyptian performances, this design showed the forces of nature, as well as culture: within, between, and around human brains.

Greek theatre emerged through ritual sacrifices, dances, and festivals in honor of Dionysus. He was the god of wine, drunkenness, sexual liberty, and spring fertility,[7] whose mythical followers included the part-human, part-animal satyrs. These creatures were often depicted in Greek art as having small horns on their heads, pointed ears, and animal tails, with otherwise human bodies and faces. Apart from such hybrids in satyr plays (and the Furies considered below), or occasional animal caricatures in comedies, the

Greeks did not show many animals or animal-human forms onstage, according to the extant dramas.[8] Yet, these ancient works also display a cognitive fluidity in self and Other consciousness,[9] emerging beyond, but still in relation to the brain's animal heritage, like the Egyptian Eye of Power in the falcon-headed Horus and the immortal corpse of Osiris, overpowering the chaotic nature of Set.

Indeed, the shamanic Dionysus was related to the fertility god Osiris in ancient thought (Wilkinson 120).[10] Herodotus, traveling in Egypt in the fifth century BCE, saw a procession in honor of Osiris, with women holding puppets having large phalluses, which the ancient historian describes as akin to the Greek parade with a phallus in honor of Dionysus (Cornford 53).[11] Yet, Dionysus also bears the disruptive force of Set — as shown in Euripides's play, *The Bacchae*, where his female followers overthrow the tyrannical King Pentheus of Thebes. Like the rebellious god, Prometheus, who gave fire to humans and was punished for it by Zeus, Dionysus represents the divine but dangerous aspirations in the evolving human brain, with its blissful trances and destructive passions. The myths of Dionysus also involve fragmentation and rebirth, like the punishment of Prometheus and the battles of Osiris, Set, and Horus — or the shamanic initiation ordeals that may have been performed in Paleolithic caves. In *The Bacchae*, Dionysus's female devotees tear apart live animals (*sparagmos*) and eat them raw (*omophagia*), through the passion of their ritual trances offstage (83, 102).[12] Ultimately, they tear apart Pentheus as well (115). Even his mother, Agave (played by a masked male actor), participates in this, seeing his head as that of a lion cub she has dismembered, when she proudly, still in a trance, displays it onstage — probably holding the full-headed mask worn earlier by the actor playing Pentheus (117). Spectators then share her horror as she comes out of trance and sees what she has done (120–21).

Offstage, earlier in this drama, Dionysus used a bull and a phantom to delude Pentheus, who thought he had imprisoned a priest of the cult, instead of the god himself (98–99). Dionysus not only creates animal, human, and spirit doubles in this way,[13] but also changes the costume of Pentheus, from masculine to feminine — ostensibly so that the king may spy on the bacchae, the worshippers of Dionysus, during their forest rites. When Pentheus appears onstage as a woman (like the males in the orchestra portraying the Chorus as another group of bacchae), he sees a double sun in the sky and a bull double of Dionysus. "Now I'd say your head was horned / or were you an animal all the while? / For certainly you've changed —/ oh, into a bull" (108). The theatre god thus represents the animal/human[14] and masculine/feminine, as well as fragmentary and whole, mortal and immortal fluidity of our cognition. Dionysus demonstrates, in Euripides's play, the illusory nature of self and Other — as cognitive anchors.

According to today's neuroscience, there is no single, unified self in the brain, but various modules or "unconscious zombies" of embodied, passionate, visceral, executive, mnemonic, vigilant, conceptual, and social selves, as well as the "filling in" of gaps to provide the illusion of a whole self (Ramachandran and Blakeslee 227–28, 247–54). According to the psychoanalytic theories of Jacques Lacan and Julia Kristeva, the ego is illusory — the Symbolic and Imaginary fiction of a whole self in the mirror, dependent upon the desires of the Other, while masking split subjectivity,[15] Real drives, and a repressed, abject, yet eruptive *chora* (a womb-like space of becoming). Likewise, *The Bacchae* warns of a tragic hubris in human higher-order consciousness.[16] Dionysus's tricks with Pentheus and Agave show the lure of cognitive fluidity, moving beyond mortal awareness and dismemberment terrors, toward an illusory anchor of godlike apprehension — of a central, whole self merging with, yet failing to become the ideal Other, as political leader or religious believer.

Self and Other consciousness in the enlarged human brain, as a theatrical gift of the gods, becomes both ecstatic and dangerous in *The Bacchae*. Dionysus is associated with the nurturing and threatening aspects of nature's power within human culture. He is like the collective Eye of Power belonging to the pharaoh as Horus: a force poured out as nourishing milk to his people, and thus "a sweet influence," yet also "a fearsome gaze" (cited above). As Tiresias, the blind prophet and follower of Dionysus, says about this god: "He it was who turned the grape / into a flowing draft / and proffered it to mortals / So when they fill themselves with liquid vine / they put an end to grief. / ... / It is the god himself / we pour out to the gods, / and so, do men some good" (88). Later in the play, when Pentheus denies the god's existence, Dionysus (disguised as a priest of the cult) says, with sly sincerity: "He is beside me now and sees my trials" (96). The Chorus of Asian bacchae, who have followed this "priest," helping him return Dionysus to Greece, then call for the god's watching eye and power against the king who plans to enslave them as well. "Do you see these things, O son of / Zeus, O Dionysus? / Your votaries on trial / Grappling with oppression / Descend, Lord, from Olympus: / Shake your golden thyrsus. / Quell this man of blood's / Brash obsession." Although performed by young men in military training (Winkler), this chorus clearly represents a potential threat to patriarchal rule.

The Bacchae was first performed in Athens in 405 BCE, a year after Euripides's death. During the City Dionysia festival, when this and other tragedies were performed, spectators also experienced the performance of animal sacrifices: the spilling of a piglet's blood to purify the theatre's orchestra circle and the killing of larger animals at the temple of Dionysus, about forty yards behind the theatre's stage, where "the smell of the dead animals must have lingered" (Ashby 123; David Wiles 58–59). During the plays, the audience wore "wreaths which helped to transform the space of the auditorium into a space dedicated to Dionysus" (David Wiles 188). Spectators were seated on the hillside above the orchestra and stage, looking down at the performance, at the *skene* representing Pentheus's palace, at the actual temple of Dionysus, and at the rest of the city, the trees, the sky, and perhaps the sea beyond (Rehm, *Play* 35–36). Thus, the singing and dancing of the bacchic Chorus, in their prayers to Dionysus, asking him to quell the blood of King Pentheus, may have resonated with spectators' feelings (in the early democracy of Athens) against the ruling order of the play. Viewers might have felt a similar appeal to the Other watching over their city and its place in nature, with their sacrifices of blood and wine to Dionysus during the festival—while praying not to suffer more themselves because of vengeful rage between rulers or gods. (The play was first performed near the end of the Peloponnesian War against Sparta, which took several decades and destroyed the Athenian Empire.)

As Tiresias describes in pouring the ritual wine and the Chorus requests through the god's descent, such cognitive anchors in the fictional performance invoked for the Greek audience a sense of communion. This aspect of theatre extended the practices of prehistoric shamans in communing with animal guides or gods, through the membrane of cave walls and echoing music. Greek actors and spectators experienced communal transcendence, to whatever degree, in a different kind of performance space from prehistoric caves: through the membranes of a scene-house and painted *pinakes* (flats), costumes and masks, a singing and dancing chorus, a crane (*mechane*) for flying gods, and other stage devices in an open-air theatre. Yet they were using similar brain networks, sculpting their art and its collective pathways out of the natural world, including the animal within human nature. We continue to use those brain areas today, making new forms of transcendent, communal experiences, through performance membranes and godlike ideals in our theatre stages and mass-media screens, which still project remnant animal drives.

In getting revenge against Pentheus, Dionysus not only watches the trials of his votaries, he also inspires their animal passions toward bloody acts. Rather than quelling their blood, through executive left-brain and cultural controls, Dionysus creates something like "temporal lobe storms" of mystical ecstasy in his maenads' brains (Ramachandran and Blakeslee 197). The intense focus on Dionysus in his followers' frontal lobes may trigger a deactivation of the left parietal lobe, merging self into the Other and causing a neural "spillover" of hyperarousal and hyperquiescence during their violent rite (Newberg, d'Aquili, and Rause 117–23; d'Aquili and Newberg, *Mystical*

25–26). Thus, the "god himself" would be poured out onto the earth, through the blood of animal and human victims, as well as wine, not just in the play's fiction, but also in the spectators' neural theatres, possessed by the god.

According to a herdsman who witnessed the offstage rite, it was at first "orderly and beautiful" (101). Young and old women wore fawn skins and snakes around their waists. "Some fondled young gazelles / or untamed wolf-cubs in their arms / and fed them with their own white milk / .... Another hit her rod of fennel on the ground / and the god for her burst forth a fount of wine." Jets of water and milk also burst from the soil and "sweet streams of honey" dropped from the thyrsus (symbolic fennel staff) of the bacchae.[17] But when they called upon Dionysus in their bacchic dance, the scene turned from quiescent pastoral to aroused blood rite. "You could see a woman with a bellowing calf / actually in her grip, / tearing it to apart. / ... / You could see their ribs and cloven hooves / being tossed up high and low; / and blood-smeared members dangling from the pines; / great lordly bulls, / ... dragged to the ground like carcasses / by the swarming hands of girls" (102–03). The women also pillaged villages and "snatched up babies out of homes" (103). Later in the play, another messenger describes a similar sacrifice of Pentheus in the hands of his entranced mother, Agave. "Gripping his left hand and forearm / and balancing her foot against the doomed man's ribs, / she dragged his arm off at the shoulder ... / it was not her strength that did it / but the god's power seething in her hands" (115). The other maenads joined in: "One woman carried off an arm, / another a foot, boot and all, / they shredded his ribs — clawed them clean. / Not a finger but it dripped with crimson / as they tossed the flesh of Pentheus like a ball."

These scenes are described after they took place offstage, thus evoking a further playing out in the various theatres of spectators' minds.[18] Dionysus brings a revolutionary ecstasy to the fictional Thebes and, in another sense, to the actual theatre audience. He shows the potential violence of the limbic brain within every human animal — and the creative or destructive force of a repressed, feminine *chora* within patriarchy (especially with all roles performed by males). Dionysus threatened ancient Athens, as well as Thebes, with his power, while being honored and evoked by the ritual of theatre. A statue of Dionysus was carried in procession at his festival and placed in the *theatron* to watch the plays, along with the audience, while animal offerings burned in the temple behind the stage (Cole 27). Theatre emerged in ancient Greece, at least in part, as a sacrificial offering to this god, extending the theatre in and between human brains. In neurological terms, this may have served to integrate the left and right hemispheres[19] of performers' and spectators' brains, through the homeopathic evocation of Dionysus — as an animal Other erupting within the human body and yet also as divine Other judging the mortal drama.

We share 98 percent of our genes with chimpanzees and bonobos, our nearest primate relatives. Wild chimpanzees go on organized border patrols, crossing deep into others' territories for violent "raids" (Wrangham and Peterson 14).[20] These raids involve a performance like that of the offstage bacchae in Euripides's play, although with groups of male chimpanzees. When they find a lone chimp not belonging to their group, they attack and kill it — holding it down, beating it to death or nearly to death, tearing its skin off, and drinking its blood (5–6, 16–17). After a group splits into two parties, chimps will even seek out and kill a fellow chimp that used to be within their own group (18) — perhaps akin to the violent realignments in human civil wars. Although at other times gentle and nurturing, with food shared and hours spent in mutual grooming, male chimps will also beat up females when they refuse to mate (7, 18, 144–46).

On the other hand, bonobos, a related species known as "pigmy chimpanzees," are organized in co-dominant male and female hierarchies, where females cooperate to control male aggression, unlike the patriarchies of chimpanzees and humans (Wrangham and Peterson 205–08). When bonobo groups meet at the edges of their territories, aggression may result, but it is usually defused by the lead females, who engage in genital rubbing with each other (209–10, 214). Through the evolution of this species and its cultural behaviors, their genitals have migrated lower on their bodies (in comparison with chimpanzees), making such "lesbian" sex more accessible. Like humans, bonobo females have evolved concealed ovulation, which also lessens the aggression of males competing to inseminate them (212). Bonobos and humans are the only primates known to have face-to-face sex. Yet, bonobos engage in sex much more often than chimps or humans, in various ways, both male and female (with both), starting "long before the onset of puberty — from about one year old" (213).[21] Most of these sex acts are just brief, pleasurable interactions, reducing tension in the troop. Also, while chimpanzees regularly hunt, kill, tear apart, and eat monkeys, bonobos merely hold and play with monkeys as if they were toys (216–19).[22]

Chimpanzees and bonobos show different elements in our evolutionary ancestry — both of which are suggested by the wildly violent and yet erotically playful women in *The Bacchae*. Today, we are still evolving the primate drives of territoriality, lust, and play in our limbic brains, revised in cultural ways by our "advanced" neocortical hemispheres and displayed in our performance media,[23] which alter our self and Other awareness, for better or worse. Non-human apes (like elephants and dolphins) exhibit some degree of self-awareness in experiments where a red dot is put on their forehead and they look in a mirror, recognizing the image as their own and touching the dot on their head (de Waal, *Our* 193–94).[24] Apes also play with their facial expressions in a mirror. Such a sense of self involves, to some degree, a "Theory of Mind" about others (see chapters 3 and 5). This evolved through group identity and deception tactics, with territorial and hierarchical struggles sometimes leading to extreme violence in the wild. It indicates a Dionysian core in our animal to human evolution, the heritage of which we still bear in our brains and cultures: the dangerous, ecstatic, communal illusions and frictions of self, with and against the Other, in the theatres of love and war.

But Greek drama helps to reveal a further tragic flaw in our big-brained species.[25] By evolving beyond animal instincts, humans achieved tremendous success in survival, reproduction, and expansion across this planet, transforming many environments. The enlarged human brain, born prematurely in comparison with other primates,[26] achieves a vast cognitive fluidity — especially with the neural plasticity of the late developing right and then left neocortex, parts of which (the prefrontal lobes) do not mature until past adolescence, in the mid-twenties.[27] We are thus set adrift with far fewer instinctual behaviors than other primates, with a "lack of being" in nature (in Lacanian terms). Our remnant animal drives — to fight, flee, feed, and fornicate, or to claim territory and nurture offspring[28] — are transformed by cultural, cognitive anchors, as shown in perverse ways by Dionysus and his bacchae, retaking Thebes through anti-maternal passions. Even today, we continue to need the Other, in the form of gods or secular memes, to shape the lacking human brain outside the womb — extending our first parental influences to give life its meaning and purpose, despite our awareness of individual death, in the past and future. From shamanic cave visions to Egyptian animal-headed gods to the theatre of Dionysus and various mass media today, humans evoke distinctive versions of the metaphysical Other, not only as watching, nurturing, and threatening forces beyond us, but also as figures for the good and evil, or complex passions, inside us. Through specific beliefs and rites, or the sacrificial questions of theatre, such cultural figures shape the remnant forces of nature within our competing and cooperating brains.[29] Ancient gods continue to structure our current embodied metaphors, as the neural foundations of secular Western culture.

Promethean Sacrifice

Another angle on the gifts of bestiality and divinity in the human brain can be found in an ancient drama from a half-century earlier, Aeschylus's *Prometheus Bound*. Probably created a few years before the playwright's death in 456 BCE, this play presents the god Prometheus as a tragic figure, unlike prior comic portrayals (5–7). In the backstory of this god's myth, as related by Hesiod, Prometheus tricks Zeus with bones instead of meat as the burned sacrificial offering of humans (Lefkowitz 18). Zeus becomes furious and takes fire away from men. Prometheus, whose name means "forethought," pities humankind living in darkness and cold. He disobeys Zeus and gives fire back to men, carrying it in a hollow fennel reed (like the Dionysian thyrsus). Prometheus, in name, feeling, and action, reflects the primal, hominid emergence of forethought and of tool-using cultures, rebelling against natural limits through sympathetic cooperation — evidenced also by prehistoric cave art, produced by torchlight.

According to the myth, Zeus punishes Prometheus by ordering that he be chained to a rock, with a bird pecking at his liver. He punishes men also by giving them females, including Pandora, who opens her box, giving birth to the world's evils. At the start of Aeschylus's play, Prometheus is chained to his rock by the characters Power and Violence, as agents of Zeus. Prometheus then tells the Oceanids (ocean nymphs and sisters of his wife, Hesione)[30] that Zeus had planned to "wipe out the whole species [of humans] / and breed another, a new one" (40). Here, Zeus embodies the ruthless force of evolution, demanding obedience (adaptation) to environmental rules, through power and violence.[31] From ancient to modern times, concepts of human reason, ethical compassion, and universal rights (as extensions of Prometheus's gift of fire) have attempted, depending on various cultural interpretations, to counter nature's cruel, selective pressures.[32] Yet, cultures also produce further cruelties, as well as liberties, shaping our evolutionary, Promethean fates.

Prometheus explains that he gave not only fire to humans, but also "blind hopes" to help them with the problem of "foresee[ing] their own deaths" (41). The god of forethought helps humans to avoid seeing too much of their mortality. But he also reflects the evolutionary gifts (and flaws) humans attained through the forethought of bigger brains: moving from the limited episodic awareness of animals,[33] to mimetic cultural learning via mirror neurons, to mythic meanings with the audioverbal expansion of the left neocortex, to theoretic technologies transforming the environment.

Prometheus describes such gifts, while lamenting the irony of his own helpless, sacrificial position. He gave humans the symbolic orders of time, numbers, and letters (50). He taught them how to tame animals to do their labor (50–51). He gave men ships for sailing, drugs for diseases, and dream interpretation (51–52). In Lacanian terms, he helped men to translate the Symbolic and Imaginary orders of natural signs and of the Real inside the brain's theatre.

> I was the first to realize what dreams are bound
> to wake up: real.
> And snatches of speech caught in passing, and chance meetings along the road,
> these too have secret meanings.

> I showed them this. And clearly analysed the flight
> of birds with crookt claws — what ones bring luck, and which are sinister —
> and the way each species lives,
> what hates it has, what loves,
> what others it settles with.
> I looked into the silky entrails, I showed them
> what color gall bladder meant the Gods were pleased,
> ...
> I gave the fire eyes, so that its signs shone through
> where before they were filmed over. [52]

As he says in summary, "All human culture comes from Prometheus."

    Non-human primates in the wild show some degree of a local "culture" with distinctive group behaviors in different geographical areas — especially while performing with specific props as tools. Chimpanzees in certain locations have invented protection from thorny branches with cushions, sandals, and gloves made of leaves (Wrangham and Peterson 9). "Elsewhere are chimpanzees who traditionally drink by scooping water into a leaf-cup, and who use a leaf as a plate for food." Other groups use bone-picks for eating marrow, sticks for eating ants, bees, or termites, and "leaf-napkins to clean themselves or their babies. These are all local traditions, ways of solving problems that have somehow been learned, caught on, spread, and been passed across generations among the apes living in one community or a local group of communities but not beyond." In captivity, chimpanzees and other apes are capable of learning and remembering human sign language, "of understanding numerical concepts, [and] of creating a simple form of art" (152). Such brain skills have enabled wild apes to have "rich and multidimensional social relationships," involving both affection and violence. "The hugs and kisses and embraces of the apes are as elaborate as their use of brute force." They show us how our own animal and divine brains, building upon those of common ape ancestors, turn "affection into love and aggression into punishment and control"— but in vastly more elaborate and brutal ways through our numerous human cultures, as exemplified by ancient Greek drama.

    In Aeschylus's play, Prometheus tells the ocean nymphs that despite his shamanic gifts to humans of fire, hope, dream interpretation, and medicine, he must suffer Zeus's punishment without such a cure. "Art," he tells the Oceanid Chorus, "is far feebler than Necessity" (53). They ask: "who brings Necessity about?" He answers: "The three bodies of Fate, and the unforgetting Furies." Prometheus represents the Imaginary and Symbolic orders of human culture struggling against our chaotic lack of natural being, like Horus and Osiris in their battle with Set. Yet he also demonstrates a pervasive sacrificial imperative in humans, with our incomplete overcoming of nature through culture.³⁴ As the Oceanids remind him: "You must have seen / how blind and weak, like prisoners of a dream, / the human beings / are. / Can the plans of things that live and die / ever overstep / the orchestrated universe of Zeus?" (55).

Each human—with a prematurely born, enlarged brain, shaped by a particular culture—loses his or her instinctual being in nature, acquiring a certain cultural Fate, while still fueled by remnant animal drives as the "unforgetting Furies." Humans gain a vast freedom to remake the world and themselves. Yet they sacrifice themselves to metaphysical ideals, whether religious or secular, in order to make meaning out of their blind, weak, dreamlike lives—with specters of pain and death always lying ahead. Such ideals often delude humans, despite their Promethean forethought, as to whether their sacrifices are good or evil. As the ocean nymphs say to the bound Prometheus: "Tell us, what's the use of doing good / when there's no good in it / for you?"

Altruism is a sticky subject for evolutionists. How would altruistic genes survive if they drive their hosts to sacrifice themselves, limiting their own reproduction and perpetuating others' genes instead?³⁵ Examples of this in insects and animals usually involve "kin selection," whereby an individual is sacrificed for the benefit of others with related genes—or "reciprocal altruism" where an organism helps another and benefits later, thus furthering its own genes.³⁶ Yet in human cultures, many individuals, from athletes to soldiers to terrorists, sacrifice their bodies and sometimes their lives for teammates, countrymen, or fellow fundamentalists, who are not biologically related to them (or might not help them to survive and reproduce). The collective, reproductive memes they fight and die for are far more powerful than the individual's genetic instincts of survival and sexual reproduction.³⁷ But within that genetic logic there is already a sacrificial drive. Individual bodies meet and mate, bear offspring, and die—at the service of genes that live beyond them. Such a sacrificial drive is exemplified, too, in the cannibalistic mating rites of certain praying mantis and spider species, where a Dionysian female eats the male while he inseminates her—thus providing food for their offspring through his death (Diamond, *Why* 11–12).

As the evolutionist and avowed atheist, Richard Dawkins, put it recently during a *Time* magazine debate with Francis Collins, a geneticist and religious believer:

> Altruism probably has origins like those of lust. In our prehistoric past, we would have lived in extended families, surrounded by kin whose interests we might have wanted to promote because they shared our genes. Now we live in big cities. We are not among kin nor people who will ever reciprocate our good deeds. It doesn't matter. Just as people engaged in sex with contraception are not aware of being motivated by a drive to have babies, it doesn't cross our

minds that the reason for do-gooding is based in the fact that our primitive ancestors lived in small groups. But that seems to me to be a highly plausible account for where the desire for morality, the desire for goodness, comes from [qtd. in Van Biema 54–55].

Collins responded: "Evolution may explain some features of the moral law, but it can't explain why it should have any real significance. If it is solely an evolutionary convenience, there is really no such thing as good or evil. But for me ... moral law is a reason to think of God as plausible — not just a God who sets the universe in motion but a God who cares about human beings, because we seem uniquely on this planet to have this far-developed sense of morality" (55).[38]

Humans have evolved to a point beyond, yet through nature's rules of instinctual behavior, transforming the sacrificial drives of sexual reproduction into various memes of self-promotion and donation. God or many gods could be behind this, starting and conducting the evolutionary experiment. Or it could have evolved by chance. Either way, humans seem to have a divine spark in their brains through their higher-order consciousness, their awareness of mortality, their Promethean technologies, and their remaking of god ideals through specific cultural laws of morality.

Collins, who led the scientific project of mapping the human genome (and later became head of the U.S. National Institute of Health), insisted that he finds a sign of God in "the sense that we all have about ... absolutes ... of good and evil," so that "our noblest acts are [not just] a misfiring of Darwinian behavior" (qtd. in Van Biema 55). I would agree that there is a moral and divine sense in all of us *potentially*— in the frontal and temporal lobes (plus other brain areas), involving executive controls, mystical experiences, and pleasure/reward systems, according to current neuroscience. But each brain's use of such inherited anatomical structures depends a great deal on how that brain is shaped and reshaped by family and social experiences, along with personal choices. ("Free will" is another sticky issue for neurologists, considered in chapter 4 here, along with the evolution of revenge and forgiveness.) Man may be made "in the image of God," with a brain evolved for divine and moral experiences. Yet, mankind remakes its gods in various images that extend cultural and personal values[39]— even at times beyond God and morality, as Nietzsche argued.

Prometheus Bound shows such a potential remaking of the divine within humans, through a questioning of sacrificial emotions. As the god who rebelled against the rule of Zeus, Prometheus is bound in punishment and realizes that his pity for mankind caused his suffering. "I'm wrencht by torture: / painful to suffer / pitiable to see. / I began by pitying people (things that die!) / more than myself, but for myself / I wasn't thought fit to be pitied" (41). Yet

Prometheus moves toward and perhaps foresees, along with his tragic audience, a catharsis of pity and fear (in Aristotle's terms), a purifying and further evolution of limbic emotions — in the next play of Aeschylus's trilogy, *Prometheus Unbound*. Only fragments of that play remain (99–105). But we know that in it Herakles frees Prometheus from his rock imprisonment and bird-attack torture, with Zeus changing, apparently, from tyrannous in *Prometheus Bound* to "pitying" in the sequel.[40] Maybe this Promethean trilogy (including another lost play, *Prometheus the Fire-Carrier*) was an ancient fable for the past and future evolution of human brains — developing more epigenetic compassion with better connections between remnant animal drives in the limbic system and godlike desires and controls in the neocortical hemispheres.[41] Thus, despite the emergence of our species through Darwinian competition, we might evolve a Promethean future, through greater cooperation and compassion, sharing resources and freedoms beyond selfish genes and superpower, mass-media, consumerist memes — as we decide what is fittest to survive, by clarifying our fear and pity through theatre.[42]

A Furious Fate and Evolving Guilt

In the plays considered so far, we can see ancient inklings of the brain areas described by current neuroscience and the corresponding realms of intersubjectivity in Lacanian psychoanalysis, related also to the evolutionary stages of hominid to human culture. Set, Horus, and Osiris — or Dionysus, the possessed bacchae, and Pentheus — show specific aspects of limbic Real, right-brain Imaginary, and left-brain Symbolic orders, reflecting the episodic animal, mimetic hominid, and mythic human layers of our theoretic culture. In *Prometheus Bound*, the offstage Zeus embodies executive, left-brain, Symbolic controls (plus prefrontal lobe ties to limbic passions, in his prior rape of Io). Prometheus, with his rebellious gifts to mankind, against Zeus's orders, displays the devil's advocate, right-brain, Imaginary desires of alternative perceptions and concepts in each of our brains. Io also appears onstage, wearing horns, as she commiserates with the suffering Prometheus. She is a human girl (daughter of the river god Inachus) who resisted being seduced by Zeus and was transformed into a cow. Fleeing the wrath of Zeus's wife, Hera, Io wanders the world as a cow, tormented by a gadfly. She exemplifies the abject Real (and Kristevan *chora*) of the human limbic system, which has lost the instinctual ordering of nature and yet continues the primal sex, fear, and flight drives of our mammalian ancestors. Prometheus, along with his pitiable human children, "blind and weak, like prisoners of a dream," also represents such abject despair, turning into revolutionary power and technological inventiveness: through the Real, Imaginary, and Symbolic areas of the brain. In

The Oresteia, a trilogy from 458 BCE, Aeschylus fleshes out these animal, human, and divine orders of the brain's theatre more fully onstage, with a variety of gods meddling in mortal affairs.

The struggle of various gods, as expressions of the evolutionary drives, desires, and controls within humans, may be seen with the monstrous Furies, instructive Apollo, and juridical Athena in *The Oresteia*. (Other gods also appear in *Prometheus Bound*— Power, Violence, Hephaistos, Oceanus mounted on a griffin,[43] and Hermes — but as extensions of Zeus's control over Prometheus.) *The Oresteia* shows how Orestes is born into a tragic fate, in the cursed House of Atreus.[44] In the first play of the trilogy, Orestes's mother, Clytemnestra, and her lover, Aegisthus, kill Orestes's father, Agamemnon, after his victorious return from the ten-year Trojan War. (Clytemnestra's violence, at least in part, avenges Agamemnon's sacrificial killing of their daughter, Iphigenia, to get favorable winds from the goddess Artemis for his battleships at the start of that war.) In the second play, the *Choephoroi* or *Libation Bearers*, Orestes returns home from exile (disguised as a messenger) and kills his mother, getting vengeance for his father's ghost.[45] He is pursued by her Furies, the Erinyes, until with the support of Apollo and the judgment of Athena, he finds sanctuary in Athens. This happens during the final play, *The Eumenides*, with that title as a new name for the Furies, as "Gentle Ones," once they are tamed.

The ancient Greeks "commonly spoke of the Erinyes of a particular individual, so that those of Cytemnestra will be different from those of Agamemenon" (Brown 28).[46] The Furies of Clytemnestra appear onstage in the third play as a collective Chorus (performed by a dozen or more young men in monstrous masks and costumes). They appear elsewhere in Greek myth (and in Virgil's *Aeneid*, Book VII) as three specific animal-human goddesses, sometimes leading a flock of Furies — with bird or bat wings, serpents in their hair, dog heads or bodies, and blood dripping from their eyes.[47] They are more ancient than Zeus, and out of his control, since they were born from the blood of the castrated Ouranos, when it hit the earth, Gaia (Apollodorus 27). The names of the three primary Furies suggest the tragic terrors and tribulations of Orestes's guilt after killing his mother and the threat of his father's Furies if he does not: Megaera (the jealous, begrudging denier), Alecto (the uneasiness of incessant anger), and Tisiphone (the blood avenger).[48]

As evolutionary theorists of "Machiavellian Intelligence" have recently argued (about humans and other primates), as psychoanalysts have long investigated, and as current neuroscientists confirm, our animal-human hybrid brains are built for self and other deception, along with self awareness. First the right brain, then the left, is pruned by early childhood and later experiences with parents and other social representatives — so that each person's remnant animal drives are shaped by the embodiment of cultural rules. While such "prosocial" connections are reasserted by the left brain's executive controls, the right brain's alternative processes sometimes differ (Cozolino, *Neuroscience of Psychotherapy* 118), even to the point of acting out limbic (id) passions against the ingrained morality (superego). Denial of such passions is a prevalent defense mechanism — for self and other deception — as theorized in the human mind by psychoanalysis and specified in the brain by neurology (Ramachandran and Blakeslee 131–56).

> The left hemisphere's job is to create a belief system or model and to fold new experiences into that belief system. If confronted with some new information that doesn't fit the model, it relies on Freudian defense mechanisms to deny, repress or confabulate — anything to preserve the status quo. The right hemisphere's strategy, on the other hand, is to play "Devil's Advocate," to question the status quo and look for global inconsistencies. When anomalous information reaches a certain threshold, the right hemisphere decides that it is time to force a complete revision of the entire model and start from scratch [136].

In neurological terms, Megaera, the jealous denier, is partly a left-brain Fury, repressing and yet increasing the right-brain force of Alecto's angry questioning and the limbic power of Tisiphone's vengeful rage.

Denial and repression may have biological as well as psychological sources. After a stroke in the brain's right hemisphere, especially in its parietal lobe, patients often suffer an extreme form of denial called "neglect syndrome" (Ramachandran and Blakeslee 113–26). The patient will not attend to something on the left side of the body (unless it is moving) and will fail to draw the left side of a picture — even when imagining it with the mind's eye, in the brain's internal theatre. But this does not occur with left-hemisphere stroke victims. The "holistic" right brain has a "broad 'searchlight' of attention that encompasses both the entire left and entire right visual fields. The left hemisphere, on the other hand, has a much smaller searchlight, which is confined entirely to the right side of the world (perhaps because it is so busy with other things, such as language)" (117). Thus, if the right brain is damaged, the left cannot make up for its lack with attention to both sides of the visual field. Such patients may not only miss what is on their left side, but also deny that the left side of their bodies has been paralyzed by the right hemisphere stroke — or that their left limb even belongs to them (119, 131). In these extreme cases where neglect becomes outright denial (anosognosia), the executive, narrative sense of self and its "belief system" in the left brain insists that nothing is wrong on the left side of the body, confabulating as to why the left limb is inactive. One of Ramachandran's patient even claimed that she was clapping her hands together, and that she could tie a shoe "with both hands," when the left one could not move at all (129, 139).

As Ramachandran suggests, such extreme cases of denial involve both neurological and psychoanalytic processes. The right-brain stroke produces the paralysis and leaves the left brain to make sense of the situation in various ways, depending on the patient's particular Freudian "'reaction formation'— a subconscious attempt to disguise something that is threatening to your self-esteem by asserting the opposite" (Ramachandran and Blakeslee 139). Without the right hemisphere's "discrepancy detector," such patients also deny that their other, non-paralyzed arm is apparently not moving — in an experiment with mirrors that makes them see it as being still when they try to move it (140–41). Ramachandran thus finds many Freudian defense mechanisms exemplified with his stroke patients: denial, repression, reaction formation, rationalization, humor, and projection (153–55). This shows the importance of considering both neurological and psychoanalytic components in understanding the brain's intersubjective, internal and external theatres.

Orestes did not suffer a right-brain stroke. But he did experience multiple traumas in the childhood pruning of that maternal-oriented hemisphere: his mother's mourning for his sacrificed sibling, her rage at his father for that killing, Orestes's own exile from home, news of his father's murder, and the patriarchal pressure for vengeance against his mother. *The Oresteia* displays not only this vengeance but also the repressive denial, incessant unease, and vengeful return of maternal Furies through the damaged neural wiring and family connections of Orestes's mental theatre. It shows the social infection and emotional contagion caused by his own perpetuation of family trauma through reciprocal violence[49] — and the ancient attempt at a cathartic cure — with the hero's internal brain theatre turned inside-out onstage. Thus, the stage actor, playing Orestes or another tragic role, may function as a cathartic scapegoat, sacrificed to the brainstem instincts, limbic passions, right-brain devils, and left-brain gods in the audience, involving their family fates and "social drama."[50]

Is Orestes guilty as a murderer or is he a scapegoat of the gods as victim of his fate? With the help of his sister Electra and his friend Pylades, Orestes plans and acts out the killing of his mother and her lover, after they killed and dismembered his father (like the bacchae did to Pentheus and Seth to Osiris).[51] However, Orestes was born into that situation and feels compelled to kill through "grief" (106) and through a duty to his father as his phantom audience. "[I] beg my father from this mounded grave / to hear me and attend" (105). He seeks the advice of Apollo's oracle, which encourages him toward violent vengeance. Indeed, the oracle threatens a furious guilt in Orestes's brain and physical symptoms in his body if he does *not* avenge his father's death: "a swarm of worries," "[l]eprosies that mount the flesh with acid fangs," and "the Erinyes, / springing from my father's blood" (117). Orestes inherits a family situation sculpting his brain through violent acts toward further violence. He also feels a territorial and patriarchal duty to recover his father's kingdom and honor his father's military colleagues: "the loss of my estates, / the scandal of my citizens, those famous ones / who toppled Troy" (118). He chooses to act heroically—given his fate, his family's sacrificial compulsions, and his own brain's pruning. But then he suffers tragically, through the conflicting gods of his neural theatre.

Orestes kills his mother with the help of his sister Electra and his friend Pylades. Yet he suffers the vengeance of the Furies alone, hounded into exile, as the Chorus says in the third play, "Like a dog on a wounded fawn / we track him by red trickles" (170). Their appetite for blood also makes them vampire-like. "I'll suck / Your limbs alive / Of scarlet chrism... / ... / Oh, to feed on you—/ You grisly elixir! / And when I've sucked you limp / I'll drag you down / Below to pay / A murdered mother's pains." Orestes cannot stay in Argos to rule, like a confident, executive, left-brain general (in Ramachandran's terms), because the maternal Furies arise through his devilish right brain, mammalian limbic system, and instinctual brainstem—with carnivorous passions that disrupt the left's control. Indeed, the mother's primal Furies are inside Orestes, in his mental theatre, as well as outside, pursuing him. As Electra says before her brother gets revenge, "For *we* like the wolf are raw / with the savage heart of our mother" (123).

At the end of the second play, Orestes tells the Chorus of Trojan women (Clytemnestra's captive servants) about his mad vision of the Furies: "You do not see them. But *I* see them. / I am driven, driven—I cannot stay" (152).[52] But in the third play, the audience shares Orestes's vision; the Furies appear onstage with him, as the Chorus. Orestes's neural theatre is turned inside-out, showing the conflicts of his left brain, right brain, and limbic system in the *agon* between Apollo/Athena, the ghost of Clytemnestra (who also appears onstage), and her Furies. Orestes seeks refuge from his mother's ghost and her Furies at Apollo's temple at Delphi. Apollo temporarily represses the limbic monsters by putting them to "sleep" (161). The oracle god also sends Orestes to Athens, where Athena will cure him, with her left-brain attributes: a rational human jury, "words to charm," and the cathartic "release for you from all this strife" (162).

Apollo then confronts the Furies, who are roused from sleep by Clytemnestra's ghost. He questions their failure to get vengeance for Clytemnestra's killing of Agamemnon. But they explain that only genetic, not social ties concern them: "Such a killing does not count as blood of kin" (168). The Furies care passionately about the animal heritage of kin selection—privileging genetic respect over the cultural meme of marriage, which Apollo relates to the executive, Olympian ideal of "Hera's consummated pact with Zeus" and even "Aphrodite's ... logic." The Erinyes seek "justice" through a different, older order of earth and blood (168)[53] — of "Tartarus beneath," as Apollo puts it (162).[54]

Scholars today disagree about whether ancient Greek spectators experienced the gods in their own lives like Orestes and other characters onstage. Jon Mikalson argues that in the popular religion of ancient Greece, people might refer, like Orestes at one point, to "the gods" as responsible for things that happen to them (18). But they usually assigned responsibility to the gods "only for what was ... good and desirable." They would blame their failures on fortune, on a particular *daimon*,[55] or on fate. Ruth Padel finds, however, that the Greeks sensed the gods everywhere: "in every part of the environment ... [and] in every activity" of human beings (Padel, *In* 139).

The audience of *The Oresteia* saw actors portraying Apollo, Athena, and the Erinyes onstage. But they could also experience such gods in the performance space around them, on the hillside where they sat and in the sky overhead. Orestes, right after killing his mother and her lover, refers to "the wide all-seeing sun," associated with Apollo, as a "witness" for him in the future when he is tried (Aeschylus 149). For the watching audience, the sun in the sky was part of the performance, too, as were other life sustaining and threatening forces. "What we think of as the natural world was for the fifth-century a divine arsenal. Gods 'send' animals, as Zeus 'sends' thunderbolts, or as Apollo, Eros, and Artemis 'send' arrows" (Padel, *In* 156). In some plays, gods or demigods might appear in the sky, flying out over the stage, as in a *deus ex machina*, or standing on the *skene* roof, which became known as the *theologeion*, "the place where the gods speak" (Rehm, *Greek* 34, 70–73). At the end of *The Bacchae*, Dionysus may have stood on the roof of the *skene*, showing his power over the palace, the royal family, and the dismembered body of the king (David Wiles 181; Rehm, *Play* 213–14). In Euripides's *Medea*, the title character made her final entry as a *deus ex machina*, in a flying chariot pulled by dragons, a gift from her grandfather, the Sun (Helios).[56] This displayed her own supernatural passion in killing her children. And it exemplified the connection between divinity, nature, and emotions in fifth-century Greece: "Gods throw at human beings the whole environment, not only weather, elements, animals, but also other daemons.... In the same way, they also 'send' emotion" (Padel, *In* 156).

Neurologist Antonio Damasio distinguishes between unconscious drives or emotions, as the body's communication with the brain, and conscious feelings, including fictional, "as if" feelings and the self-reflective "feeling of a feeling" (*Feelings* 231–32). Primal emotions arise in the limbic system, but thoughtful feelings also engage the neocortex (*Descartes* 131–34). "Emotions play out in the theater of the body. Feelings play out in the theater of the mind" (*Looking* 28). Such a double sense of theatre is shown in Athenian performance space, as Nietzsche found with his famous distinction between the Apollonian actor onstage and the Dionysian chorus in the orchestra. This becomes apparent in the final play of *The Oresteia*, with lead actors portraying the higher-order feelings of Apollo and Athena onstage, while a chorus of Furies performs underworld emotions on the orchestra floor.[57] Such emotions also circulated through spectators' bodies in the *theatron*, evoked by the music, chants, and dancing of the chorus, as well as the drama onstage, while feelings played out in audience minds. This created a neural network of divine aspirations and conflicts, with the thoughtful feelings of Olympian sky gods and the animal passions of underworld daemons playing in the reflective, "as if" spaces of spectators' minds — while emotions performed in their bodies, as if "sent" by the gods.

In *The Bacchae*, the theatre god Dionysus manifests Apollonian authority and Dionysian passions, thoughtful feelings and daemonic emotions. He displays his shape-shifting tricks against the repressive left-brain rule of the king and shows his madness infecting the animal brains of the bacchic chorus and Agave. Likewise, the performance of this drama circulated from the spaces of stage and orchestra to individual and communal spaces in spectators' minds and bodies, involving "as if" and self-reflective, neocortical feelings, plus primal limbic passions. Perhaps it also created a catharsis of emotions into more thoughtful feelings, through the fictional sacrifices on- and off-stage, reintegrating neural pathways in new ways. Rather than just the good or evil outlook of Pentheus, stressing left-brain, binary abstractions, there might be, with this and other tragedies, originally or today, a more complex view. New sensitivities may evoke a greater awareness of how animal passions, structured by parental and cultural gods, fuel our higher-order consciousness and yet also our tendency to scapegoat others as devils to be destroyed.

Regarding *The Oresteia*, Padel finds that Aeschylus used the "daemon-discourse of epic, popular cult, and the visual arts" to summon the Furies as "part-human, threatening female figures" (*In* 179). She also relates the Erinyes' "underground sense of anger" to the fifth-century theatre's "consciousness of its own underground," the channel under the stage "through which an actor playing a ghost, for instance, might crawl up" (171). This may have occurred when Clytemnestra's ghost arrives, provoking the sleeping Furies to wake up and pursue her murderous son. The abject *chora* of the stage's underground channel and the ghost's eruption from it, into the Furies' dream, would thus reflect a potential within each spectator as well as the lead character. For the Erinyes "work punitively in the inner world, in the mind of a person who has hurt someone else. They are activated from the external inner world, the underworld." They are thus akin to the animal spirits in Paleolithic cave art — as underground externalizations of the inner worlds of prehistoric minds.

For the ancient Greeks, emotions were external, as well as internal (Padel, *In* 157). They were often personified as gods, such as Fear and Terror, like other conditions of the human body and of nature. Recall, for example, that Power and Violence are characters in *Prometheus Bound*. And yet, emotions were also "daemon, animal." As shown by the hound-like Furies and Io's gadfly, emotions maul, bite, and sting (119–21). They may act of themselves or be sent by a god. Indeed, the gods use daemonic animals to contact humans and humans use "animals, or parts of them, for getting in touch with gods," due to animals being in the natural world and thus having an "extra daemonic dimension" (144). Like Paleolithic shamans envisioning their spirit guides as animals in cave rituals, the ancient Greeks saw "particular gods, permeating and disturbing all things, acting through the world's solid fabric, in 'natural' elements..., and in animate nature, in the animals" (140).[58]

Here we see a connection that makes sense regarding current neuroscience and our biological evolution. We share with animals the primal instincts and emotions of our "reptilian" brainstem and paleo-mammalian limbic system — where drives and emotions arise. But emotions are also intersubjective, since we are highly social animals. With over 90 percent of the brain acting unconsciously, behind the scenes (in Baars's terms), emotions may perform like external and internal daemons coursing through us, sometimes beyond our conscious, cognitive control. The Furies embody this, as daemonic, animal-human, underworld goddesses inside Orestes, yet also seen externally, spurred to hunt him by his mother's ghost, arising from under the stage.

Today, the ending of *The Oresteia* may strike us as unbiological and grossly chauvinistic. Apollo argues at Athena's court in Athens that Clytemnestra's murder of Agamemnon is more heinous than Orestes's of his mother because a woman contributes nothing to her child but the space of the womb; only the father gives a generative "seed" (186). And yet, however wrong the science here, we can still see in this ancient argument[59] the different brain areas that we share with Aeschylus and his original audience. The executive, audioverbal, Symbolic, left neocortex is pruned by the Name/No of the Father — by the parent beyond the mother and by further representatives of society's rules — because of its "growth spurt in the middle of the second year," when the child is moving beyond its primal bond with the mother (Cozolino, *Neuroscience of Psychotherapy* 107).[60] The right hemisphere's intuitive, visuospatial, Imaginary order is sculpted primarily in the first two years of life, through the child's "attunement" with the mother[61] — or, in Lacanian terms, through the mirror-stage desires of the (m)Other. Therefore, the "seed" of Orestes's Symbolic identity comes from the father, according to Phoebus (radiant) Apollo, because this sun god exemplifies left-brain processes, repressing right-brain anomalies and limbic passions (although the Furies may also operate through left-brain defense mechanisms).[62]

Athena makes the final judgment in this case, after the human jury splits in its decision on whether Orestes is guilty or justified in killing his mother. Athena also admits her alliance, like her brother, with the patriarchal. They are both children of Zeus and she was born without a mother, straight from her father's head — or his left brain perhaps.[63] "No mother ever gave me birth: / I am unreservedly for male in everything / save marrying one —/ enthusiastically on my father's side" (190). Of course, it was a male actor who said this onstage, underlining the patriarchal bias of this goddess.

Orestes is free then to return as a king to Argos, reborn through his god-communing, somewhat like a Paleolithic shaman after his cave-art ordeal, involving a psychic vortex, rock membrane, and animal spirit visions. The bestial Furies insist, however, that they have been wronged and will continue to "haunt" Athens as a "contagion ... / [of] ... / Leafless and childless Revenge" (189, 192). They thus represent a primal, competitive territoriality — a Real *chora* erupting through the maternal ghost of Orestes's right brain, as devil's advocate to the left, disrupting the patriarchal culture of Athens. Yet, Athena promises the Erinyes a cave-art shrine, "a cavernous deep place" where they will be "abundantly ... worshipped" by her city (192). They complain that they will become an "old wisdom under the earth / Displaced like dirt" (194). Yet they agree to have a "home" in Athens, "under the ground," and thus become the Eumenides, the Gentle Ones (196, 199–200).[64]

In psychoanalytic terms, a cure involves not just discovering the family source of one's symptomatic repetition compulsions, but also taking responsibility for that fate, rather than blaming others. Athena's judgment of Orestes does not absolve him of guilt. She uses a jury, representing the human society of Athens, for the decision about accepting the outcast Orestes. This would justify his Symptom[65] and allow him to return home. When their vote is split evenly, for and against, she decides in Orestes's favor, but then negotiates with the Furies to make a better place for them in Athens. Apollo's excuse for Orestes's violence, and Athena's agreement with that argument, may ring false today — biologically and politically. Yet, the play's final cure makes sense as neurotheatre. The Furies are not just repressed. Clytemnestra, Agamemnon, and other monstrous predecessors in the cursed House of Atreus do not become excuses for Orestes's fatal crime. Instead, a new tragicomic understanding is reached, showing the further, personal and cultural development of left-brain, right-brain, and limbic connections within Orestes and in Athens. But the drama does not end there. We are still evolving our neurotheatres — for better or worse — as we continue the struggle of reintegrating animal passions, ego anxieties, and divine ideals in our brains and communities.[66]

As mentioned in the previous chapter, the nerve fibers connecting our left and right hemispheres "cannot convey complex thoughts and perceptions, only nuances" (Newberg, d'Aquili, and Rause 21). The hemispheres each solve problems independently, using different processes, which may conflict or converge, thus involving emotions from the limbic system. Such problem solving includes the spiritual challenge of our mortal awareness and yet immortal hopes, in the higher-order consciousness of self and Other. The left brain uses "causal," "reductive," "abstractive," and "binary" operators to make meanings through perceptions, memories, and projections of good and evil ideals in this world and beyond (47–51, 63–64). The causal operator considers causes and effects. The reductive operator (in the left temporal-parietal region) analyzes component parts. The abstractive operator (in the left inferior parietal lobe)[67] creates categories and links. And the binary operator produces polar opposites (48–51, 63–64, 196n22). However, the right hemisphere uses the "holistic operator" in its parietal lobe, connecting with the "existential operator" and "emotional value operator" in the limbic system, to give a sense of depth and authority to certain images and intuitions — especially when they agree with the logical causes, component parts, abstractions, and binaries developed by the left brain (48, 51–53, 69, 195–97n22). Thus, in a specific cultural context, divine figures of good or evil "become emotionally charged convictions ... [of] a visceral belief" (69).

Aeschylus's *Oresteia* shows a battle between these brain areas, through its various gods and their interactions with Orestes. Yet it ends with a convergence of divine ideals — with causal, reductive, abstractive, binary, holistic, and emotional operators resolving most of their differences in a meeting of minds between Apollo, Athena, and the Furies. This play shows rational communication between human and divine characters, with left-brain operators persuading the right. However, Euripides's *Bacchae* reveals the joys and horrors of mystical communion with Dionysus, as the limbic *chora* triumphs in the madness of the possessed — through emotional and holistic operators in the right brain overcoming abstractive, causal, and executive controls in the left. In another way, Aeschylus's *Prometheus Bound* displays the tragic suffering of the divine within the human, with the god of "forethought" and cultural gifts rebelling against (and being punished by) Zeus's repressive, left-brain, hierarchical power.

There are many ways that Greek plays and myths show human interactions with a diversity of gods — in a culture very different from our own. Yet, the three examples summarized above fit with Nietzsche's famous generalization about Apollonian and Dionysian modes in the "birth" of tragedy.[68] *The Oresteia* depicts vulnerable and yet triumphant Apollonian aspirations toward rational, balanced, communication between humans and their contending deities, harmonizing the conflicting forces in our brain architecture. This relates not only to ancient but also to modern art and society, when left hemispheres and parasympathetic (quiescent) nervous systems appear to rule, at least officially, in our post–Enlightenment culture.[69] On the other hand, Dionysian ecstasy in mystical communion with the gods — through animal, limbic drives, sympathetic (arousal system) passions, and right-brain, holistic, devilish intuitions — may erupt in the brain as temporal-lobe presences, individually and collectively, as in *The Bacchae*. Prometheus, a shamanic medium bound between the divine and human realms, embodies both the sublime Apollonian gifts of our higher-order consciousness, with its cultural technologies, and the rebellious Dionysian passions of our fiery neural structures that lack the natural being of our animal ancestors, yet make new worlds of meaning and survival.

In other plays, the gods are depicted as comical buffoons, for example the demigod Herakles in Euripides's *Alcestis*, or Prometheus in Aristophanes's *The Birds*.[70] Sometimes the gods are helpful to humans, as when Herakles mediates between Neoptolemus and Philoctetes, while also promising to cure the latter's perpetual wound, in Sophocles's *Philoctetes*. But more often, the gods are cruel to those who offend them — as in the plays explored in this chapter. Likewise, in Euripides's *Hippolytus*, Aphrodite manipulates the limbic and right-brain passions of love and honor in Phaedra and Theseus to get vengeance on Hippolytus, after he favors Artemis over her. In the end, Artemis promises Theseus that she will get revenge, too, against Aphrodite, using another human as a scapegoat. "I'll wait until she loves a mortal next time, / and with this hand — with these unerring arrows / I'll punish him" (289). Even Athena, who exemplifies left-brain control of the Furies in Aeschylus's trilogy, becomes furious herself in Sophocles's *Ajax*, deluding that heroic warrior with Dionysian madness, for the sake of her favorite, Odysseus.[71] (Ajax attacks, slaughters, and dismembers a herd of cattle, seeing them as the Greeks he wants vengeance upon.) In Euripides's *Trojan Women*, Athena becomes "mood[y]" against the Greeks, after she helps them win the Trojan War, because they "dishonored" her temple while sacking the rest of Troy (463). She collaborates with Poseidon, who favors the Trojans, to take revenge on the Greeks in their journey home — through the violence of the sea.

The ancient Greeks tried to make sense of natural and human destructiveness, as well as the contending processes within their skulls, by extrapolating such forces as gods watching, punishing, provoking, and possessing mortals on earth. A thousand years earlier, the Egyptian coronation drama also expressed the brain's animal, human, and divine powers in relation to political rulers and the threat of social or natural chaos. The ancient Egyptians did not value the brain materially — preserving other organs, yet discarding the brain, when embalming their dead. But their depiction of the sky-god Horus (and the new pharaoh) as dominating the rebellious, desert creature Set shows their personal and social experience of left-brain controls vying with right-brain and limbic passions. The symbolic props in the coronation play, involving Horus and his Eye of Power, but also the Osiris myth about death, dismemberment, and resurrection, reflect primal, shamanic, initiation rites of physical ordeals, psychological fragmentation, and spiritual rebirth. Such rites relate as well to the animal spirits in Paleolithic cave art, tens of thousands of years before. And they are echoed in the identity-shattering ordeal of Orestes with the Furies, plus his rebirth in Athens, as well as the tragic reshaping of Pentheus by Agave, through the spirit of Dionysus.

The Egyptian coronation drama, as an environmental, processional performance, represented the pharaoh and his father's mummy, as Horus and Osiris, at the top of a pyramid of social and cosmic power, repeatedly containing the threat of Set. This involved various performance spaces along the Nile River, which had a natural rise and fall that sustained Egyptian civilization, its agricultural gifts returned symbolically to the pharaoh, as Horus, and its pyramidal power structure — against the desert dangers beyond. But Greek drama demonstrates a further stage in Western culture's externalization of its neurotheatre in relation to the natural and cultural environment. The Greeks reveal a crucial distinction between rational, left-brain communication and intuitive, right-brain (or mystical, limbic) communion with the divine, which is still inherent in our neural equipment today.[72] With a performance space of hillside seating, earthen orchestra circle, and fixed stage, plus the temple, city, and sky as surrounding background, Greek tragedies showed human heroes (or the human-oriented god, Prometheus) suffering at the bottom of an inverted pyramid of power — unlike the triumph of the Egyptian pharaoh, as Horus over Set. While these Egyptian gods were represented in animal-human forms, involving sky power and desert danger, the Greeks showed their Olympian sky gods as human figures onstage, reserving animal-human hybridity for older earth goddesses (Erinyes) or the mythical creatures of nature (Dionysian satyrs). Yet, Greek gods and daemons also signified the cultural shaping of the animal brain within the human neural theatre, as they helped, punished, or possessed the human characters. This occurred through remnant animal drives, anatomical operators, specific family fates, and a vastly transformed social environment of ego memes and superego morals (beyond genetic patterns in a fixed habitat) competing or combining to survive.

Music and dance might also be considered, along with myth and words in ancient tragedy, as expressions of limbic emotions and right-brain mimetic systems, deriving from earlier stages in our species' evolution. (The mimetic stage of our hominid ancestors involved a kinematic image of self, through rhythm, building on the prior, episodic stage — as mentioned in the introduction here.) Ancient theatre continued to involve its primordial performance element and space: the dithyrambic chorus in the orchestra, out of which Greek theatre was "born" (according to Nietzsche). In the plays considered above, such primal communications of music and dance, involving deeper neural structures, reveal inner and outer danger, ecstasy, or communal reflection — through the monstrous Furies (and captive Trojan women), the mad bacchae, and the questioning Oceanids.[73] Yet, these female choruses, while critical of the patriarchy, were performed by males, perhaps beardless *ephebes*, young men in military training (Winkler).[74] They sang, played instruments, and danced between the human and divine heroes onstage and their watching audience, in an open-air theatre built into the hillside, near the temple of Dionysus. Thus, the primal, disruptive *chora* of limbic emotion and right-brain mimesis, in and between performing brains, was still contained within and shaped by the musical orders, performance conventions, and architectural spaces of Greek theatre.[75]

The ancient Egyptians and Greeks had very different cultural views from ours today on political rights, the afterlife, male and female reproductive roles, and the gods' interactions with humans. Yet, their plays show specific externalizations of the human brain's evolving equipment, which we also struggle with today. They extended the sculpting of self and Other within the brain's theatre toward myths and plays about the gods as watching and interacting with humans, in performances of animal-human gods along the Nile or of Olympian sky gods, tragic human heroes, and underworld monsters. Their plays demonstrate how our human neurotheatre continues to remake the cosmos by evolving new cultural worlds. And yet, their metaphysical figures form the basis for the Western cultural unconscious, as embodied metaphors of the divine within the human animal.

In subsequent centuries, as Christendom took hold of Egypto-Greco–Roman Europe, these and other varieties of divine experience became consolidated into one God — and his opposite. Mysterious natural forces in the environment, within and between bodies, and in the brain's anatomy, were figured as the binary battle between God and Lucifer/Satan. Yet angels or devils on each side (and various saints) were also watching, nurturing or threatening, framing the meaning of, and acting in human struggles to survive and reproduce, through mortal awareness and immortal ideals. As considered in the next chapter, the medieval stage of Western theatre shows the human brain being externalized and pruned through specific permutations of biblical stories, miracle figures, and moral allegories. These paradigms continue to influence our god ideals today, through Judeo-Christian culture, more directly than Egyptian and Greek deities. The monotheistic projections of good and evil ideals may still be seen in the theatres of politics, war, and terrorism, sculpting new generations of competitive and cooperative egos, through neural circuits of sacrifice, even in our secular, mass-media rites of stage and screen.

3

One Medieval God — with His Angels and Devils

Humans bear the animal heritage of archaic neural structures that have, for millions of years, assessed natural signs as nurturing or threatening: the immediate reactions of brainstem instincts, the learned behaviors of mammalian limbic emotions, and the greater, yet conflicting perceptions and reconceptions of higher-order cortical hemispheres. We also inherit specific cultural evolutions of the human lack of natural being — involving projected desires and drives of the Other — in our (post)modern Western society and our enlarged, prematurely born, socially sculpted brains. A crucial catalyst for how we relate to the Other today, as believers in various faiths or as secular humanists, came through the cultural dominance of Christendom. A profound change in world views emerged during the Roman Empire and the Middle Ages through Judeo-Christian monotheism (and its binary tendencies): from numerous animal and human spirits, as gods and daemons, to one God in control of it all, or as a dominant force of good, through a continual cosmic battle with evil.[1]

Under Constantine and Theodosius in the fourth century, the Roman Empire converted to Christianity, aligning itself with an anti–Semitic sect of the Jesus movement. After the fall of Rome, yet survival of the Catholic Church's power in Western Europe, theatre re-emerged in the Middle Ages (in the 900s to 1500s) as a way of acting out religious stories and morals. The complex nourishments and threats of both Mother Nature and human culture, shaping our species' evolution, were personified as multiple gods in ancient times. But these forces became consolidated in the medieval, patriarchal system of one God, as ultimate king in heaven, with Lucifer or Satan as the opposite in hell, and many levels of powerful beings between them: angels, saints, clergy, secular rulers, and devils. Prior "benevolent pagan deities" were associated with saints and those "deemed sinister with the Devil and his cohorts" (Lima 84). Certain human groups were also stereotyped as evil, especially the Jews, as biblical killers of Christ (along with the Romans). This solidified the spreading politics of Christianity, creating a transnational allegiance, by demonizing the "Chosen People" at its roots, as well as various indigenous cultures with their pagan gods and certain heretics with alternative ideas of God. The cruel success of such an enterprise, with its imperialistic crusades and inquisitions, shows the collective power and yet ethical danger of Western culture's particular evolution of our human neurotheatre. (Such demonizing also occurs in other cultures, including those with polytheistic beliefs.) The left brain's binary, reductive, and abstractive operators, combining with holistic and fear/rage functions in the right hemisphere and limbic system, may produce a monotheistic unity of many brains and subcultures. But it can also stage horrific violence, through the righteous alignment of our "good" against their "evil." This is shown in the religious and secular developments of anti–Semitism, from medieval persecutions and crusades to modern genocide.[2] It is also seen in the current clash of Islamic terrorists and Western imperialists, involving another "holy war" about the "holy land."

Bloody Jews

In the Croxton *Play of the Sacrament*, probably performed in the late fifteenth century, five Jewish characters are presented as villainous, yet comical scapegoats[3] — to show the power of the consecrated bread and wine as the body and blood of Christ, in the Church's ritual theatre of symbolic human sacrifice.[4] In this English play, set in Spain, a Christian merchant, Aristorius, steals the sacred bread (the Host) from a church tabernacle and sells it to five Jews, who then experiment with it, testing its holy powers and thus giving Christ "a newe passyon" (240). They stick daggers in the Host and it bleeds (257). They try to throw it in a cauldron of boiling oil, but it sticks to the hand of one of the Jews, Jonathas (258). They nail the Host to a post, symbolizing the nailing of Jesus to the cross, and then the other Jews pull on Jonathas's arm (258–59). But his hand sticks to the Host and becomes detached from his arm — a cruel, comical dismembering, probably using a fake hand and a hidden blood bag. (The loss of a hand onstage occurs in a very different context a century later in Shakespeare's *Titus Andronicus*.) The Jews boil the Host and hand in oil (264). Miraculously, blood overflows the cauldron. When the Jews use pincers to take the Host out, it remains whole, while the hand is reduced to bone (265–66). They cook the Host in an oven to stop its bleeding. But blood continues flowing from the oven (266). Even more miraculously, Jesus appears with his bleeding wounds to show not only their foolish villainy, but also his body and blood: present in the Host as holy prop and redemptive of their mistaken sacrifices (267). Jesus then restores Jonathas's hand to his arm (by telling him to put it in the cauldron), converts the Jews to Christianity, and changes himself back into "brede" (268–70).

A more bloody show occurs in a similar miracle play, *Mistere de la Sainte Hostie*. According to a witness in Metz in 1513: "a great abundance of blood ... shot upwards from the aforementioned host, as if it were a child pissing. And the Jew was all soiled and bloody and played his part very well" (Philippe de Vigneulles, qtd. and trans. in Enders, *Medieval* 194).[5] The Jew stabbed the Host several more times and it "spit out blood until the whole center stage glistened with blood and the whole place was full of blood" (195). This profusely bleeding Host, as Body of Christ, like the dismembered hand of the Jew in the English play, shows a primal connection between blood, body, and bread — and wine also since these dramas reflect the priest's re-enactment of the Last Supper during mass. As Jody Enders points out, such dismemberment and bloody sacrifice was familiar to medieval audiences, with public displays of torture and death used to teach and enforce the law, especially during the Inquisitions in Spain, France, and Italy. While God was often a character in medieval drama, and believed to be watching from above, religious and secular rulers also watched the bodies performing in the play and in the audience. The violence onstage served not only as entertainment, but also as a warning about further pleasures and punishments — shaping the neurotheatrical networks of subjects in the audience.

Such bloody shows reveal the continued cultural development, in the Western tradition, of the human brain's attempt to transcend its mortal nature and environment. Ever since the agricultural revolution, humans in Europe were dependent on grain and bread as a staple food source. The biblical story of the Last Supper (also drawing on the Jewish Passover ritual) turns the nourishing bread and dangerous Dionysian wine into Christ's Body and Blood — with a rite that believers are commanded by Jesus to repeat, in memory of him. This sets up a ritual of holy cannibalism: consuming God's body and blood in the bread and wine. Thus, "communion" reflects ancient rites of blood sacrifice that Christianity transforms from Abrahamic[6] and Dionysian traditions. The Croxton play shows this with its Jewish bacchae and the "madness of the Host ... [as] contagious" (Scherb 75). In this miracle play and the holy sacrifice of the mass, Christians tried to demonstrate that they, not the Jews, were God's "chosen people." They were building upon, yet going beyond the promises of Yahweh to Israel in the "Old Testament." They were also displaying the continued influence of ancient Greek and Egyptian gods: Dionysian *sparagmos* and the Set-Osiris-Horus myth, involving bloody body parts, wine, bread, and other symbolic props.

In the bleeding Host miracle plays of England and France, Jewish villains are explicitly scapegoated.[7] They become the internal Other that must be purged, as in the Spanish Inquisition. They are "extimate" (in Lacanian terms), too intimate and therefore externalized, in order to purify Christian identity, purpose, and salvation, while consolidating the earthly rulers' controls. Jews were expelled from England two centuries earlier (in 1290). Yet, the exiling of Jewish villains in the Croxton play not only reminds the audience of that prior banishment. It also turns the Jews into vehicles for an "apocalyptic teleology," which seeks to "transcend both universal and local historical narratives" (Nisse 108, 113). Jesus tells them that their experiments with the Host have given God "a newe tormentry," even though He died for them, too, on the cross (*Play* 267). The Jews repent and convert; they are healed, forgiven, and christened. Yet, they must leave then on a diasporic "vyage" (275). Aristorius, the Christian traitor who sold them the Host, repents, but he is sent away as well. The Spanish city where the play takes place (Heraclea in Aragon) is thus symbolically evacuated, in a shift from historical to apocalyptic time, when all will be judged and converted by God (Nisse 122–23).

Such scapegoating and expelling of villains, through symbolic bread and blood, reflects the cultural changes in human evolution, more than ten thousand years ago, when nomadic hunter-gatherer groups settled into territorial, grain-growing communities with altered perceptions of survival threats and benefits, as they transformed their environment. The flow of blood had been crucial to the feeding exchange between hominids and nature for millions of years. But with the advent of settled human communities, bread became a vital commodity in Western culture—with grain grown, stored, and used, along with domesticated animals, inside boundaries that had to be defended, or could be captured. Wine also represented the transformation of nature into culture, yet pointed to the power of animal passions still within the human body and brain, which might be triggered by this Dionysian drink or shown in ancient theatre. Thus, the Christian transubstantiation of bread and wine into sacrificial flesh and blood evokes not only the power of God—to become man, to suffer for the sins of mankind, to bleed, to die, and to rise again—but also the power of Western culture(s) to shape the natural world and redefine its survival and reproduction drives.

The producing and protecting of human, hyper-territorial, transformative, and transcendent wealth involved the distinction of certain people as cultural threats, or "terrorists," outside and within the community. It also involved a new sense of God as a symptom of expanding sacrificial demands, ego desires, and imperial appetites. Consuming God's body and blood through the bread and wine at mass, or experimenting with it in the *Sacrament* play (and losing/regaining the hand that touches it), shows a distinctive cultural extension of the nurturing and threatening forces in human evolution. Nature and the brain are altered by medieval culture and its sacrificial values of good and evil, heroes and villains, communion and alienation. But the Jewish villains, experimenting with the Host, also point to a new era of neurotheatrical expansion (considered in the next chapter here): Renaissance science as a divine power in human hands.

According to theatre historian Donnalee Dox,[8] a miracle play such as the Croxton *Play of the Sacrament* demonstrates how medieval theatre represented imaginary space onstage as reflecting the creative mind of God, beyond the real world of nature, from earth to the heavenly spheres. The performance spaces of the play move from chaos to order (168), with realistic special effects involving animal blood and fire under a cauldron (172), for the miracle of the Host to be presented, producing the Jews' torture, dismemberment, and then conversion. This also reflects a theological debate in the prior, thirteenth and fourteenth centuries, about God's power to "create worlds beyond those observable by the human senses"—with an "extracosmic void" beyond the natural world as the divine "space of imagination" (185–86). Thus, the performance of the Croxton play in the fifteenth century brought into the imaginary space of theatre, with its realistic material effects, "the power of an unseen God to effect miraculous transformations from a realm inaccessible to human sight and mind" (187). God was viewed as "omnipresent and capable of altering material conditions" in the "bounded cosmos" of the natural world, as this miracle play showed (186). God worked in "an extratemporal 'time' of creation and judgment," just as the transformation of the Host and conversion of the Jews took place "in *durati*, or God's time, ... and [in] the imaginary space of Christian belief in an intervening deity" (188).

In a neuro-psychoanalytic sense, the right neocortical, Imaginary spacetime of the medieval spectator's inner theatre (also involving left hemisphere, Symbolic orders and subcortical, Real drives) became extrapolated as the divine dimensions of God's ruling view of the human world and His miraculous interventions in material reality. Real terrors erupted onstage with bodily fragmentation and blood spurting (even if comical). Then a miraculous Symbolic order restored Imaginary wholeness, while revealing another sacrificial body, with its bloody offering transcending time and mortal nature. Like the human brain using its perceptual circuitry to create fictional possibilities for the self and others within the mind—and like the stage's special effects showing violent transformations—God was an extracosmic playwright and yet an immanent presence: watching, directing, and acting in the natural world.[9]

But with one God in all-knowing control, how does the medieval mind account for suffering, mortal threats, and temptations of evil? Is God evil as well as good? Does God give a special identity and immortal salvation to Christians alone — despite the beginnings of such monotheistic principles in Judaism? In order to purify God of such inconsistencies, medieval theatre shows Lucifer, Satan, and other fallen angels as evil rivals, while demonizing certain humans, especially the Jews, as external to divine goodness and Christian salvation, even if they were God's favorites before. Yet, it also shows the guilt of all men, through Adam and Eve's original sin, as well as every man's potential to join with God, through Jesus and the miraculous Host.

A particular cultural evolution of the human brain occurred in Europe's Middle Ages, as manifested in theatrical performances: shaping the Symbolic beliefs and executive controls of the left neocortex, the Imaginary fantasies and holistic perceptions of the right hemisphere, and the Real emotions of the limbic system. This helped provide a collective order on the social level, with potential cosmic meanings for joy and suffering in this life and the next. Yet, it required the submission of limbic, animal passions and right-brain, alternative views to orthodox beliefs. It also produced the persecution of those deemed antithetical to Christian goodness (such as Jews and witches)—through the brain's binary, causal, reductive, and abstractive operators, as well as its temporal and parietal lobe devices for spiritual experiences. Hence, the "binary distinction between God and the devil became the model for a series of parallel oppositions that profoundly influenced thinking about science, history, religion, and politics" (John Cox, *The Devil* 2). Like the Jews of the Croxton play, devils in English medieval theatre incarnated villainy, enacting "whatever opposed individual wellbeing and the sacramental community"[10]—through the left brain's binary control of the right's alternative intuitions and limbic passions, as diverse (or perverse) forms of spirituality. Rather than honoring a tragicomic trickster, like the ancient Dionysus, as god of theatre and of nature's fertile, transformative drives, medieval performance focused on God's scripting and ruling of the world versus Lucifer's rebellion and Satan's temptations — as melodramatic forms of good versus evil. But tragic moments still emerged with these and other characters, offering the cathartic potential of more complex connections across neural and political divides.

God versus Devil

In medieval cycle dramas, the biblical history of the world was shown, from God's creation of it to the Last Judgment of all humans by Christ in heaven, through a series of short plays staged over many days, organized and acted by clergy and townspeople. Such plays were presented for hundreds of years, from the 12th to 16th centuries, in various cities around Europe. The Wakefield (or Towneley) Cycle, staged in England in the mid 15th century, begins with God on a throne, declaring His "Ego" as "alpha et omega" (*Wakefield* 59).[11] From the start of *The Creation*, the first play in this cycle, God is shown as a heavenly ruler, as the ultimate ideal of self awareness and creativity, who then creates man in that "likeness" (64). Prior to man's creation — yet symbolizing the tragic hubris of his higher-order consciousness — Lucifer, in self-reflective admiration (and a bright costume), sees himself as "a thousandfold / Brighter than the sun" (61). He mounts the heavenly throne as a "king of bliss," after God exits the stage (62). Other angels debate whether Lucifer deserves such a status, with his brightness or vanity, as if echoing primal human arguments about which alpha male should rule and how many gods to believe in.

When Lucifer flies upwards "above God's throne," he and the other bad angels fail in their presumptive evolution and fall toward hell (63). They then emerge from the "hell-mouth," a cave-like medieval set piece, howling that they are now "as black as coal, / And ugly...." For medieval spectators, who were mostly illiterate, not allowed to read the Bible, and yet believers in a book-based religion, the Wakefield Cycle displayed the mystery of good and evil, of nurturing and threatening forces in nature, in the brain, and in human culture. It presented these many mysterious forces as related, yet polar opposites: from God's ultimate creative power to Lucifer's similar solar brightness, but then vanity, fall, and embodiment of evil.[12] The cycle's opening drama involved regal rivalry, collective orders, conflicting desires, and yet also an admirable striving for greatness. It showed the competition of divine ideals, for good and ill, as extensions of humans' evolving higher-order consciousness. Lucifer's tragic fall into darkness, along with other devils, may give meaning to the cruelties of nature and culture, with God as the ultimate ruler of heaven. But it also suggests to a medieval or modern audience that such cruelties — and our own evil temptations — are inherent in the workings of the cosmos, theologically, historically, and biologically.

Aspects of medieval devils may be traced back to ancient Greek theatre, to the rebellious energy and "anti-character" of the Dionysian chorus (Soule). Devils were God's enemies and yet also carnivalesque folk figures, more physically involved with humans than the elite character of God in heaven (76). "[T]he Devil was present not only as a character or theatrical persona, but as a natural agency, acting within (i.e., possessing) the performers playing these figures" (77). Like Soule, but using current neuroscience, I would view the devils in the Wakefield *Creation* play, and elsewhere in medieval drama, as representing the "natural" heritage of the human brain. They show the animal drives of the limbic system, plus the intuitive desires and anxieties of the right hemisphere, conspiring against the dominant cultural beliefs and executive controls of the left cortex (with God as its supreme, regal extrapolation).

One might also see devils as representing the psychoanalytic *chora*, disrupting the thetic and symbolic orders of medieval cosmology (Kristeva). Or in neurological terms, such figures onstage reflect limbic emotions producing conscious feelings in the right brain, as a "Devil's Advocate," contradicting the war-room General in the left (Ramachandran and Blakeslee 135–47). This is shown in medieval performance practices, with evil assigned to the left side of the stage, since perception and action on the left side of the body are connected with the right cortex (and vice-versa). Indeed, the medieval hellmouth and its devils were traditionally placed on the left side because "all things considered evil were sinister, that is, of the left-hand path" (Lima 30). The right brain's greater ties to limbic, mammalian emotions can also be seen in the medieval tradition of devils being presented collectively, as "a threat from without," running in packs, "like wild dogs" (Soule 86)— or like the *Oresteia*'s Furies. Their masks were probably "animalized," as with the horned satyrs of ancient Greek theatre (88). Yet, they were shown, too, as "a threat from within, expressing the rampant bodily eruption of psychic forces" (86).

Both the single and collective anti-human character of the Devil, as external and internal threat, can be seen in the Wakefield *Creation* and in a much earlier Genesis play, the Anglo-Norman *Mystery of Adam*, performed in the twelfth century. In that earlier play, the Devil conspires not only with other demons, but also with the audience, by making an "excursion through the square" (outside the church), before returning to his temptation of Adam onstage (*Adam* 307). The Devil and his demons make several such excursions, "involving carnivalesque interplay between the performers and the audience" (Soule 78). Then, after God banishes Adam and Eve from paradise, the devils put "chains and iron shackles" on our primal ancestors' necks (*Adam* 314). They push and drag Adam and Eve toward hell, where other devils "make a great dancing and jubilation over their destruction." Devils also thrust them into hell, causing "a great smoke to arise." And they "dash together their pots and kettles, so that they may be heard without." With this "rejoicing," some of the devils "run to and fro in the square." Through such carnivalesque entertainment, the noise and spectacle of devilish lures may have evoked spectators' rebellious desires as well as animal fears, even beyond the left-brain authority of God's presence. In sixteenth-century England, the Puritans spread warnings about actual devils appearing in such plays, dragging the human actors toward hell (Elliott 10).

Likewise, at the start of the Wakefield Cycle, spectators are engaged in a dialectic between God's gifts and the Devil's temptations. According to the envious Lucifer, now "Satan," God "fashioned man [as] his friend" and gave him "bliss without an end," which the rebellious angels have lost (*Wakefield* 67).[13] Humans are created to take the "fallen angels' place," by God's "will." In revenge, Satan tempts Adam, through Eve, to eat from the tree of the knowledge "of good and ill," to "Bite on boldly" and "As gods ... [to] know everything" (69–70). Banished by God for these divine aspirations,[14] Adam and Eve fall out of paradise. They are forced by God's angel, Cherubim, to "middle-earth," a lower level of the drama's staging, below heaven and paradise, but above hell (72).

Through these performance moments, medieval spectators might have glimpsed in a biblical way something of what we know today about human evolution. As if eating from a tree of knowledge, our hominid ancestors gradually acquired from their environment, through selective adaptations across many generations, the powers of a bigger brain, knowing more "good and ill." The brain expanded further with humans' survival in new environments (of middle-earth) and transformations of nature and themselves through mimetic gestures, symbolic languages, and creative cultures. But the powers of the human brain, with its godlike aspirations, ego anxieties, and animal drives, also produced tremendous destructiveness, as well as creativity. Today, we continue to experience, like our medieval ancestors, a lack of being within nature's patterns of "paradise," the mortal loss of "bliss without an end," and the temptation to bite boldly into divine technologies of competition, cooperation, and change.

The devils of medieval drama, like Set in ancient Egypt, threaten the dominant order with their chaotic, rebellious, and possessive energies. Like Prometheus in ancient Greece, they tempt humans toward new powers, yet also bring about great sacrifices. Like the Furies, they haunt the action of the drama even when not onstage. Like a Dionysian chorus, they move collectively between the stage and audience areas, connecting spectators to the show, yet also tempting them with further evolutions of animal drives and ego desires. Yet, medieval devils represent a distinctive problem in monotheistic religions, as mentioned earlier in this chapter. If there is only one almighty God, how is evil allowed to exist? Did God create it? Are the many sufferings of humans, even if caused or inspired by devils, ultimately permitted by God? For medieval plays also showed the cruelty of Yahweh, when He demanded that Abraham sacrifice his son and provided an animal substitute only at the last moment.

As Karen Armstrong points out, this Old Testament God is "essentially male ... due to his origins as a tribal god of war" (*History* 50). He appears to be "dependent on man when he wants to act in the world.... There are even hints that human beings can discern the activity of God in their own emotions and experiences, that Yahweh is part of the human condition" (57). Eventually, however, the negative side of this emotional, experiential, human-embodied God was split off as an evil Satan (or Lucifer), in the medieval Christian revision of Jewish theology — and thus connected with anti–Semitism and the witch craze. "Satan had emerged as the shadow of an impossibly good and powerful God. This had not happened in other God-religions.... Satan became a figure of ungovernable evil. He was increasingly represented as a vast animal with a priapic sexual appetite and huge genitals" (275).[15]

As shown at the beginning of the cycle dramas, Lucifer rebels against God and loses, as competing tribal god of war. Yet Lucifer wins the second round (as Satan), in the form of an intimate, animal figure: the serpent tempting Adam and Eve to disobey, through their similar aspirations to become like God. This ultimately leads to Jesus as God's way of becoming, once again, an intimate friend of mankind — suffering (like Prometheus) for the sake of human knowledge and freedom. Satan plays an increasingly important role as the evil side or counterpart to God,[16] as a separate rival force that the Son of God will battle and overcome. But Jesus is not just a melodramatic hero, countering Satan's villainy. Fully human, he is also a victim of our evolution, through the brain's persistent drives of territoriality, ego rivalry, scapegoating, and sacrifice — as Jesus inherits the Father's conflict with Lucifer over heaven's throne and human souls, in this divine tragicomedy.

Despite Adam and Eve's evolving rebellion (like Lucifer's), God does not lose interest. They eat the fruit of higher-order consciousness and fall to middle earth, with mortal awareness and sinful desires. Yet God continues to watch mankind as a divine spectator and sometimes acts upon the earthly realm. In the second play of the Wakefield Cycle, *The Killing of Abel*, Cain shows the devilish side of mankind's dominance of nature. While plowing, he yells at his horse — "Up bitch! Ye'd scarcely pull a straw"— and then fights with his servant, Pickbrain (*Wakefield* 74). He also tells his brother to "kiss the devil's tail," when Abel comes to remind him about "the customs of our law" that they tithe one-tenth of their agricultural yield as a "sacrifice" to God (75). Cain and Abel thus exhibit different relations between left-brain Symbolic, right-brain Imaginary, and limbic Real orders. But God, along with the medieval audience, oversees this drama of various brain biases toward sacrifice — like an extra cortical layer to the intersubjective theatre of Cain and Abel's evolving consciousness.

Abel's insistence upon gratitude to the God "of our law" reflects an appreciation for the geographic fortune of human groups in the Fertile Crescent, where advanced agriculture developed (Diamond, *Guns*). But his brother sees things otherwise: "Go to the devil.... / What gives God thee to praise him so? / To me he gives but sorrow and woe" (*Wakefield* 76). Cain also says that his sacrifices, in the past, ended up in "the priest's hand" instead of God's. Abel claims, however, that all their "goods" are merely on "loan" from God, by His "good grace." Yet Cain continues complaining that the God of the natural world has not nurtured him well: "he has ever been my foe.... / ... / When all men's corn was fair in field, / Not a needle would mine yield...." The apparent unfairness of nature's sustenance, in the competition between humans to fit within or master their environment, challenges Cain's right brain to question the sacrificial rationale of his brother's left.

Evolutionary psychologists argue that the great ape line split from monkeys about 12 million years ago — with primates eventually having a "Theory of Mind" (ToM) about others' views in social relations (Byrne, "Primate" 87–88). This was a key step toward the Machiavellian (or social) intelligence of great apes and humans today. Like, but far beyond apes in the wild, humans after the age of 4 show an awareness that other individuals have different perspectives from their own (Corballis, "Evolution" 133) — using the inner performance spaces and "representational skills" of their brains to imagine others' views (Bjorklund and Kipp 46–47). Chimpanzees may use this ability to demonstrate tool use, or to deceive others by inhibiting their own hunger or sex drives, pretending a predator is near, and then getting delayed gratification while others are distracted (Bradshaw 65, 73). In humans, this probably involves the dorsolateral prefrontal cortex as "the substrate of fluid problem-solving intelligence of an abstract nature and the cingulate and orbitofrontal cortex [as] the substrate of Machiavellian (or social) behaviors or intellect" (74). It also involves the higher-order development of "mirror neurons" in human brains, as different from other primates, in the language and motor skills module (Broca's area) of the left hemisphere. This produces not just "motor empathy" as observable with monkeys, but also the "intricate mental perspective taking, often with several levels of recursion, that makes human relationships so complex" (Corballis, "Evolution" 134).[17]

Chimpanzees "perform on tests of mental attribution and self concept at about the level of a 2-year-old child, but lack the sophisticated, recursive mental capacity ... termed *metamind*— that emerges [in humans] around the age of 4" (Corballis, "Evolution" 134). The bipedal locomotion of our hominid ancestors freed their hands for the development of manual gestures as a mimetic language, also serving "to sharpen and expand the internal representations of objects and actions," through the inner theatres of their brains (135–36). This then allowed the body to assume a more and more "theatrical role"— especially through the relative prematurity of human birth, "allowing much of the growth of the brain" to take place after birth, with increased cultural or "environmental input ... critical to the development of hierarchical structures in language ... and perhaps to generativity in general" (136–37).

Abel and Cain exhibit different aspects of the human metamind and its generativity, as they extend the internal theatre of their own brains and their theories of others' minds to discern the desires of a divine Other, as the ultimate Metamind. Abel sees nature's goodness as a "loan" from God, who demands a one-tenth tithe in return. But Cain projects nature's unfairness as coming from God's Metamind. Cain responds by using his mirror neurons and internal theatre to turn the Other's desires against Him — through a bodily gesture of Machiavellian deception.

Both Abel and Cain climb to a higher level of the stage to offer their sacrifices to God, as divine spectator and consumer (*Wakefield* 78). Abel burns a sheep as his tithe to God. But Cain tries a sleight of hand as he offers his sheaths of corn, miscounting to give less than a tenth. When Abel catches him and warns, "come not in God's hate," Cain responds, "I will not give my goods away" (79). Cain even insists it is "Reason" that "rules" him (80). He uses his prefrontal, abstract reasoning to inhibit limbic fear of God's hatred — after his Machiavellian achievement of mirroring and yet deceiving the Other's desire for ritual sacrifice. "If I need not dread him sore, / I were a fool to give him more." But then the theatrics of his altered rite backfire: Smoke comes from the offering as it fails to burn any more "than snow." Reflecting on the disaster, Cain complains that when he tried to blow on the sacrificial fire, the smoke almost "split" his lungs and "stank like the devil in hell" (81). Earlier, the brain of Cain's brother inhibited his hunger drive and ego greed to offer the tithe of a burning sheep to the Other's desire. But Cain's brain inhibited his fear of God to cast a Machiavellian, mind-reading trick. The natural sign of his fire signals Other-wise.

The watching Other appears then "above" them, verbalizing His Metamind with a rule mirroring Cain's choices. "If thou tithe right, thou gets thy meed; / But ... if thou tithe ill, / Repaid thou shalt be thy evil" (*Wakefield* 81). Seeing and hearing is not believing for Cain. After God withdraws on the uppermost level of the stage, he reacts sarcastically to the divine spectator's left-brain logic: "who is that hob-over-the-wall? / ... / God is out of his wit." Yet Cain also shows his fear of the Metamind's Symbolic order on the upper level. He tries to find a place where "God's eye shall not see." His fear of God's watching eye then turns to jealousy of his brother's "clean" sacrifice, which he recalls in contrast to his own that "foully smoked" (82). This memory, both visual and smelly (for the medieval spectators, too), triggers a limbic rage in Cain, through his right-brain Imaginary order and ego rivalry with Abel. He acts out suddenly, in a perverse mirroring of God's desire for sacrifice, killing his brother with an animal's "cheek-bone."

The politics of sacrifice were displayed on three levels in the medieval staging of this drama (whether on processional wagons or on platforms in the town square). Nature's unfairness to Cain becomes reinterpreted by Abel, on the lower level, as just desserts. Abel's devotion to God, on the middle level, or sacrificial "hill," is shown with his burning of a sheep, in contrast to Cain's trickery with corn (*Wakefield* 78). Then God frames the view by defining His justice as a vocal spectator from the upper level.[18] When the brothers descend again to the lower level from the hill (81), terror and rage emerge in Cain. Thus, the three levels of the stage show Cain's primal passions, ego rivalry, and divine aspirations — externalizing the limbic Real, right-brain Imaginary, and left-brain Symbolic of his brain's inner theatre.

After Abel dies, crying to God for "vengeance," Cain involves the medieval audience in the crime scene, as complicit with his expanding evil (*Wakefield* 82). "And if any of you think I did amiss, / I shall amend it, worse than it is, / That all men may see it...." This echoes Cain's reaction to God's watching eye and judgment. If the divine Metamind reflects guilt upon him, he will compound the perversity. "Much worse than it is / Right so shall it be," he says to the theatre spectators. This presages the perverse logic of John Milton's Lucifer (and Mary Shelley's Frankenstein) who finds that evil becomes his "good," after the fall from grace. Cain is also like Lucifer or Satan in the earlier plays of this cycle, displaying rebellion and deceit against the authority of God. Yet Cain moves from divine hubris to animal panic, immediately after his threat to the audience. "Into some hole I fain would creep; / For I fear I quake in so sore dread, / For be I taken I be but dead."

But the left-brain Metamind, on the upper level of the stage, rouses Cain from his limbic terror and lower-level withdrawal, demanding from him: "Where is thy brother, Abel?" (*Wakefield* 83). Cain avoids responsibility here, replying that his brother is "in hell" or just "a-sleeping" and asks, "When was he in my keeping?" Yet God is not deceived. The divine spectator hears the "voice of thy brother's blood / ... / [which from] earth to heaven vengeance cries." God tells Cain that he is "caught in a fierce flood [and will] ... / Under the flood of my fury drown." Cain continues to rebel against God's "curses," saying he will "give them back"— like his earlier perversion of sacrificial desire. But then he expresses shame. "That I may not thy mercy win / ... / I shall hide me from thy face." Having failed in his right-brain, devil's advocate rebellion against the left-brain Metamind, Cain now sees his death sentence: "if any man me find / Let him slay me and not mind." He states his will for his burial, specifying a site in Wakefield, "Goodybower at the Quarry Head"— bringing this Old Testament drama into the medieval audience's neighborhood. But he still shows the desire to escape his fate there, in comical terms. "If safe I can this place depart, / By all men set I not a fart."

Once again, the left-brain Spectator makes a rule. He forbids the reciprocal vengeance that Abel's ghost desired. God dams the flood of His own divine fury, contradicting Cain's view of his doom. "No man may another slay, / For he that slays thee, young or old / He shall be punished sevenfold" (*Wakefield* 83). God "withdraws," but gives the impression that He continues watching, invisibly. Cain takes out his stress on his servant as scapegoat, who mocks him in return. This forms a comical sub-scene of the master beating upon his rebellious servant, which also reflects, in microcosm, Cain's own battle with God and his brother. Then the servant, Pickbrain, addresses the medieval audience with a moral lesson, showing his glimpse of a higher order. He tells his anachronistic (or synchronic) contemporaries to cherish God's blessing, which was also granted to his master, "through God's grace" (86). But Cain orders the servant to come down from the "hill"— the middle level of the stage — back to his own lower level, in a fall akin to Lucifer's. "Come down yet in the devil's way." Finally, Cain says "farewell" to his audience, admitting that he is still enthralled with the devil: "World without end." Like Satan, Cain shakes his fist "at God's throne" above and tells the human audience: "Accursed I needs must hide" (86–87).

Like Set and Dionysus in ancient drama, or Lucifer and Satan in this cycle, Cain rebels against authority and tries to trick others with his Machiavellian Intelligence, through his theory of what the Other's mind desires. Like Set and Lucifer, he fails and falls, becoming a villain through jealousy, ego rivalry, and hubris. Like Orestes, he is hounded by his fate — and by his victim's ghost, who calls for vengeance against him. Like Satan using Eve and Adam in his vengeance against God, Cain uses Abel and his servant as scapegoats. Yet, as the villain, Cain becomes a scapegoat, too, for the medieval audience, like the Jews in the miraculous host plays. His right-brain questioning of God's left-brain demand for sacrifice, and of Abel's submission to it, distinguishes Cain as alien to the Christian context of the play's performance. His errors are laughable and his suffering enjoyable — as a negative example for medieval spectators. But is he a scapegoat of God, the ultimate spectator? Was he already fated for sacrificial punishment as the first murderer, like the sheep his brother kills, in God's omniscient scripting of mankind's fate?

Evolutionary psychologist David Barash recently took a line from Stephen Sondheim's musical, *Sweeney Todd*, to claim that the "dark and angry God" served by the title character, a London barber and serial killer,[19] is "universal." According to Barash, humans have always had "a peculiar need, as insistent as it has been tragic," deriving from our animal ancestors, to make others "suffer the pain we feel." He gives evidence from studies of laboratory rats that were stressed with repeated shocks and thus developed oversized adrenal glands and sometimes stomach ulcers. But if rats given the same treatment were allowed to chew on a stick, they could endure the stress longer (before becoming apathetic) and suffered less severe physical damage. If they had other rats to fight with, they survived even better, enduring the stress without any adrenal gland enlargement or stomach ulcers.

> When animals respond to stress and pain by redirecting their aggression outside themselves, whether biting a stick or, better yet, another individual, it appears that they are protecting themselves from stress. By passing their pain along, such animals minister to their own needs. Although a far cry from being ethically "good," it is definitely "natural." (Barash B6)

Barash relates this to the scapegoating that occurs in many, if not all human cultures, as a way of redirecting aggression, akin to "retaliation and revenge," but directed at an innocent party. From Bosnian genocide to the U.S. invasion of Iraq, Barash sees many recent examples of violent scapegoating as redirected aggression. He refers to physiologists who have found that both humans and animals experience "subordination stress," when attacked or threatened, and may suffer adrenal and ulcer damage, unless they display aggression outwardly. "By passing along their pain, they modulate their own internal distress while generating trouble for the next ones down the line."

Barash does not mention that a play such as *Sweeney Todd* may also be a form of scapegoating — perhaps a better evolved (or less costly) type of redirected aggression. Like "rough-and-tumble" play, which has also been studied in rats (Panksepp 295), violent theatre is emotional exercise. Like dreams, according to a current theory in neuroscience (Revonsuo 90), plays may also be a form of "simulating threat-perception and rehearsing threat-avoidance" for the audience. The personal traumas and subordination stress of spectators' own lives might find a redirected aggression through fictional characters suffering onstage, while actors make sacrifices to embody them. But such theatrical sadomasochism, rather than repeating actual harm, may produce a greater awareness about the tragic causes and consequences of violence. This catharsis could reshape Sweeney's dark and angry God, through the internal theatres of his audience, instead of repeating that "natural" need to pass along stress to a victim or an object.

Scapegoating continues today in many real life situations, far beyond the cathartic power of stage and screen performances. Perhaps subordination stress and redirected aggression will always be necessary for social order, through the shifting hierarchies of human groups. In the ancient Greek ritual of Thargelia and other *pharmakos* (scapegoat) rites a poor or ugly person, innocent but at the margins of society, would be chosen, dressed in a special costume, well fed, and then beaten and exiled — appearing to remove communal evil and disease in a collective catharsis (Bremmer 273). The tragic actor in ancient theatre may have functioned in a similar way, particularly when playing an exiled character such as Orestes or Oedipus. As René Girard points out, the ancient Greek word for a symbolic object used in shamanic rites, pulled out of the patient to show the evil being expulsed, was *katharma* (*Violence* 286). It was an alternate term for *pharmakos* and related to Aristotle's notion of *katharsis*—with the scapegoat rite as an attempt to defuse reciprocal violence with a substitute victim (287).[20] *Pharmakos* is related also to *pharmakon*, which meant both "poison" and "remedy" (288), and from which we get today's term "pharmacy." In the most extreme cases, the scapegoat would be killed, sacrificed to appease the anger of the gods and to embody social evil. According to Hipponax, in the 6th century BCE, the *pharmakos* was eventually burned and his ashes thrown to the sea, in order to purify the city (Bremmer 272–73). In the medieval Wakefield Cycle, there is a similar projection of collective illness and evil on the actors playing Lucifer, Satan, and Cain. But, as in ancient stage drama, they are only fictionally abused—with such redirected aggression relieving social fears and subordination stress, through mirror-neuron sympathy and tragicomic catastrophe, in traditions shaped by both biological and cultural evolution.

The term "tragedy," from *tragoidia* or goat song, may suggest the vital role of sacrificial scapegoating in ancient, medieval, and modern performances. Yet there is a difference in the characterization of the scapegoat in some plays as a tragic hero (such as Orestes) or villainous devil and rebel (Lucifer/Satan and Cain). The gods' anger or a trickster's whims, as with the Furies and Dionysus, may be refocused as coming from the scapegoat instead. This purifies God, as ultimate spectator and judge, from bearing an ambiguous, good and evil nature (like Zeus in relation to Prometheus). However, traces of that tragic ambiguity remain, despite the appeal of melodramatic stereotypes to the left brain's binary operator. God tricks Cain in return, with the choking smoke of his sacrifice, and proclaims the "flood" of his fury against this first murderer, while also prohibiting his being killed by others. Cain's initial complaint against nature's unfairness with his crop suggests that he may have been picked from the start as the *pharmakos*—like the laboratory rat made available for others to abuse, or provoked by stress into fighting with others, by the watching scientist. This may also be akin to the omega wolf, placed by nature at the bottom of the pack's hierarchy and thus serving a vital communal role as continual scapegoat.

The first two plays of the Wakefield Cycle demonstrate various forms of villainous scapegoating as foundational to Judeo-Christian biblical history with God as a medieval king, watching and ruling, even at times in a dark and angry way, like Sweeney's god. When Cain tries to trick the divine spectator with his tithe and then kills his brother, the wrath of God emerges, spurred (like the Furies by Clytemnestra) through the ghostly voice of

Abel's blood. Cain also beats on his servant, who is then even more the underdog, as omega scapegoat, drawing the laughter of the audience. He connects a familiar, medieval hierarchy to the biblical, cosmic order represented onstage.

All of this was scripted, probably by the clergy, under the aegis of earthly rulers, who were also watching the show. Religious and political rulers were able to shape their subjects' animal drives for surviving subordination stress — with redirected aggression and sacrificial substitutes shown onstage as moral warnings. Through theatre, medieval authorities could prune their subjects' brains, to some degree, using a sacrificial Christian paradigm that is still influential in today's global, mass-media culture. But medieval writers, actors, and spectators expressed alternatives, too, by twisting the biblical script in tragicomic ways, raising questions about human and divine rulers as moral authorities.

Spectator, Helper, Avenger, and Trickster

According to primatologist Frans de Waal, chimpanzees have a "revenge system" to punish negative actions within their groups (*Primates* 18). Likewise, a macaque monkey, when attacked by a dominant member of its troop, will "redirect aggression" to a vulnerable member who is a relative of the attacker. And yet, positive emotions are also shown as "retributive" among primates. Chimpanzees "kiss and embrace after fights," with such reconciliations preserving communal peace (19). Protection of others against aggression, through coalitions and alliances, is also a "hallmark of primate social life." Primates, such as capuchin monkeys, involved in food sharing and cooperative groups show a dislike for inequity while in captivity, with "emotionally charged expectations about reward distribution" (48). Apes also exhibit empathy, consolation of others after fights (35), and "targeted helping"— at times taking "great risks on behalf of others" (32). These and other aspects of current primate behavior provide evidence for the evolution of human morality from our common ancestors millions of years ago, through the group benefits of both revenge and forgiveness (McCullough).[21]

But morality and forgiveness did not evolve at first as universal ideals, with the same definitions in and for all human beings. They probably developed as "within-group" phenomena, when social pressures toward the common good increased through external threats (de Waal, *Primates* 53). They thus have "evolutionary ties to our basest behavior — warfare" (55). With humans, moral sentiments and social pressures eventually reached a more abstract level of self-reflective, logical judgment (168). And yet, such rational, left-brain morality is still based today on primal, primate drives: right-brain holistic intuition and limbic emotions (Haidt, "Emotional").[22]

It may be used for social control, but theatre, at its best, assists human groups with their self-reflection about moral and immoral situations, toward a cultural evolution of Symbolic, Imaginary, and Real brain areas. In the second play of the Wakefield Cycle, Cain expresses his dislike for inequity — like capuchin monkeys — concerning God's distribution of food rewards. He attempts revenge with his tithe trick and turns his redirected aggression toward his brother, suddenly killing him. Given no chance then for reconciliation with his brother or with God, Cain becomes a prime example of immorality to the medieval audience. The judgment of him comes from an actor playing God on the upper level of the stage, as a higher form of self-reflective spectatorship. Yet the Cycle also shows that God does not intervene to save Abel's life or to redeem Cain's. He allows humans to err and suffer for it, through His wisdom and justice — or His targeted help and vengeance.

In *Noah*, the third play of the Cycle, God gives selective aid to one family and to certain pairs of animals, while punishing others who rebel against Him, like Adam, Eve, and Cain. At the start of the play, the 600 year old Noah lives in "dread" of God's potential "vengeance" for the sins he sees around him, which are being performed "without repentance" (*Wakefield* 90). God appears on the upper level of the stage, where He has been watching "as a lover," who gave mankind angelic status but received "sins" in return (91). Because people are not repenting, God says, "I repent full sore that ever made I man," and decides to destroy "both beast, man and woman," except for those on Noah's ark. God gives as his reason that Noah and his wife "offered no strife, / Nor me did offend." And yet, the rest of the play shows Noah and his wife in a continual, comic battle, as she argues with him about getting into the ark. Noah tells the men in the medieval audience to "chastise" the tongues of their wives when they are young (101). He threatens to beat his wife, "back and bone, and break all." At last, his wife submits to this subordination stress: "I am beaten so blue / And wish for no more strife."

God's targeted help to Noah and his wife, saving them from a vengeful flood, seems misplaced, especially to modern eyes, unless the rest of mankind is much worse than this warring couple. Divine wrath against mankind for its sinfulness may offer a moral reason for nature's periodic destructiveness. But God's special aid to Noah and his wife, saving them from the flood, despite Noah's violence against her and her striving against him, makes the ultimate Spectator/Actor appear unjust, almost a trickster, playing favorites. In this way, though, God the Father may also be a whimsical Mother Nature.

God on the medieval stage takes the place of nature and various ancient gods, as both nurturing and destructive. Human characters treat God with reciprocal nurturing through sacrificial exchange, or reciprocal violence through rebellion and redirected aggression, responding to the subordination stress coming from their divine spectator. But the question remains — as humans extend their Theory of Mind from primate awareness to higher-order consciousness and posit God as the ultimate, watching Metamind: what does the Other desire? The Wakefield Cycle's ironic farce of a hen-pecked Noah beating on his wife to get her into the ark shows one possibility. God apparently enjoys, along with the medieval audience, not only seeing His favorites being saved, but also their scapegoating of each other under stress.

In a French production of the same story, a different spectacle was stressed. At Mons in 1501, a continuous rain of five minutes was shown, with water coming from wine barrels at the edges of rooftops above the town square. Such a "flood" demonstrated God's wrath against sinners and his enjoyment of a disaster scene. Yet it also displayed man's increasing power to imitate the forces of nature. This exemplifies a crucial step in the evolution of Western gods: from ancient tricksters with limbic meta-passions to one God as left-brain ruler (but still with right-brain trickster tendencies) — reflecting human artistic and technological powers, developing more and more in His image. In the next half of the second millennium after Christ, during the early modern, modern, and postmodern eras, this god-man ideal evolves further. Man becomes more godlike or even replaces god, with moral, political, and technological achievements, yet also tragic flaws.

Another image of the double-sided (or Janus-faced) God occurs in the fourth play of the Wakefield Cycle, *Abraham*. Here God is both helper and trickster, both nurturing and destructive in His sacrificial demands, revealing tragic traces of evil desires. First, Abraham, the founding father of Judaism gives the medieval audience a quick review of Old Testament stories before his time. These involve men displeasing God and being punished by Him, or pleasing God and being saved: from Adam and Eve to Cain, Noah, and Lot, plus the Sodomites who were destroyed by divine wrath when Lot escaped (107–08). Then God appears on the stage's upper level and tells the audience that He will "help Adam and his kin" if they are "loyal within," if they tender Him love and truth, "Shunning pride and showing ruth" (108). But God also explains his plan to "test" Abraham's faith, to prove if he "be true of love." God then speaks with Abraham, in response to his "devout prayers" (109). He directs him to take his son, Isaac, "As a beast to sacrify, / To slay him ... [and] / In offering burn him as a brand."

The economic ring in God's sacrificial demand, to "brand" Isaac, is echoed in Abraham's response: "No profit is God's will to shun" (*Wakefield* 109). Yet Isaac's initial words to his father, when called by him, evoke the play's emotional irony as well: "I love you greatly, father dear." In effect, Abraham must prove his love to God, as the ultimate Father and consumer, by doing the worst thing a father can do: kill his son, who loves him dearly, and turn him into a sacrificial beast. As father and son journey to the place of sacrifice, perhaps on a donkey between stage platforms (164–65), Isaac asks why they have no "beast for burning" (112). Abraham admits to his son that he must die. At first, Isaac is as dutiful to him as Abraham is to God: "To do your will I am ready...." But then he asks to "over-take" his father's will and wonders what he has done that his "flesh [must] be rent." In an aside to the audience, Abraham acknowledges that Isaac draws on his sympathy, speaking "so ruefully." Abraham foresees the slaying of his son as a "great sin" and worries about explaining Isaac's absence to his mother. Thus, the audience is put in a supernatural position, as Other to the dutiful, yet guilt-ridden Abraham, sympathizing with him and his potential victim—against God's bloody demand. This evokes limbic compassion and right-brain devil's advocacy in the watching spectators versus God's left-brain testing of Abraham's "love" and control of his progeny.

Yet God, on his upper level, then sends an angel to tell Abraham about a change in plans (*Wakefield* 114). The angel almost arrives too late, seizing Abraham's hand just as he is about to kill his son (115). Abraham is given a "beast" to sacrifice instead of Isaac, because God perceived his "meekness" and "goodwill." Grateful for this happy ending, despite the traumatic demand, Abraham calls God a "well of goodness." But God's targeted helping of Abraham also reflects a darker side to his patriarchy. In many species of mammals, from lions and bears to gorillas, a male taking control of a female with offspring from another male may commit infanticide—to clear the way for his own reproductive agenda. This is a significant source of infant death in 35 primate species (Hrdy 34). According to a study of gorillas in the wild, about one in seven infants dies this way (Wrangham and Peterson 22, 148–50). God's demand for Abraham's "love" almost forces a similar result, branding the founding father of the Chosen People with an exemplary test of total submission to divine, alpha-male rule.[23] Abraham thus proves his willingness to sacrifice what is most valuable, emotionally and biologically, to the ultimate Father above.

According to cognitive anthropologist Pascal Boyer, such elements can be found in many religions around the world—with dead ancestors and gods often functioning as intentional agents, exemplars, moral witnesses, legislators, exchange partners, and protectors—due to "our minds' evolved dispositions" (146, 170–72, 200–01, and 321). Primate researcher Barbara King cites Boyer's cognitive approach, but stresses also that empathy and belonging are crucial to all religions, as they "evolve god" through emotions that go back to our primate ancestors and are shared with other primates today (206).

In the first four plays of the Wakefield Cycle, God appears on the upper level as the ultimate intentional agent or Metamind, as exemplar of creative love and yet alpha-male control, as moral witness and judge, demanding costly sacrifices in exchange for His (and Mother Nature's) nurturing goodness and protection — or giving punishments to those who go against Him. God becomes a helper, avenger, and trickster: conveying patriarchal benefits when given the correct sacrifice, expressing nature's fury against the disobedient, and challenging Abraham, as a human father, to contradict his own reproductive instincts.

The obedience required for divine empathy and the cost of disobeying are displayed with the fall of Adam and Eve, the doomed rebellion of Cain, and the offstage massacre of many in the flood that Noah's family escapes. The sacrificial cost of belonging to God and his Chosen People is also shown in *Abraham*, despite its happy ending. But the measuring of divine favoritism, or sacrificial violence, is given further twists in the subsequent plays, *Isaac* and *Jacob*, where the human struggle with God becomes more explicit onstage. When Isaac, as an old man, tries to give the proper blessing to his eldest son, Esau, the younger one, Jacob, steals the "benison" (*Wakefield* 117). Ironically, this is like Isaac himself, who benefited as a child from a switch at the sacrificial altar, as shown in the previous play. Here, Jacob becomes a luckier trickster than Cain, gaining God's blessing as well his father's (120).

After Jacob flees his brother's vengeance, with the sympathetic aid of his parents, he goes to Aran, near "Jordan's stream" (118–19). God appears on the upper level of the stage, giving Jacob his own territory and a vast progeny — thus rewarding the primal drives of his mammalian brain and Machiavellian Intelligence. "This land that thou sleepest in, / I shall thee give, and thy kin; / I shall thy seed multiply, / As thick as powder on earth may lie / Thy generation shall spread wide..." (120). In his dream vision of this promise, Jacob sees God on a ladder — as the medieval audience may have — and responds with his allegiance if God brings him "home to kith and kin" (121). But when Jacob returns home from Aran, he learns that Esau still wants a violent revenge, bringing 400 men against him (122). Through his successful challenge to birth-order obedience, and God's apparent failure to protect him from his brother, Jacob begins to question God's identity, as intentional agent and judge, as spectator and director of sacrifices.

After learning of Esau's army, and considering an offering of "gifts" that might "his wrath slake," Jacob wrestles with an Angel, as God's double on the lower level of the stage (*Wakefield* 122, 167). God calls down from above: "now let me go." But Jacob refuses, demanding another blessing. So God gives him, via the Angel, a "touch" on the "thigh," which will make him limp, but not feel sore. God also gives him a new name, "Israel" (123). And

God says: "Since thou to me such strength made known / All men on earth thy might must own." God touches Jacob in all three Lacanian orders: right-hemisphere Imaginary, with the wound that makes him limp, yet live beyond pain; left-brain Symbolic, with a new name, which becomes a national identity; and limbic/brainstem Real, with survival "strength" over all men on earth. The touch on the "thigh" also suggests phallic castration, and yet potency. It involves body image, patriarchal power, and animal drives, through a renewed promise of success in the fight for alpha-male identity and reproductive territory. Jacob then makes peace with his brother, since, as Esau says, "it is God's holy will" (124). Jacob also humbles himself, as a "servant" to his brother. But Esau calls him, instead, "my lord through destiny." The lesson here for the medieval audience is paradoxical. It involves tricking the father and wrestling with God's Angel, yet submitting to the Other's difficult-to-know desires, in order to gain multiple blessings and powers, through family empathy and tribal belonging.

From medieval to modern times, Christianity extended Judaic monotheism across tribal, national, and racial groups — spreading God's sympathetic help to mankind and yet also the blessing tricks of sacrificial demands. Christianity expanded both cooperation and competition between "believers," on a vast, imperialistic scale. Many sects and orthodoxies formed shifting political hierarchies, evolving godlike figures of "moral" power within their groups, by demonizing others as internal scapegoats or external enemies. Some of this is foreshadowed in the Wakefield play, *Pharaoh*. The imperial leader of the ancient Egyptians shows his alpha-male, infanticidal drive, ordering that male children of the Hebrews be killed, at the hands of Egyptian "midwives" (*Wakefield* 127). The Jews are internal scapegoats to Pharaoh because their population has grown amazingly fast in 400 years and because "learned clerks" predict that one of them will "strike us dead" (126–27). Pharaoh's pre-emptive strike, a brutal massacre of those who believe in a different god, seems justified on his side.

But then Moses appears onstage and announces to the audience that he was saved from the massacre — by God's narrowly targeted help (*Wakefield* 127). God also speaks to Moses from a bush, "burning full bright," which still has "leaves [that] are green," showing nature's destructive and creative powers, beyond mortality and human consciousness (128). God tells Moses to bear a message to Pharaoh about letting his Chosen People go to the "wilderness" to worship, and then to show him his "wand" turning into a "serpent" as proof of God's power to manipulate nature (129). Moses warns Pharaoh about God's "vengeance"; but Pharaoh calls him a mere "warlock with his wand" (131). So God sends various plagues to change Pharaoh's mind. The Egyptian ruler lets Moses and his people go. But soon he changes his mind and leads his troops in chasing the Jews across the miraculously divided Red

Sea. Strips of fabric were probably used onstage to show the sea, its division by Moses's wand, and its closing again upon Pharaoh and his soldiers (136, 169).

As they are about to drown, Pharaoh anachronistically prays to "Mohammed" (as do Romans later in the Cycle), suggesting a medieval rivalry between monotheisms, which still haunts us today (*Wakefield* 137). Another detail toward the end of the play points to the medieval scapegoating of Jews by Christians, which eventually led to the modern horrors of the Holocaust. In the Bible, God repeats his infanticidal drive—this time with targeted help for Abraham and Jacob's progeny—massacring first-born sons of the Egyptians, as the tenth plague of vengeance against the Pharaoh. However, in the Wakefield *Pharaoh*, the last plague is a "pestilence," a familiar horror to medieval audiences (135). The audience was encouraged to identify with the Jews fleeing Egypt, but may also have identified with the Egyptians as victims of God's wrath. In fact, Jews were often blamed for pestilence when it struck medieval cities, scapegoated to make sense of nature's otherwise whimsical destructiveness. They were massacred at York in 1185 (and in various German cities during the Middle Ages) in revenge for the pestilence there—inverting the hero and villain identities of *Pharaoh*, but paralleling its logic of human sacrifice (Rose in *Wakefield* 168).

In evolutionary terms, the Jews' escape to the wilderness, like Jacob's move with his family to Aran, reflects the numerous hominid migrations that extended our species' adaptation to and alteration of many environments, through neural and cultural expansions—whether or not with the targeted help of God and his control of nature's powers. The Wakefield Cycle's selection and staging of stories from the Hebrew Bible shows particular medieval interpretations of the ideal Other of nature's powers in the environment, the human brain, and its cultural extensions. God, as the Metamind in and beyond Mother Nature, watches, helps, tricks, or punishes humans, while also demanding sacrifices for their purposeful, moral belonging. But such a connection between God and mankind is incarnated in a new way as the Cycle turns to Christianity's distinctive stories about the divine as an actor on the human level of the stage. Ironically, Jews become the prime scapegoats—the internal Other as *katharma* (evil object to be purged) and *pharmakos* (both poison and cure)—when God is shown coming to earth in the form of a Jew. Thus, the tragic flaws of God's image in man, and of evil within the good, as displayed in ancient theatre, reemerge in a medieval Christian context.

The Son and Other Scapegoats

Neuroscientists Jean Decety and Thierry Chaminade have used empirical research to theorize the self-other interaction as a "driving force" in self development, with the right hemisphere (especially its inferior parietal and prefrontal lobes) having a "special role" in distinguishing self from others and representing the other ("When" 577). They use Colwyn Trevarthen's notion of "intersubjective sympathy," which is inherent in human infants, who are predisposed to be sensitive and responsive to the subjective states of others. This is shown with the imitation of facial gestures, even by newborns, and with infants at 15 months who seem to have "beliefs about the goal of human actions" (580). Such functions involve areas of the right brain, which develops earlier than the left and is dominant in children up to the age of 12. Also, in both hemispheres of the brain, there are "shared representations," with the same circuitry being used, including mirror neurons, while we watch or perform an action (582–83).[24] These circuits are not only motor, but "semantic and affective," as ways of understanding others' "intentions, desires, and beliefs" (582–84).[25]

For example, watching someone smile activates the same facial muscles, "at a subthreshold level," as smiling, and this creates the "feeling of happiness in the observer" (Decety and Chaminade, "When" 584). Viewing facial expressions trigger similar expressions in the observer's face even without "conscious recognition of the stimulus"—using right brain areas that are "necessary in both expressing and recognizing emotions." This process includes pain sensing neurons in the anterior cingulate cortex, which are involved in both the experience and observation of pain. Such shared representations of the brain's theatre, "in action, pain processing, and emotion recognition," demonstrate the neural basis for sympathy as a fundamental social interaction—through the specific "mental code" of a given "cultural group."

Then why do we not imitate and resonate with everything others do? According to Decety and Chaminade, the prefrontal cortex plays a key role in "self-ownership" and "self-agency," distinguishing first versus third person, and self versus other representation ("When" 585). This involves executive functions, including inhibition, which may be lost with lesions in the prefrontal cortex, resulting in "environmental dependency syndrome" (with the person automatically mimicking a role in a given environment, like a human chameleon). Also, feedback and feed-forward mechanisms, which predict the consequences of one's own actions, may be used to estimate the intentions of the observed other. The left hemisphere is dominant for the execution of an action, and thus "over-activated" (especially in the inferior parietal lobe) when the intention to perform "comes from a third-person via the visual modality" (588). In Lacanian terms (not used by Decety and Chaminade), this type of left-brain performance might be called Symbolic via Imaginary, with the intention of the Other becoming a demand for performance.

But the right hemisphere is activated more when "the intention of the self" is reflected in another person's imitative performance (according to Decety and Chaminade), with the Imaginary ego confirmed by the Other's mirroring.

As the Wakefield Cycle moves into New Testament territory, God speaks directly to the medieval audience, explaining His desire for humans to mirror His creative awareness and divine intentions. Despite Adam's fall from grace, God plans to show "mercy" and "make redemption" because, He says, "I will not lose what I have made" (*Wakefield* 175). God blames Adam's fall on Eve and the serpent, who "wrongfully beguiled" the first male and thus "doomed him to a life full sad." God's sympathy for Adam involves the right-brain desire for a better mirror in mankind's performance, and the redemptive plan of a left-brain, commanding Other, who was foiled by the Machiavellian Intelligence of animal and female characters. God as spectator, director, and actor thus extends the left-brain, right-brain, and limbic theatrics of human social interaction, of personal sympathy and the drive to belong. God reflects the human brain's sense of self-ownership and self-agency, expressing His intentions as divine Other. And yet, unlike the Greek gods, who took possession of humans, sexually or through madness, this Judeo-Christian God offers humans the choice of an interpersonal relationship, albeit with sacrificial demands, such as giving up forbidden fruit or offering the life of one's son.

This god also likes symmetry. He plans to create a parallel triptych of a "man, a maiden, and a tree," in the New Testament world, to replace the failed man, woman, and "tree of life" in the Old (*Wakefield* 175–76). He will evoke from the medieval audience their redemptive sympathy for His son, who will "take on human form," through a maiden, and then be sacrificed, like Isaac almost was, but on a tree (176). Later in the Cycle, spectators will experience these stage images in their neural theatres — through mirror neurons and action, pain, and emotion pathways — as synchronic with the divine viewpoint.

The Christian God is more of a left-brain, audio-verbal lover of his human "maiden" than the licentious Greek divinities.[26] He sends the angel Gabriel to "greet" her and tell her that God has "chosen" her as a "maiden sweet" (*Wakefield* 177). Through the angel's "word and her hearing," God will then "In her body ... come, / ... cleanly ...." Gabriel also tells Mary that her child, though the son of God, must submit to the bloody law of Jewish sacrifice: "He shall take circumcision." God uses His "word" to make a fully human god, generating Jesus cleanly through Mary's ear, which incarnates the intersubjective sympathy of divine being and mortal flesh, as actor and audience. Mary's son is God's equal, not a mere demigod. Yet he must die as a man, suffering for mankind.

Mary has a more immediate bodily concern: "How should it be? / I slept never by man's side..." (*Wakefield* 178). She says she has "taken a vow" to remain a virgin "Unshaken." But the angel explains that the "holy ghost shall ... / ... in his virtue ... enshroud" her, and thus "infuse" her, without a loss of "maidhood." And yet, while Mary retains her virginity, the making of a God-man still involves sacrifice — at least from a feminist view today. This God usurped the primal goddess powers, positive and negative, of maternal creativity and rage, which are apparent in many other traditions (Neumann), as well as in the Furies of the *Oresteia*. God keeps Mary an intact human virgin, entering cleanly through her ear via the angel and Holy Ghost, because in monotheism there is no room for a rival mother goddess. Yet there is room in this play, *The Annunciation*, for a comical scapegoat.

Unlike the ancient Greek Io, Mary submits with her body to God's sympathetic, clean, and redemptive plan. She tells the angel: "My lord's love will I not withstand. / I am his maiden at his hand, / And in his fold" (*Wakefield* 179). But then Joseph appears onstage and confesses to the audience how his wife "amazes" him. "Her body is great, and she with child! / By me she never was defiled, / Mine therefore is it not." Joseph expresses an age-old fear of males and a flaw in their patriarchal power: the offspring that they provide for might not be their own progeny. "I am irked full sore with my life, / That ever I wed so young a wife." He fears that he has lost the genetic reproduction contest, not to God, but to "a younger man." With a boy playing the role of Mary in medieval England, Joseph's scapegoating may have been even funnier in performance. "Youth and age are poorly paired; / ... / Some other she dotes on / ... / Who could trust any woman now?" When Joseph confronts Mary, demanding to know the father of her child, all she can say is: "God's and yours" (180). She tells him she "know[s] no other man." But Joseph insists that her bodily "state" clearly shows she has done something "amiss."

Joseph speaks again to the audience, while alone onstage, complaining of Mary's "villainy" (*Wakefield* 181). He recalls, anachronistically, how their marriage was arranged by "bishops" in the temple. Although old in relation to other suitors, his "wand" bloomed: "it flourished in my hand" (182). Yet, the old man suffered from a common phallic illusion of patriarchal power, now undermined by his wife's apparent infidelity. "Young women ever yearn to play / With youths and turn the old away, / Such is the world's report" (183). Admitting, however, that Mary may be pregnant with "God's son," Joseph fears that he is not "worthy" to be near that "blessed body" and must flee to the "wilderness."

As if Gabriel had been watching *through* the audience, the angel then appears to Joseph with clean, left-brain answers, ordering him not to abandon human culture and family (*Wakefield* 184). Gabriel reassures Joseph that Mary "has conceived [by] the holy ghost" and will bear God's son. Joseph is told by this sympathetic Other to be "Meek and obedient" and to "take good care" of his wife. After the angel departs, Joseph repents to God, as divine spectator, for his "ungracious" slurs against Mary's "matchless maidenhead." And he repents to Mary for his "unclean" thoughts. He asks forgiveness for his "trespass" against both her and God. Though not a goddess herself, Mary, as a medium between the human and divine, gives her own sympathetic forgiveness and God's as well, "With all the might I may" (185). Thus, Joseph realizes that he still belongs within the divine plan and its Symbolic order of Imaginary wholeness through Real sacrifice. Joseph expresses his *jouissance* in natural, deciduous terms: "Lo, I am as light as a leaf!" His left-brain performance now dominates his right-brain's shame, anger, and fear—as he accepts the intention of the third-person Other through the angel as visual (and verbal) modality.

The medieval audience was also a medium for Joseph's confessions, enjoying a comical view of his scapegoating by God, as the ultimate trickster. Yet, the males in the audience might have sympathized with him, too, as uncertain patriarchs, and the females with Mary, who submits to patriarchal demands and subordination stress. Joseph was as much a medieval fool as a Jew, using contemporary terms like "bishops" for ancient Hebrew priests, at a time when Jews were already exiled from England. Hence, he sets up the synchronistic redemption of all mankind through God's sympathetic plan, with Joseph's foster son, Jesus, becoming the ultimate alien (and alienated) *pharmakos* on the cross.

The Cycle foreshadows that scapegoating climax by showing two comical versions of the birth of Jesus, with medieval shepherds complaining about their poverty and stealing a sheep. They then go to ancient Bethlehem, bringing presents for the Lamb of God. Herod also hears about the new King of the Jews and responds to this threat against his alpha-male power with a vast demonstration of the infanticide drive in male mammals.[27] First, he tells the audience how such news torments him: "This boy burns my brain" (*Wakefield* 263). He says his anger is growing stronger and that a "devil" ails him (264). Eventually, he orders his soldiers to take "vengeance" on the potential usurper by slaying all male children in Bethlehem "and all the coast about" (270). The audience then watches three such killings by the soldiers, with mothers fighting and yet failing to protect their boys (271–73). The soldiers return to Herod and report the killing of "Many thousands" (274). Pleased by this, and swearing anachronistically by "Mohammed's renown," Herod promises each of them "a maid to wed," good wages, and "castles and towers" as his "knights" (275). Medieval spectators could thus identify

Herod's evil with the threat of Islamic rulers who vied with Europeans for control of the Holy Land in various crusades for hundreds of years prior to the performance of this play. Yet, *Herod the Great* also shows the dangerous insecurity of a powerful ruler, at the top of the totem pole, sacrificing many lives as scapegoats, because of news about a potential terrorist force — which may resonate with the geopolitics, pre-emptive strikes, and collateral damages of our own time as well.

God does not intervene in this massacre of children, like He did in the sacrifice of Isaac. But He sends an Angel, as targeted help for Joseph, warning him to flee with his family to Egypt (*Wakefield* 254). In fact, God does not appear again onstage in the Cycle, except through Jesus (or when He is with him in the Garden of Gethsemane). Here, the boy Jesus escapes the scapegoating of thousands of others in and near Bethlehem, while God gives no aid to those others. Neither God nor Jesus explains to the audience whether this mass sacrifice, like many others in human history, is part of His divine plan — or, if so, why.

As an adult in a later play of the Cycle, Jesus shows his power over death with Lazarus. When Jesus tells his apostles they are going to Bethany to visit their sick friend, Peter and John warn him that "the Jews" there see him as their enemy and that he should trust "not one Jew" (*Wakefield* 324). Here, Jesus and his apostles are aligned with the Christian audience against the Jews. It is even suggested that a former Jewish friend in Bethany might be a "foe." Thomas reminds Jesus: "When we were last in that country / ... / We thought thou there should have been slain." Then Jesus tells them that Lazarus "has fallen asleep" and that they must go to "stir that knight and bid him wake." (Here again a term is used that connects the ancient setting to the medieval audience.) When the apostles continue to resist, Jesus says that Lazarus is "dead" and they will go into enemy territory for his sake (325).

With the faithful support of Martha and Mary, Jesus prays to his Father in heaven to bring Lazarus out of "hell's fierce blaze," i.e., out of cosmic enemy territory and its pains (*Wakefield* 326). Jesus also orders those around him to take the "band[s]" off the four day old corpse, freeing him from immobility (327). Ironically, the resurrected Lazarus then presents himself to the audience as a sign of death's inevitability — with a long monologue about mortality to conclude the play. He says that none can escape death, displaying his body as proof, in a final mirror-stage reflection of nature's alien recipe and cyclical scripture. "Your death is worm's cook, / Your mirror here ye look, / And let me be your book, / Example take from me; / Though charms for death ye took, / Such shall ye all be." He describes flesh falling away and worms gnawing on knights' bodies — at their lungs, "lights," and hearts (327–28). He gives further details of worms breeding "in you ... as bees breed in a hive" (328).

This description of biological decay reminds the medieval audience not only of their mortality but also of the continued power of nature within their animal bodies — even with one God as ruler of all. Lazarus warns spectators about the "dreadful day" of God's judgment and tells them — like Death in *Everyman*— to "amend" their ways while they still can, because their "goods however great" will not help them when they die (329). Lazarus's resurrected corpse and revealing experiences of death offer the audience a glimpse of the Real, usually masked by the immortal illusions of the right brain's whole body image and the left's self-preserving rationalizations. He becomes a wise, yet abject scapegoat, foreshadowing the suffering and triumphant Christ, later in the Cycle. But Lazarus, while getting a second chance at life, must also die another time with even greater awareness of his mortality.

Sacrificial Villains and Heroes

Eventually, Judas and Jesus are shown as opposite types of scapegoats, as sacrificial paradigms of satanic villainy and godly heroism, at the climax of this melodramatic epic about the world's cosmic history. In *The Conspiracy*, Judas tells Pilate that a woman rubbed expensive ointment on Jesus's head and feet, which Judas saw as a "waste" (*Wakefield* 339). He wanted to sell the ointment for "300 pence" and keep a tenth of that for himself, as he usually did with the "treasure that to us fell." He uses this as a left-brain rationalization in demanding 30 pence now for Jesus "to be sold." Pilate encourages Judas in this financial and personal vengeance: "Since he tricked you with such a sleight, / Repay your wrong, your hurt now heal...." Later, Judas betrays Jesus with a kiss, while claiming to offer nurturing affection: "I have come to succour thee" (351). But Jesus sees the threat behind his enemy's falsely theatrical nature: "Judas, thy part is overplayed!" Thus, the danger of human role-playing is shown, as a transformation of nature's creative nurturing, when it warps the brain's care system through competitive greed and deceit — or Machiavellian Intelligence.

In *The Hanging of Judas*, this vengeful villain even aligns himself with the monstrous Oedipus of classical times, as he confesses to his audience: "I slew my father, and after lay / With my mother; / And later, falsely, did betray / My own master" (*Wakefield* 390). But this may also suggest a tragic dimension to Judas. Did he try to avoid his fate of betraying the Son of God, like Oedipus fleeing from his supposed parents, so as not to fulfill his oracle, yet find that his own goodness produced evil? Judas's monologue does not reveal such a tragic flaw. Yet in it he relates his mother's dream of him, before he was born, as a "loathly lump of fleshly sin / Of which destruction should begin / Of all Jewry." This might have evoked some degree of sympathy in the medieval audience for Judas as not just a purely vengeful, greedy villain, but as traumatized by his mother's vision of him. It might also remind a modern audience of the destructive villainy of anti–Semitism in the centuries since.

Judas tells how his parents abandoned him as a baby in a basket (like the biblical Moses) on "the deep salt sea," until he was washed ashore on an island and adopted there by a "queen" (391). Unfortunately, the rest of the speech has been lost, so we do not know how much sympathy the play evoked for this paradigmatic villain, relating his life story and confessing his sins to the audience, as his Other, while hanging himself. But the monologue does suggest that Judas's rivalry with his "master" might have emerged through his abject poverty and alienation, through his parents' horror at his birth, as well as his rise to power as an adopted prince. For he describes his birth as "a cursed thorn / That spread full wide" and says he was "born without a grace." This might have evoked cathartic pity and fear in the audience, raising an awareness that even this villain, who betrayed the Son of God with a kiss, also became a victim of fate, with an earlier, heroic struggle for survival that somehow turned evil—through his loss, recovery, and then repeated perversion of nature's nurturing.

On the other side, Jesus is not just a melodramatic hero, triumphing over Satan with his sacrificial death and subsequent resurrection. To be sure, the main arc of the Cycle shows such a victory of good over evil. But that is not the whole story. While praying in the garden, in *The Conspiracy*, Jesus resists his sacrificial fate, and yet accepts it as his Father's will (*Wakefield* 346–47). This shows the conflict within him between abject fear as a victim and yet courageous struggle as a flawed hero. God also reveals that He has been watching, like the sympathetic audience, by appearing to Jesus and giving the "reason" for His Son's painful mortal fate (347–48). And yet, in the subsequent plays, Jesus must suffer alone—as he is interrogated by human rulers, is scourged, carries the cross, and is crucified.

While Jews are often the villains in such scenes, their reasons for fearing Jesus are also shown. One of his four "Torturers" mentions Jesus's "witchcraft ways" in raising Lazarus from the dead (*Wakefield* 359). Caiaphas sees a "devil's dirt" in his beard (361). The torturers even become somewhat comical—as they tie Jesus to his cross and complain about one another "slacking" in their work (398).[28] They joke anachronistically of Jesus jousting in a tournament, of setting him in a saddle and seeing how he can wield a lance, since he calls himself "king" (395–96). Such cruel humor might have aroused a tragicomic awareness of complicity in the watching medieval audience.

Yet, the Wakefield Cycle repeatedly stereotypes Jews as historical villains who tortured and killed Jesus, apparently justifying anti–Semitic vengeance for centuries to come. As Mary Magdalene puts it, during *The Scourging*, "To death these Jews this day lead him in unbelief" (386). She then calls for "vengeance" on his torturers (387). Jesus himself predicts "misery" for the daughters of Jerusalem, as his sacrificial "blood that guiltless spills will bring my foes despair" (386). His mother in *The Crucifixion* says: "These Jews with him have striven, their evil he withstands" (404)—allowing Christians to forget that she and Jesus are also Jews and encouraging them to repeat the sacrificial violence in real-life vengeance against similar villains as scapegoats of evil.

On the other hand, Jesus tells his mother it is "for mankind's miscarrying" that he must suffer and die, not just because of Jewish villainy (*Wakefield* 407). While on his cross, Jesus speaks to "the people that pass," including the medieval audience, standing around the stage or walking by it (400). He asks what he has done to them and what further kindness he could do, after making man in his "likeness" and then being "mocked by all men" in return, through their "wickedness" (401). Here, Jesus not only gives his audience a guilt trip, but also asks his Father's forgiveness for the men mocking him, who "know not what they do." After his death, Jesus goes to hell and debates with Satan about freeing certain souls there, including Jewish characters (Moses, David, and Isaiah), which he then does, through "love" of mankind (453).

And yet, in the final play of the cycle, *The Last Judgment*, Jesus reasserts a moral lesson for his audience, praising those souls who were good to the poor and rewarding them with heaven, while condemning those who were not to eternal hellfire forever. This stresses the value of compassion for all mankind, under God's watching eye, rather than genetic survival and competition between humans with diversely reproducing gods. It also points toward the modern expansion of Western monotheistic culture, with its values and judgments, for good or ill.

The Wakefield Cycle characterizes its one God in various ways: as artist of Creation, as demanding ruler and trickster through various sacrificial scenes, as heroic victim at the scourging and crucifixion, as righteous victor over Satan, and as judge of souls at the end of the world. God watches, to some degree directs, and then, through Jesus, acts in the sacrifices of the human realm (and in hell) through tragic love, melodramatic vengeance, and yet also compassionate control. This God, as ultimate Metamind and ruler of the universe, offers personal connection, targeted help, and purposeful belonging to those who believe loyally—like Noah, Abraham, Jacob, and Moses—even if that involves becoming, like Jesus, a sacrificial scapegoat or trickster for others.

Everyman's Death and Redemption

In medieval drama, nature's nurturing and threatening forces are often presented in orthodox, binary, and melodramatic ways—to reinforce a moral lesson. At times, however, such figures reveal more complex temptations and tragic paradoxes. In *Everyman*, a late fifteenth or early sixteenth-century morality play,[29] the allegorical figure of Death is not a villain, like the Jews or Satan. He is a servant of God and a messenger helping mankind. Death was a frequent presence in the lives of the audience, through terrifying plagues and feudal cruelties in the Middle Ages. Yet here, every man's inevitable meeting with death is given divine meaning. God, as the ultimate spectator and ruler, offers Everyman a choice to reform his life before it is too late. This offer comes through Death and other allegorical characters that are symptomatic of a new stage in the Western development of human world-making: an early modern confidence in self-transformation.

In ancient Greece, "Olympian gods and the dead have nothing to do with one another; the gods hate the house of Hades and keep well away" (Burkert, *Greek* 202). But the Christian God of *Everyman* summons Death as his servant, showing the audience that their mortal destiny has a reason. First, a Messenger onstage conveys this theme directly to the spectators: "How transitory we be all day" (207). He also reminds the audience that sin is their enemy, not death. "Ye think sin in the beginning full sweet, / Which in the end causeth the soul to weep, / When the body lieth in clay." He introduces God as "Heaven King," who will call "Everyman" to a "general reckoning" or judgment. God then appears onstage, like in the cycle plays, ruling rationally and justly.

As a divine spectator, God "perceive[s]" that humans ignore their spirituality; they are "Of ghostly sight ... so blind" (207). Humans are immersed in "worldly riches," rather than fearing God's "righteousness, the sharp rod" (208). He reminds the audience how much He has done for them: showing His law of sacrifice by shedding His "blood red" and suffering "to be dead." God sees people getting worse every year. If He leaves them alone, "they will become much worse than beasts." So, He will have "a reckoning of every man's person." And yet, such a view of God as the ultimate spectator and actor, saving mankind from its beastly tendencies, by watching in judgment and offering Himself as a merciful sacrifice, evolved from that animal nature — from primal morality and forgiveness in apes to the Christian sense of cosmic justice.

To the medieval mind, nature's artistic experiment with humans, giving them the powers of a bigger brain to become both better and worse than beasts, requires God's continued attention, as director of the world's theatre and judge of every man and woman, as they move toward their afterlife reckoning. This belief in a divine judicial theatre enabled medieval civilization, with its alpha males and feudal hierarchies, to survive for hundreds of years, despite the subordination stress of such social controls. Yet it also helped those in the lower classes to make sense of their remnant animal drives, of nature's calamities (such as the Black Death), of many social evils, and of their brain's higher-order awareness that everyone's life will end someday.

As mentioned above, ethologists have discerned a proto-human demand for justice, for a fair reckoning in this life, in capuchin monkeys. "Capuchins refused to participate [in experiments where they would get a food reward] ... if their partner did not have to work ... to get the better reward but was handed it for 'free'" (de Waal, *Primates* 48). These monkeys "seem to measure reward in relative terms, comparing their won rewards with those available and their own efforts with those of others." But primate species differ in this respect. "As opposed to primates marked by despotic hierarchies (such as rhesus monkeys), tolerant species with well-developed food sharing and cooperation (such as capuchin monkeys) may hold emotionally charged expectations about reward distribution and social exchange that lead them to dislike inequity."[30]

Yet even the despotic hierarchy of the rhesus monkey involves an aspect of human "morality." In experiments they refuse to pull a chain that delivers food to themselves when they realize it also gives an electric shock to companions (de Waal, *Primates* 29). One monkey did this for five days, another for twelve. "These monkeys were literally starving themselves to avoid inflicting pain upon another. Such sacrifice relates to the tight social system and emotional linkages among these macaques.... [T]he inhibition to hurt another was more pronounced between familiar than unfamiliar individuals." But it is unclear whether this self-sacrifice is due to: "(a) aversion to distress signals of others, (b) personal distress generated through emotional contagion, or (c) true helping motivations."[31]

The Judeo-Christian paradigm of filial sacrifice, from Isaac to Jesus, extends both the despotic hierarchy of rhesus monkeys and their surprising sacrificial instinct. The despotic Yahweh demands the ultimate offering from Abraham. But God also sacrifices his own beloved son, thus suffering and dying Himself, in the person of Jesus. This theological reversal of Jesus's political defeat as a preacher, at the hands of the Jewish priests and Romans, structures the primate morality of Christian monotheism: aversion to others' distress via "agape" (self-sacrificing love), emotional contagion through "holy communion," and true helping motivations as "charity." Such ideals fueled the vast expansion of Christianity, from a small cult to the dominant religion of Europe, in the late Roman and medieval periods. Christianity extended reciprocal altruism and genetic kinship sacrifice, as biological drives in primate groups, toward a monocultural logic of human compassion, far beyond familial or ethnic ties, to believers who were Jewish or Gentile, slaves or free men.[32] In practice, however, there continued to be vast and bloody persecutions of those deemed to be alien or heretical to the Church's com-

3. One Medieval God—with His Angels and Devils 117

munal body—with many centuries of pogroms, crusades, inquisitions, and witch burnings.

In *Everyman*, the character of God reminds the audience of His bloody sacrifice for them, involving His own suffering and death through Jesus. Yet, as Heaven King with a "sharp rod," He insists on justice, extending into the cosmic realm both a despotic hierarchy and a sense of fairness, akin to that of rhesus and capuchin monkeys. Indeed, this one God sits at the top of the Great Chain of Being—even if He humbles Himself to suffer and die as Jesus, or to appear in the Holy Eucharist, or to allow His character to be represented by an actor onstage.[33] The medieval theologian, Thomas Aquinas, argued that God left His signature in nature, as its creator, through its signs of order: motion, causation, contingent beings, human values, and "traces of intelligent design" (McGrath, *Science* 92–94). Such rational arguments in the Middle Ages laid the groundwork for the scientific revolution of the sixteenth and seventeenth centuries—involving Newton's new laws and a Deist concept of God as a "clockmaker," who designed the natural world and then let it run on its own, without further involvement (2–3, 102–03). *Everyman* foreshadows this idea of a divine clockmaker, while encouraging its audience to find signs of Him in the natural world. God does not appear directly to Everyman. He sends Death as His servant and messenger. This shows what Aquinas argued: God, as the primary cause of nature, works in it through "secondary causes," but only indirectly, which helps to explain why suffering and pain are sometimes involved, due to the "frailty" of those causes (McGrath, *Science* 104–05). "God, so to speak, *delegates* divine action to secondary causes within the natural order" (105). Thus, Aquinas tried to provide an answer for the human awareness, evolved from our primate ancestors, of unfairness in the cosmos. This delegation of God's power to secondary causes is shown, too, in the chance that Death gives to Everyman to change his afterlife sentence.

When Death appears, as a delegate from God, he tells Everyman that he "must take a long journey" and bring his "book of account" to the divine judge, showing his good and bad deeds, for which he must "answer" (210). Everyman offers Death money to "defer" his reckoning for twelve years (210–11).[34] But Death says all must obey him, because of God's "commandment," thus demonstrating to the medieval audience the divine signature in their own mortal bodies (210). He tells Everyman to "prove" his friends, if he can, because "each living creature / For Adam's sin must die of nature" (211). He also warns that Everyman's "wordly goods" were only "lent" to him. Everyman has the current day to "make ... ready."

This gives the medieval audience a view of their own mortality as cosmic fairness and divine will. Death is not an evil force or a devil acting counter to God, but rather His delegate helping Everyman to see his choices, as an indirect yet divine influence. Everyman goes to Fellowship and asks for help at his life's end (213). Fellowship pities him at first, and promises to get vengeance for him, before learning why Everyman is distressed. Fellowship says he will eat, drink, and be merry with Everyman, or "haunt to women the lusty company," or help him to "kill," but not die and go to the final judgment with him (214–15). Everyman then meets Kindred and Cousin, who also refuse to go with him, the latter giving a comical excuse: "the cramp in my toe" (217).

Everyman invokes "Jesus" briefly, through his existential fears — "is all come hereto?" — showing a sign that he is stepping in the right direction (217). But then he goes to Goods, trying to get that allegorical character to help him with his death sentence. Goods is "in chests ... locked so fast" and "sacked in bags" so much that he cannot move, suggesting how he may have appeared on the medieval stage (218). Goods also says that he is "too brittle" to continue with Everyman, admitting that his "love" is not everlasting (like God's) and that his "condition is man's soul to kill" (219). Punning on the terms Good and Goods, he admits that he is a "thief" to Everyman's soul and will deceive others "in the same wise" (219–20). Next, Everyman goes to his Good Deeds. He finds that character lying "cold in the ground" and not able to stand because of Everyman's sins (221). Yet, Good Deeds sends him to "a sister," Knowledge, who then takes him to Confession, "that cleansing river" (222). Stressing the significance of Church rituals in Everyman's redemption, Confession gives him "a precious jewel ... / Called penance," to chastise his body with "abstinence and perseverance in God's service" (223). Reminding Everyman of Christ's suffering for him, Confession orders him to scourge himself, too, in order to find the "oil of forgiveness."

These details may suggest props that were used during this scene on the medieval stage. Such visual or verbal associations show a careful consolidation of various theological ideas into a theatrical demonstration of Church rules for redemption, during this "moral play" (207). God's signature in nature and in the mortal human body, with symbolic props and allegorical actors onstage, is tied to the Church's rites and teachings. Everyman also prays to God through Mary as intercessor, for help at his "ending," though he sees Death as his "enemy," not God's servant (224). He craves to be a "partner" in her Son's "glory," by means of Jesus's "passion," which he then shares a taste of, through his self-scourging. Everyman punishes his body onstage, whipping himself, "for the sin of the flesh" (225). He blames his body for delighting in "the way of damnation," while trying to save his soul from "purgatory, that sharp fire."

This splitting of Everyman's character, between soul and body, moves toward an early modern psychology of the tragic hero in existential conflict with himself, as in Shakespeare's *Hamlet*. Such a split subjectivity also appears at times in ancient Greek theatre, as when Euripides's Medea fights with herself and with specific body parts, about carrying out her vengeful plan of infanticide. Yet Everyman's split-self aligns with the medieval theology of a hierarchical Chain of Being, with God as ultimate spirit at the top, then angels below him, and mankind having both spirit and flesh, leading down to other material levels of nature. Everyman's scourging of his flesh exemplifies a specific cultural evolution of primate sacrifice, akin to that observed today in rhesus monkeys, although he sacrifices his body not to save others, but to free his soul from hell or purgatory.

This also enables Good Deeds to rise from the ground, to be free of "sickness and woe" (225). Knowledge then gives Everyman a "garment of sorrow" to wear, the costume of "Contrition" wet with his tears, for his appointment with God. It will help him to get "forgiveness," Knowledge promises, because it pleases the divine spectator and judge — as well as the medieval audience and rulers. First, however, Everyman and his audience have more things to learn about their mortal bodies. Good Deeds and Knowledge tell him to take his mental and physical faculties with him: Discretion, Strength, Beauty, and his Five Wits (226). Everyman meets them and gets their promises to go with him. This inspires him to give half of his goods to charity in order to avoid the peril of the "fiend of hell"—a rare mention of Satan in this play (227). Knowledge also advises Everyman to go to "priesthood" for another sacrament, extreme unction, before his death. Five Wits also gives him and the audience a brief, doctrinal lesson, naming the seven sacraments, as a counter to the "seven deadly sins," which God lamented mankind's involvement with, at the start of the play (208, 228). Despite such orthodox teachings, however, Knowledge hints at future Reformation critiques. She says that sinful priests have "Jesu's curse" when they sell the holy sacrament, engender children that "sitteth by other men's fires," or become lecherous (229). An aversion to religious hypocrisy becomes apparent here, like the outrage of capuchin monkeys at unfair rewards, when their social order is unsettled.

With the notion of trusting in God that he will find a good priest as his shepherd, for his soul's "succour," Everyman leaves the stage and then comes back rejoicing at the sacraments of "redemption" he has received (229). But when he reaches his grave, as a "cave" in which he must "turn to earth, and there to sleep," he finds that his mental and physical faculties betray him (230). Beauty, Strength, Discretion, and Five Wits all take leave of him, like his earlier, external friends (230–32).[35] Knowledge must "forsake" him, too, when he "sinks into his grave"—again suggesting a crucial use of levels in the medieval staging. However, according to Knowledge, Good Deeds will go with Everyman to the Last Judgment. An Angel appears, announcing that Everyman's soul, as "excellent elect spouse," will arise to Jesus "Hereabove"

(233). The Angel promises this also to the medieval audience: an ascension "into the heavenly sphere," if their souls are likewise "crystal-clear," as evolving spirits in the Great Chain of Being. A Doctor concludes the play, recapping its moral and offering spectators the further promise of their "body and soul together," reunited by God in heaven, when He "brings us all thither" (234).

Such a paradoxical vision of humans rising to angelic status, of their souls and bodies being reunited through the sacrificial scourging of the flesh and its temptations, performed onstage with abstract characters embodying ideas, rituals, and physical or mental faculties, shows a particular twist in the collective externalizing of our evolving brains, a half millennium ago in England.[36] Sorrow, as Contrition, becomes a garment to be worn, wet with tears. Penance is called a "jewel." Goods is too brittle and Good Deeds initially too weak to join Everyman in his pilgrimage toward judgment. Actors play these abstract ideas and also Fellowship, Cousin, Kindred, Knowledge, Confession, Beauty, Strength, and Discretion. An actor performs Everyman's "Five Wits" as well: sight, hearing, smell, touch, and taste. Such an elaborate allegorical display illustrates medieval theatre's potential for staging many complex metaphors. But how does the brain's inner theatre enable such embodiments of human social relations, emotions, and physical attributes, aspiring to the divine or weighed down by the flesh, as performance metaphors onstage?

Recently, neurologists have speculated that the primal evolution of metaphor usage in human prehistory emerged in ways akin to the surprising phenomenon of "synesthesia." There are many forms of synesthesia, experienced as certain colors appearing with letters or numbers, time units, musical or general sounds, phonemes, tastes, odors, pains, personalities, touch, temperatures, and orgasms (Day 15). It can also be experienced as smell, sound, taste, temperature, touch, or various aspects of vision and personality — crossed with other specific senses. Although such "cross-modular" experiences are unusual in adults, we may all experience them in some ways as infants, due to "less specificity of cortical areas during early infancy" (Maurer and Mondloch 201). Distinct cortical areas with different sense perceptions develop during the first three years of life, reducing such synesthetic cross-over between the Five Wits. And yet, this cross-talk of the senses can also be found in adult language, through the use of multi-sensual metaphors, such as "soft light," "loud colors," "bright thunder," "dim coughs," and "quiet sunlight" (204–05). There is a widely shared propensity for this "metaphorical mapping," with associations between high tones, light colors, small objects, and high positions, or between low tones, dark colors, large objects, and low positions — or, more specifically, of hot tempers and the color red, or calm and a green shade (Treisman 241).

3. One Medieval God—with His Angels and Devils

The synesthetic roots of metaphor and allegory can also be found in universal associations between visual shapes and language sounds. V.S. Ramachandran and Edward Hubbard have found that 98 percent of subjects in their American experiment associated a hard-edged shape with the new term "kiki" and a rounded shape with the term "bouba," instead of vice-versa ("Synesthesia" 457–58). Similar results were found for Tamil speakers in India and for Chinese speakers, whose written language does not use letters matching such shapes, as English does ("k" with harder edges, "b" with rounder). There are cross-cultural, embodied constraints to our creatively free associations, revealing ties between different sense-perception areas in our brains. We feel certain visual and verbal shapes together, through the embodied experience of language (Lakoff and Johnson)—despite the arbitrariness of many cultural signifiers.

Ramachandran and Hubbard examine the universal "disgust response" to a foul smell or taste, shown by a scrunched nose and palms turned up, which Darwin noticed even in newborn infants. They ask why this is something we relate to moral issues, when calling someone "disgusting" in English or in other languages, such as Tamil and French. They explain that the brain's "olfactory bulb projects to the orbito-frontal cortex, and olfactory and gustatory 'disgust' is almost certainly mediated by this part of the frontal lobes" ("Emergence" 172).[37] In evolution, as early mammals became more social, "the same regions were taken over for mapping moral and social dimensions (which makes sense given the close link between territoriality, sexuality, aggression, and smell); hence the same facial expressions and the same terminology" (172–73). They also relate this to the "near-universal use of sexually loaded words for aggression" and to the Freudian sense of anatomical drives and "destiny" (173). Yet, they consider as well the "arbitrary associations" that are reinforced in the brain's neural circuits, beyond genetic ancestry, through cultural practices and contingencies, which "humans also excel at."

Moral and social dimensions are mapped in *Everyman* through various cross-modal associations in audience brains, between sensual perceptions, personal memories, and allegorical ideas. These include mortal premonitions with Death onstage, camaraderie with Fellowship, greed with Goods,[38] ideal Christian acts with Good Deeds, bodily powers with Beauty, Strength, Discretion, and Five Wits, and self-reflection with Knowledge and Confession. Penance becomes a jewel and sorrow a garment wet with tears—in specific "cognitive blends" of ritual, emotional, visual, physical, and fashionable values (Fauconnier and Turner). God as a character onstage is also a synesthetic blend of divine power and authority (extending that of humans over animals), of sacrificial compassion, and of the sensual aspects in performance, spectatorship, and judgment.

Yet, such metaphorical blends in historical texts and performances build upon the much earlier hominid evolution of "cognitive fluidity" between brain areas and faculties. Today, as in our hominid and medieval ancestors, this involves the animal drives of territoriality, sexuality, and aggression. These are linked not only to smell and taste (as Ramachandran and Hubbard point out), but also to the visual, auditory, and tactile aspects of performance, through the Five Wits of actors and spectators. The inner synesthetics of the brain's theatre are thus projected as collective stage practices in the "moral play," *Everyman*. As Everyman loses his territory of life on earth (and fears purgatory), while also losing his Goods and bodily attributes, he barters with Death, seeks help from his companions, and fights with himself over how to make new neural connections of metaphoric meaning and give sense to his mortality.

Our synesthetic potential is located in the angular gyrus of the brain's parietal lobe. Lesions in the left side cause difficulties with verbal metaphors, which are then perceived literally (Ramachandran and Hubbard, "Emergence" 174). Thus, the left angular gyrus specializes in "analogical reasoning" and the right in "spatial and artistic metaphors" (176).[39] Growth in these areas occurred in the evolution of our ancestors from *Australopithecine* to *Homo habilis* (175). Visual synesthesia also involves the fusiform gyrus and auditory synesthesia involves the superior temporal lobe. Ramachandran and Hubbard propose that the metaphor power of human verbal language evolved from primal mimetic communication through a "synergistic bootstrapping" of three effects: (1) visual to auditory mappings, as in the kiki-bouba experiment, with fusiform, superior temporal, and angular gyrus "cross-activation," (2) visual and auditory representations via manual-to-orofacial mirror neurons in Broca's area, and (3) "synkinesia" with the mouth mimicking hand gestures through the "ritualization of real actions" (181; "Synesthesia" 463). Such synkinesia was also observed by Darwin: the mouth moves unconsciously when scissors are used ("Emergence" 180). Neurologists explain this as a "spillover" effect between the hand and mouth areas of the brain's body map in the motor cortex. Synkinesia was probably key to the gradual shift from mimetic to mythic cultures in human evolution, as outlined by Merlin Donald. It might also be applied to theatrical performance today when gestures influence the effects of verbal utterances — from the Real and Imaginary realms to the Symbolic order, within and between brains.

Consciousness itself may have evolved as a way to focus on present priorities, as animal nervous systems became more complex (Gregory 108–09). Eventually, the "cross-modal abstraction" ability of the angular gyrus led to the vast creative connections that humans make, through language and artistic metaphors, recomplicating the brain's inherited structures (Ramachandran and Hubbard, "Synesthesia" 459). And yet, the appeal of abstraction in performance, as with allegorical characters in *Everyman*, can also be observed in animal behavior. Seagull chicks will peck at the red spot on their mother's beak to get food from her, but will peck even more at several red stripes on a long thin stick, as a "super-beak" (444). In the chick's brain, "the receptive fields for 'beak detection' ... are actually more optimally stimulated by this weird pattern than by a real beak." Ramachandran and his colleagues use this evidence to explain the appeal of cubist art.[40] But it might also apply to the super-figures of God and the devil in medieval cycle drama, to the stereotyped Jewish scapegoats in the Croxton miracle play, and to the various allegorical identities of *Everyman*.

After millions of years of evolution, the human brain became conscious in a unique way. With a growing awareness of death's inevitability and of the powers of nature to sustain or destroy human communities, our ancestors began communing with animal and hybrid spirits that they perceived through — and painted upon — the interior walls of caves tens of thousands of years ago. And then, thousands of years ago in ancient Egypt and Greece, theatre emerged as an art form exploring the troubled relations of humans with their gods. This reflected the conflicts of the brain's inner theatre — between higher-order consciousness, animal emotions, and remnant drives — as well as the nurturing and chaotic forces of nature and culture. A thousand years ago in Europe, through the Greco-Roman and Judeo-Christian traditions, monotheism with its hierarchical Chain of Being dominated the stage. Theatre focused on biblical myths and related stories of God's creative power, watchful judgment, and yet sacrificial compassion in the person of Jesus. Such a supreme Metamind gave a complete cosmic rationale to all of nature and society, tied to the soul of every man and woman, and their eternal rewards in the afterlife, as long as they obeyed religious and political rulers in this realm. But this subordination stress also relates to God's scapegoats onstage: from Lucifer to Judas in the cycle dramas, or the Jewish villains in the Croxton play, or Jesus in both as inverting the sacrificial rite.[41] These rebels may have evoked a righteous sense of melodramatic vengeance, or the ultimate victory of good over evil. Yet they also evoked, at certain points, a tragicomic sympathy and fear, with the eventual catharsis of subcortical emotions, through the right and left cortical networks in the audience.

Synesthetic, allegorical characters in a "moral play" like *Everyman*, as with gods embodying the forces of nature and culture in ancient drama, became ways for medieval spectators to interpret signs of divinity (or devilishness) in their own bodies and environment, through cross-modular mixtures of various senses and abstract social relations. But they also show the need for new collective orders in the brain and society, given the evolutionary freedom of humans beyond instinctual patterns of behavior, through our bigger brains, premature births, and heightened degrees of neural and environmental plasticity. Although not yet a Renaissance play, *Everyman* suggests an early modern sense of the hero's conflict with himself, showing different aspects of his split subjectivity. Here, the focus is on his soul versus his body. With a warning from God's servant, Death, about his mortal reckoning, Everyman's ego tries to collect various ego ideals, as superficial friends to help him, but they, like most of his bodily attributes, betray him. Yet Knowledge, by directing him toward the Church's rites, including the scourging of his flesh, strengthens his Good Deeds. This enables his soul to rise beyond the body's sinful temptations and mortal limitations — becoming the "spouse" of Jesus.

As a lesson for the medieval audience, this "moral play" shows God's signature in the various synesthetic metaphors of social and physical relations, from Fellowship and Goods as deceptive antagonists to Beauty, Strength, Discretion, and Five Wits as limited allies, plus the immortal help of Knowledge, Good Deeds, and the Church. Yet, *Everyman* begins to reveal cracks in this cosmic vision, with God ruling at a distance through Death, with a good versus evil struggle inside Everyman, with his body's ambiguous value in that battle, and with complaints about sinful priests. All of this hints at future revolutions in cultural consciousness, from the Renaissance to Enlightenment, Romantic, Modern, and Postmodern eras — as considered in the next chapter. These further stages of cultural evolution show an ever more distant God, with pagan spirits returning as reflections of good and evil forces in nature, the human psyche, and society, with the bodily values of scientific Knowledge outpacing the Church's rites, and with new godlike egos as ideal figures for the divine aspirations and yet tragic flaws of mankind.

4

From Renaissance Rebirths to Postmodern Experiments

The personal mystical visions evidenced in prehistoric cave art eventually became consolidated into normative myths — under collective ideals, identities, and leaders — as shown in ancient Egyptian and Greek theatre. Medieval monotheism continued this consolidation, through collective left-brain causal binaries and right-brain holistic intuitions, focused on the Judeo-Christian myth as inscribed, not on cave walls, but in the Bible. Access to God and fears about evil beings were mediated by the book, by the priests who could read and interpret it, by stained glass windows and theatrical pageants depicting biblical scenes, and by the developing politics of anti–Semitism. The evil Other of Judeo-Christianity was figured not just as a devilish snake in Eden, or as Satan moving between realms, but also as the Jewish people in their European Diaspora. Ironically, the primal source of Christianity was externalized as its enemy, like the devil's advocate in the human right-brain rejected by the left's war-room general and projected upon others as evil influences — using limbic emotions of panic, fear, rage, and sacrifice. This stereotyping of certain humans as melodramatic villains, who should be conquered or sacrificed by righteous heroes, then became acted out, not only in dramas such as *The Play of the Sacrament* and the Wakefield Cycle, but also by the Inquisition, which burned heretics at the stake. Yet, as considered in the previous chapter, such medieval plays also bear tragic edges of potential cathartic insight, which may have evoked new pathways between the neocortical hemispheres and limbic systems of spectators — affecting the evolution of Western culture and its politics of the supernatural.

Jews, often scapegoated as villains in real life, not just onstage, were forced to leave England in 1290 and Spain in 1492. But Jewish culture survived in Europe, despite the dominance of Christianity and later, secular attempts at genocide. Pagan beliefs and rites, though repressed by the Church, also survived, especially in rural areas, as benevolent deities became associated with Christian saints. As with ancient pagan fetishes, the relics of martyrs — any object or body part associated with a dead Christian hero, even bits of a fingernail — were considered so holy that these props performed miracles for believers, through the force of their *virtues* as a cognitive blend of immortal soul and mortal remains (Robert Scott 215–19). Saints became mythic characters and ideal egos. As role models for the living, their remains became holy props. They were signs of Heaven, existing beyond death, above the changing moon and amid the stable stars, yet also as spirits attached to their leftover relics on earth (215). The cults of the saints, with pilgrimages to their holy sites, with miracles involving ordinary people in divine cures, revealed again and again, as in medieval drama, that the Great Chain of Being was a cosmic theatre, with God as the ultimate director, spectator, and actor.

However, a change gradually occurred in the dominant collective consciousness during the 1500s to 1600s, as Europe moved from the Middle Ages toward a neo-classical "Renaissance." The melodramatic, binary warfare between good and evil forces — along a vertical spectrum between heaven and hell, with Everyman's soul and body on earth as the main battleground — shifted toward the horizontal exploration of human evolutionary potential. Binary projections of good and evil persisted, along with more complex, tragicomic realizations, as Western theatre moved from medieval and Renaissance views, through Enlightenment rationalism and Romantic passions, into the age of "melodrama" per se, which dominated popular theatre, film, and television, from the nineteenth to twenty-first centuries. We now bear the legacy of all these eras, and of ancient and prehistoric precedents, in our brains' inner theatres, through current worldviews and performance values, however much our gods, angels, demons, martyrs, and props are transformed.

Faustus as the New Everyman

First performed in the late 1580s or early 1590s, Christopher Marlowe's *Doctor Faustus*, like *Everyman* a century before, explores the human relationship to death, God, the devil, knowledge, body, soul, and possible afterlife territories. Yet, while angels, devils, the "Seven Deadly Sins" as allegorical characters, and Lucifer himself appear onstage, God does not. Marlowe's play reveals a crucial Renaissance step, beyond the Great Chain of Being[1] and divine theatrical control of the cosmos, toward a greater human freedom: to direct, act in, and interpret the theatre of nature, with its signs of God or other spiritual and physical forces. But it also shows the dangers and losses in that step toward secular, humanistic freedom — even as gods, angels, and devils reemerge from the unconscious.[2]

Like Everyman, Faustus gains a greater awareness of his own mortality and potential immortality. Unlike Everyman, Faustus starts the play with philosophical, medicinal, and theological knowledge, but becomes dissatisfied with such power.[3] He then turns to the dark arts, conjuring the devil, Mephistophilis, as his advisor and servant. Rather than showing Death as a servant of God and messenger to mankind, who will not be bartered into delay, as in *Everyman*, Marlowe presents Faustus as avoiding mortality and gaining magic powers for two dozen years, through a deal with Mephistophilis and his master, Lucifer.[4] *The Tragical History of Doctor Faustus* is thus an immorality play. Yet it also shows mortal decisions, emerging from and changing the human relationship to nature, through immortal projections of meta-nature.

Mephistophilis first comes in a "Dragon" shape too horrible for Faustus to work with, so he commands the devil to change form and return as an "old Franciscan friar" (1.3.19, 25). Here, Marlowe takes up the note that *Everyman* struck about sinful priests and extends it into a devilish masquerade. The play also offers the spectacle of the body's life force being turned into a supernatural, synesthetic power, not with allegorical characterizations of Strength, Beauty, or Five Wits, but in the blood from Faustus's arm, which he uses to sign his contract with the devil.[5] Whereas Everyman frees his soul for heaven, by scourging himself in imitation of Christ's suffering, Faustus draws his blood and writes with it to sell his soul to Lucifer, gaining godlike powers temporarily. Everyman finds God's signature in nature, from Death to his own bodily faculties, pointing to salvation through the Church's rites. But Faustus uses his signature and body — after conjuring the devil with heretical rites, with holy names "anagramatiz'd" — to test and defy the divine spectator, against the authority of His Church (1.3.9).

Faustus rearranges left-brain language orders and orthodoxies to raise a right-brain devil's advocate with limbic, animal-human (dragon-friar) drives. And yet, Mephistophilis tries to reason with Faustus — even to dissuade him from his drive to disobey God and gain illusory power. "O Faustus, leave these frivolous demands, / Which strike terror in my fainting soul" (1.3.81–82). Perhaps this is reverse psychology. But the devil claims he came to Faustus of his "own accord," not charged by Lucifer (44). He explains that Faustus's conjuring of him was "the cause, but yet *per accidens*" — thus a secondary cause, involving free choices by the human conjurer and evil spirit (46). He tells Faustus that "hell" is everywhere for him, because he "saw the face of God / And tasted the eternal joys of heaven," but then lost that "everlasting bliss" when he and other angels "fell with Lucifer" (70–80) — as shown in the first play of the Wakefield Cycle. Mephistophilis presents a morality-play warning to the Renaissance audience that the animal drive for survival, turned into greed for superior power, can bring endless suffering, especially through the higher-order conscious of humans (or angels), twisted by language and culture into tragic passions.

Faustus boasts, however, of his "manly fortitude," and insists on a deal with Lucifer, since he has already "incurr'd eternal death / By desperate thoughts against Jove's deity" (1.3.85–89). Through this competitive reasoning and classical reference, a tragic hubris appears, countering Mephistophilis's profound regret at the loss of a Christian heaven. Repeatedly during the play, Faustus's manly fortitude and growing ego connect with reasons and allusions, across the corpus callosum of his brain's inner theatre, between the right-brain Imaginary and left-brain Symbolic orders, to produce a flawed aim (*hamartia*) in his desires and actions — despite his knowledge, courage, and divine aspiration.

Yet, as Faustus contemplates his next meeting with Mephistophilis to sign the bloody contract, Marlowe shows his split subjectivity. "Now, Faustus, must thou needs be damn'd? / And canst thou not be sav'd" (2.1.1–2). He tells himself to stop thinking of God or heaven, as tempted by the devil in their previous meeting. "Away with such vain fancies, and despair; / Despair in God, and trust in Belzebub" (4–5). But he also questions his own indecision here: "Why waver'st thou?" (7). He hears a sound in his ears and repeats the lines of his left-brain superego or Good Angel: "'Abjure this magic, turn to God again'" (8). Then he doubts this voice: "To God? He loves thee not" (10). Against such left-brain, good-angel pathways, Faustus tries to redefine his sense of God as brainstem and limbic drives (akin, perhaps, to the ancient Furies). "The God thou serv'st is thine own appetite," he says to himself, "Wherein is fix'd the love of Belzebub" (11–12). Faustus evokes in his audience the Christian fear of villainous Jews and their rites of infanticide, according to folk myths. He promises to offer Belzebub the "lukewarm blood of new-born babes" (14). This relates also, in evolutionary terms, to primate fears about infanticidal instincts in competitive males — as considered in the previous chapter.

Then the Good Angel appears, as does a Bad Angel, extending Faustus's inner theatre of split voices (and temporal-lobe spirit experiences) into the shared performance space. The good one reminds Faustus of the *Everyman* lesson that contrition, prayer, and repentance "are means to bring thee unto heaven" (2.1.16–17). But the Bad Angel says these are "illusions, fruits of lunacy" (18). Faustus sides with the latter view — with evil against the good — through the binary operator in his left brain's inferior parietal lobe. For he summons Mephistophilis and writes with his own blood a "deed of gift" to

Lucifer (60). During the writing, however, his blood "congeals"—as if nature within him were rebelling against the supernatural contract (62). "Is it unwilling I should write this bill?" (65). The rebellious, congealing blood also makes him wonder whether his soul is his "own" to give (68). Yet Marlowe's play demonstrates that this Everyman does have a free will, or its illusion, as secondary cause in the Great Chain of Being—with the devil's help, as Mephistophilis brings him fire to warm the blood and make it usable again for evil.[6]

After he has written the bloody deed, Faustus's arm also rebels, offering him a good-angel message inscribed in the flesh: "*Homo fuge*" (2.1.77). At first, he doubts the vision, fearing that if he flees, God will catch him and throw him "down to hell" (78). Then the tattoo vanishes: "My senses are deceiv'd" (79). In the next moment, he sees it again. So Mephistophilis distracts him with a courtly masque of devils, giving him "crowns and rich apparel" and then dancing away. The doctor's limbic flight system, triggered by panic about eternal damnation, is thus inhibited. He is charmed by this devilish interlude, which has no meaning, according to Mephistophilis, except to "delight thy mind," letting Faustus and his audience "see what magic can perform" (84–85).

The devil promises Faustus that he will be the artist of such performances and "greater things than these" (2.1.87). Here and elsewhere, Marlowe's play reflects the seductive dangers of theatre, even while appealing to its audience in a similar way. In ancient times, theatre emerged, at least in part, as a seductive spectacle for social control, although in Greece the political and religious orders were also questioned by comedies, tragedies, and satyr plays. In the Middle Ages, theatre emerged primarily to reinforce Christian ideologies, while drawing upon the popular appeal of rebellious devils as wild, satyr-like, interactive, and sometimes comical figures.[7] This comical aspect can be seen in Marlow's tragedy as well—in its subplot of Wagner and the "clown" Robin conjuring devils. Yet in the contract-signing scene, the devils are serious and sophisticated, like Lucifer and his cohorts at the start of the Wakefield Cycle. With Mephistophilis's metaphysical knowledge, with the devil-actors' gifts of upper-class costumes and cultured dancing, and with the foolish conjuring of devils elsewhere in the play, temptations are displayed for various classes in Marlowe's audience. Thus, the devils here become more ironic than the friendly antagonists, Goods and Fellowship, as detours for Everyman's soul.

If Faustus's goal is to become powerful through his contract with Lucifer, then Mephistophilis and other devils are his allies—while God (offstage) is his main antagonist, along with the Good Angel, the Old Man (who appears later), and himself. Mephistophilis helps Faustus commit to a certain course of action, in "body and soul," and eventually to recognize his tragic flaw, by becoming his "servant" and guide (2.1.98, 109).[8] This makes God, watching from somewhere above, and the Good Angel, appearing from inside Faustus, his antagonists. Marlowe's audience must choose how to identify with Faustus on his pilgrimage toward death and a certain afterlife territory. They may sympathize with his drive for knowledge and power, or with his existential doubts about God's love and about hell as "a fable" (126). They may likewise fear his tragic temptations and ultimate punishment. And they may identify with the other side of Faustus's mind — externalized by another actor playing the Good Angel. If they see that left-brain soul as the protagonist of the play, spectators may yearn for a melodramatic victory of good over evil, as with Everyman's success in empowering his Good Deeds by scourging his body. Yet Marlowe's play repeatedly lures its spectators toward a more ironic, tragicomic view.

Mephistophilis says he is "in hell" while on earth, because he lost the joys of heaven (2.1.136). But Faustus's hubris reverses that warning: if "sleeping, eating, walking, and disputing" on earth is the same as hell, then he shall "willingly be damned" (137–38). Perhaps Mephistophilis is showing his own inner conflict, based in a prior tragedy. Even while luring Faustus toward hell, as an agent of Lucifer, he also warns Faustus away from such a catastrophic end — aware of the fundamental flaw in flying too high and falling. Both Faustus and Mephistophilis demonstrate the aspiration for greatness, in humans as higher-order animal spirits, evolving through survival and reproduction drives toward bigger brains and many cultural powers, with critical neural flaws.

Faustus asks Mephistophilis for "a wife," describing himself as "wanton and lascivious" (2.1.140). His supernatural servant first tries to dissuade him, and then fetches him "a wife in the devil's name" — with fireworks accompanying her as she appears onstage, originally played by a boy (144). Faustus dislikes her, calling her a "hot whore" (146). But Mephistophilis uses Faustus's lustful, reproductive drive to lure him beyond the Church's sacrament of marriage, beyond the idea of a wife, toward a promise of "the fairest courtesans" — if Faustus will love this devilish servant who makes the wildest fantasies appear real (149). As the play progresses, Faustus does fall in love with the powers that the devil gives him. He feels "immortal" when kissed by an ancient supermodel, whose face "launch'd a thousand ships, / And burnt the topless towers of Ilium" (5.1.97–98). Then, however, he senses how the miraculous image of Helen is a virtual succubus: "Her lips suck forth my soul, see where it flies" (100). And yet, he wants more of her: "come give me my soul again. / Here will I dwell, for heaven is in these lips" (101–02). Marlowe presents the cumulative figure of an ageless demigoddess, built with Mephistophilis's magic, Faustus's classical knowledge, and his animal lust.

She nurtures human aspirations for an ideal world, while also drawing the "soul"—the godlike, higher-order consciousness of the brain's theatre—toward a tragic fall. This Aphroditic demigoddess exemplifies Mother Nature at work within the human species: driving survival greed, stimulating reproductive pleasures, and showing the power of beauty in sexual selection, while nurturing the evolution of higher-order consciousness and its fatal flaws.

Yet, most of the play reinforces a patriarchal cosmology, from ancient myth and medieval magic to the Renaissance science of man, godlike Father with a controlling Mother Nature. After Faustus rejects the "Devil dressed like a woman, with fireworks," Mephistophilis offers the doctor some great books (2.1.144).[9] Faustus not only gets a book for casting his own spells and raising spirits. He gets his request for the ultimate telescopic book: to "see all characters and planets of the heavens," including their "motions and dispositions," and to "see all plants, herbs and trees that grow upon the earth" (166–70). Here, Marlowe displays the Faustian wager of modern science. Humans have evolved the knowledge and power to manipulate nature's forces and fruits—while surviving and reproducing on this planet, beyond other species. But will we develop the "soul" to use science and technology wisely? Or indulge in pleasure, mischief, and future doom, as in Faustus's case?[10]

Faustus himself starts to glimpse this tragic temptation of magic and science, saying to Mephistophilis: "When I behold the heavens then I repent, / And curse thee...." (2.2.1). His inner angels reappear, buzzing in his ears that if he repents, the divine Spectator will or will not "pity" his tragic plot (12–13). This anticipates a future question of European culture as it moves toward the eighteenth-century Enlightenment notion of nature as a scientific machine, with God as a "clockmaker" who winds it up and lets it run. Will such a God continue to watch and work in the human world, pitying Everyman or the Faustian scientist and changing his fate? The Bad Angel's answer is anti-tragic and existential: "God cannot pity thee" (13).

Faustus senses terrifying medieval tortures when he considers repenting like Everyman. So he takes succor with his friend Mephistophilis through scholarly "dispute," about "reason of divine astrology" (2.2.33–34). But when Faustus asks his Fellowship devil "who made the world," they reach an end to the enlightened detours of the brain's higher-order powers (67). Mephistophilis, the intellectual nurturer, refuses to answer with God's name and instead reminds Faustus that he is simply "damn'd" (73). Yet, like Descartes a century later, Faustus wants God to exist as a Metamind in the cosmic theatre, as a guarantee for his *cogito*. Speaking to himself and sounding like the Good Angel, he says: "Think, Faustus, upon God, that made the world" (74). He tells Mephistophilis to go to hell and asks the cosmos, "Is't not too late?" (76–78). The two angels reappear then, with their left- and right-brain answers. One says it is never too late if Faustus repents. The other says that devils will tear him "in pieces" if he does (81). While fearing this *sparagmos*, as sacrificial scapegoat in the battle of good and evil, Faustus shows here that he is leaning, like Everyman, toward salvation. He calls on "Christ" to save his "distressed" soul. But then Lucifer appears with other devils to turn the tide back to tragedy (83–84).

At first, Lucifer taps into the doctor's left-brain logic: "Christ cannot save thy soul, for he is just" (2.2.85). The contract, signed in blood, has clear, legal, and cosmic consequences. Then Lucifer and Belzebub, his "companion prince in hell," try to draw on Faustus's limbic and right-brain sympathies (88). Like God at the start of *Everyman*, they say Faustus "injure[s]" them by not appreciating what they offer — and that he is not keeping his "promise" (90–91). They offer him a courtly masque, a devilish "show" of the Seven Deadly Sins, which are also mentioned by God in *Everyman* (107). Although this reminds Faustus of Adam in Paradise enjoying God's gifts, they quickly dissuade him from such heavenly thoughts, by bringing onstage the allegorical characters of Pride, Covetousness, Envy, Wrath, Gluttony, Sloth, and Lechery. Faustus interacts with these spirits, asking them questions and getting the stories of their origins and personalities (unlike the allegories in *Everyman*). But it is the theatrical "sight" of these figures that delights Faustus's "soul" and perhaps also tempts the souls in his audience (155). Lucifer promises him even more such delights in hell, and will allow him to visit there, yet return to earth — like the Renaissance audience glimpsing hell's spectacle with this play (156). Lucifer also gives Faustus a "book" that will enable him to change his "shape" into whatever he wants, thus appealing to his left-brain language areas and his right-brain holistic functions, while extending his body image toward supernatural, theatrical powers (159–60).

Current neuroscience demonstrates that the human body image is not as stable as the physical body, but is malleable and continually updated by the brain's somatosensory and motor cortices, in the parietal and frontal lobes, involving top-down projections, not just bottom-up stimuli (Ramachandran and Blakeslee 58–68). "Your own body is a phantom, one that your brain has temporarily constructed..." (58). The body image (or Freudian "imago") draws on motor and emotional mappings in the cerebellum, anterior cingulate, and insular cortex, while utilizing mirror neurons and intuition neurons to mimic the actions and feelings of others, often at an unconscious level (Blakeslee and Blakeslee 24, 32, 187–89). Building upon a "body mandala" (the brain's network of body maps) and a "body schema" (the felt experience of a body, usually in motion, as constructed by these maps), one's body image also draws on a "lifetime's library of personal experiences and memories" (32, 42). Embedded in body mappings and autobiographical memories, the brain's body image is an "amalgam of beliefs," which involves "attitudes, assumptions, expectations, with an occasional delusion thrown in" (42). There is even a new academic discipline, sensory anthropology, which "focuses on how cultures stress different ways of knowing through the body maps and the senses. Notions of sight, sound, touch, taste, smell, balance, proprioception, and personal space are all conceived or even mapped differently in people from various cultures" (127).

Lucifer's gifts to Faustus are not as miraculous or unscientific as they might appear to be. Faustus's mirror and intuition neurons—and those of the audience—are unconsciously transformed by the spectacle of the Seven Deadly Sins. His body image is also alterable, as his beliefs, attitudes, assumptions, expectations, and self-delusions change under the influence of various devils and their theatrical wizardry. In the second half of the play, he flies to the heavens with Mephistophilis and plays ghostly tricks on the Pope in Rome. He acts mischievously with others, too, making his leg reappear after it is pulled off. He even comes back to life after his enemies chop of his head (in the B-text). Yet such wonders, in the theatres of spectators' brains as well as onstage, relate in many ways to neuroscientific investigations of out-of-body experiences, which can be stimulated with electrical currents to the right angular gyrus (Blakeslee and Blakeslee 121–23), or of phantom limb syndrome, where perception of a limb persists after amputation (Ramachandran and Blakeslee 21–58).

All the tricks and spectacles that entertain Faustus and his audience culminate in his tragic reckoning with the other side of his contract, when its time limit has ended. Despite the godlike power he enjoys, and his soul's flight to the heights of knowledge and experience, Faustus ends up with a double terror: his body's *sparagmos* by devils and his soul's descent into hell. Thus, his body image may be malleable through his pact with Lucifer, and the ideal ego of his soul becomes all powerful, through that perverse ego ideal. But he must pay for it in the end—with eternal damnation. Or must he?[11]

Marlowe's play touches upon a key issue in neuroscience today: do we have "free will"? Repeatedly, the Good Angel and Old Man entreat Faustus to repent. But he becomes distracted by the devils' theatrical charms, luring both his high-minded soul and lustful, mischievous, animal body. Even at the end, he has several chances to turn to God before his soul is damned forever and his limbs are "torn asunder by the hand of death" (5.3.7). A fellow scholar tells him to "call on God" (5.2.49). Yet Faustus says he cannot weep because "a devil" draws his tears (51–52). He cannot pray or raise his hands to God because Lucifer and Mephistophilis stop his tongue and hold down his arms. Even so, the Bad Angel offers Faustus and the theatre audience a cautionary vision of damnation, with a tie to ancient terrors, as with Orestes's underworld problem. Hell is revealed ("discovered") onstage as the Bad Angel points out "the furies tossing damned souls / On burning pitch forks" (5.2.108–12). Faustus realizes, with the clock striking, that he has but one hour left to live (127). He yearns for more time to "repent." He sees "Christ's blood ... [as it] streams in the firmament" and realizes that "half a drop" would save his soul (139–40). But then he sees "where God stretcheth out his arm, / And bends his ireful brows" (144–45).

This vision of a watching, ireful God above and devilish furies below, in contrast to the sacrificial Son's blood, shows Faustus's struggle to believe in a New Testament promise of divine love and forgiveness. He cannot drink of Eucharistic salvation, due to his terror at "the heavy wrath of God" (5.2.147). Yet he asks, through Christ's ransoming blood, for a limit to his sentence in hell, a thousand or a hundred thousand years, "and [then] at last be sav'd" (163–64). He wishes he were "a creature wanting soul," that he could be "changed into some brutish beast" and not be immortal in hell (166–70). He curses "the parents that engender'd" him—along with himself and Lucifer—promising in his final breath to burn his books (174, 184).[12] Faustus apparently chooses to go to hell, without accepting God's forgiveness through Christ's blood. But that choice may be forced by his fateful engendering, his seductive devils and books, or his skepticism as an early modern everyman.

Psychologist Daniel Wegner has recently marshaled evidence from several decades of experiments by himself and others to show that we do not have a "conscious will," only the "illusion" thereof—like the illusion of a stable body image. Our brains make automatic or unconscious decisions which we then think are conscious choices because of other neural pathways that produce similar thoughts, along with the desire for a persistent sense of self. For example, a "readiness potential" was explored by Benjamin Libet in the 1980s. He discovered that a bodily movement involves the firing of unconscious brain circuits a fraction of a second prior to the "conscious wanting" and choice to move (Wegner 52–55).[13] Also, experiments by Ramachandran with phantom limb patients show that they can experience a "willful movement" through the visual cue of a mirror image or someone else's hand, appearing in place of their missing one (Wegner 42–44). Such experiences—of one's own will causing an action when it is actually caused by someone or something else—have been evoked in other subjects through a Ouija Board or a similar device and through posthypnotic suggestions (74–78, 149–51). Freud also noticed this phenomenon with patients he hypnotized. They would confabulate to explain a strange action, making up a reason for it as their own choice, without remembering the prior hypnotic suggestion—as a "defensive rationalization" to support the ego as willful (Wegner 151).

Lacanian theory has extended this discovery of ego defensiveness and rational confabulation to describe the ego as illusory, and as paranoid in its insecure Imaginary projections. Wegner's cognitive evidence supports this psychoanalytic theory — though he discounts Freud's exploration of the "darker side" of human nature and does not consider Lacan (151). Yet, Wegner's evidence helps to demonstrate a key idea in Lacanian psychoanalysis (*contra* Descartes and American ego psychology): the true subject is not in the conscious ego or *cogito*, but in the unconscious, masked by fundamental fantasies and structured by chains of signifiers (Lacan, *Four*). This is an inevitable, tragic flaw in the human condition of being alienated from nature through culture and language. And yet, some degree of cathartic "cure" is possible by crossing fundamental fantasies, embracing the loss of self, and accepting the non-existence of the Other as a godlike ideal watching and guaranteeing one's ego.

Faustus fights with himself (like Euripides's Medea) because his unconscious brain is making choices that his conscious brain is trying to match or change. "Devils" inside him prevent him from weeping, raising his hands to God, and moving his tongue in prayer. The Bad Angel gives him a vision of furies in hell, tossing damned souls, and he sees Christ's blood in the sky, knowing that half a drop can save him. Yet his limbic fear of devils — and of God's Old Testament wrath — prevents him from accepting the sacrificial blood of Christ and saving his soul, like Everyman, through repentance. His unconscious neural pathways, experienced as devils in his inner theatre, pull his illusory ego-soul down to hell. So he wishes that he had no soul, as just a "beast," that he could undo the tragic flaw of human evolution into higher-order consciousness and its immortal cultural dreams.

According to Wegner, the illusion of conscious will works the other way as well — not just as a persistent sense of self, but also with projections upon an outside, willful agent, as affecting or possessing the body, or producing actions in the world. As Wegner puts it, the ultimate "ideal agent ... is God" (148). In terms of today's attempts to replicate human consciousness with the Artificial Intelligence of computers, God is the most ideal agent because He knows all ("has perfect sensors"), can do all ("has perfect effectors"), and always acts correctly ("has a perfect processor"). Wegner mentions further aspects of God's hypothetical character as spectator and actor, or of other ideal agents, that relate to current attempts to build a conscious machine. He cites Leonard Angel's notion of "basic interagency attributions," which are qualities that agents (possibly even artificial ones) find in themselves and look for in others: presence, focus of attention, belief, desire, plan, and movement (148–49). Wegner finds that an ideal agent — whether virtual, human, or spiritual — should have all of these qualities and more, but at least "intention and conscious will" (149).

In theatre, an actor finds these qualities in a character, even an allegorical figure, and performs them specifically, so that spectators may perceive an ideal agent onstage—as conscious will, unconscious drives, and perhaps an eternal soul. God, as perfect sensor, effector, and processor, is only present offstage during *Doctor Faustus*—or in drops of Christ's blood that Faustus sees in the firmament. Yet God provides the ultimate cosmic framework, along with Lucifer, for the play's interagency conflicts. The audience of the play also helps to create the ideal agencies of the Good Angel and Bad Angel, externalized from the inner theatre of Faustus's brain. Mephistophilis, the doctor's companion agent for theatrical magic and evil temptations, despite giving him initial warnings, eventually leads him toward his tragic catastrophe. Lucifer also appears, as the ultimate negative agent and deceptive theatre director. Faustus experiences certain devils, including Mephistophilis and Lucifer, as external agents that possess him, restraining his body—and his soul's conscious will to repent. "People often describe their 'weakness of will' as a matter of succumbing to habits or impulses or demons, along with a concomitant failure to think the right thoughts at the right time that might have guided their actions in a desired direction" (Wegner 92–93). Faustus also succumbs to the "will of the group" (215)—with the Seven Deadly Sins and many other devils that he desires or fears as interagency forces. He follows their will instead of the Good Angel's within him or the Old Man who visits him. These good and evil figures, along with the doctor's own sense of self, may be illusory projections of the unconscious as conscious will, according to today's cognitive science and Lacanian psychoanalysis.[14] They may even reveal what we, like the Renaissance audience, experience through our brain's inner theatre: not only the conscious illusion of self, but also the "zombies" and "multiple selves" of our unconscious neural pathways (Ramachandran and Blakeslee).

Marlowe's *Doctor Faustus* presents a riddle for its spectators to willfully and unconsciously solve—depending on their predilections of belief (in God, Lucifer, or other ideal agents) and their changing perceptions during the show.[15] Is Faustus damned beyond salvation when he signs the contract with the devil, as he later fears? Or does he still have a chance for heaven despite his many sins, as he glimpses at the end? His contractual doom points back to an Old Testament God with sacrificial demands and strict judgments, as supreme spectator above the stage in the initial Wakefield plays. The latter possibility relates to the New Testament stories later in that cycle,[16] yet also points toward a Renaissance sense of mankind as having the "free will" to create a rational, civilized, heaven-like world on earth, by the grace of higher-order consciousness and culture. Faustus is thus a tragic antihero and neoclassical anti-saint, yet he demonstrates a new hope as Renaissance everyman—even with his will for evil.

Shakespeare's Neoclassical Spirits and Psychic Spaces

The most famous, indecisive "will" in Shakespeare's work appears in *Hamlet* (1600–01), written about a decade after Marlowe's *Doctor Faustus*. Yet in *Hamlet*, the ideal agency of God's will is referenced only briefly and God as a character does not appear. Is a watching God, as divine author and spectator, still framing the play and drawing Hamlet's willful soul — as in the earlier Wakefield Cycle and *Everyman*, as well as *Doctor Faustus*? Is there another cosmic will, as negative ideal agent, luring him toward hell?

Hamlet's first line in the play is an aside, in response to his uncle, the new king, calling him "cousin" and "son" (1.2.64). Here, Claudius offers Hamlet, in front of others at court, a promise not to do what happens in many other species when a new alpha-male takes over a female and the group hierarchy: kill the offspring of that female sired by a previous male leader. But in response to this offer of being adopted, Hamlet turns to the theatre audience, putting it in the role of personal confidant or godlike spectator, to understand his inner feelings about the current cultural order, about his physical and social relations to Claudius. "A little more than kin, and less than kind" (65).

Hamlet's initial reference to God occurs in the same scene, after all the others have left the stage. His first soliloquy to the audience, wishing that his flesh would "melt," involves the fear of God's prohibition of suicide — of the "canon" law that is "fixed" against "self-slaughter" (1.2.129–32). The pun here on "cannon" also suggests the wrath of God, which Faustus feared so much that he could not believe in God's mercy and his own potential salvation. Hamlet then exclaims, "O God, God," as he describes his despair at the world's uses being "weary, stale, flat, and unprofitable" (132–34). He feels a lack of being or meaning in human culture, expressing the fundamental alienation in all of us, according to Lacanian theory.[17] Yet he is not free to will his own death, due to God's law and possible wrath. Hamlet is caught between a medieval view of God ordering all of human life and a Renaissance sense of every-man as free to will a new social order, to cultivate "the unweeded garden / That grows to seed" without him (135–36).

Of course, Hamlet's melancholy derives specifically from his father's mysterious and sudden death, with the rest of the court not honoring the old king as much as his son expects. The prince associates his father with the ancient sun god, "Hyperion," and Claudius with a "satyr" (1.2.140). He recalls how his father cared so lovingly for his mother that he, like a force of nature, would not allow "the winds of heaven" to roughen her face (141–42). This idealization of the prior king as godlike reveals how the son is split — even before meeting the Ghost — between desire and doubt about both his father and mother. His father failed to survive immortally and his mother failed to be loyal as the regime changed, and perhaps even before that. Yet, Hamlet's terms show the appeal to his Renaissance audience of ancient ideals of gods as willful agents — extending parental authority, social power, and natural forces toward a higher order of sensors, effectors, and processors.

There are many references to heaven and hell in this play. When Hamlet first sees the Ghost, he says it may be a healthy spirit or a damned goblin, and may bring "airs from heaven or blasts from hell" (1.4.40–41). When the Ghost beckons Hamlet to follow it, Horatio cautions that it might tempt him toward madness. But Hamlet reasons that it cannot harm him since it is a "thing immortal" like his own "soul" (66–67). This immortal thing comes from purgatory, which is also a key cosmological space, framing the play and its stage edges. Despite a change in the ruling theology of England, from Catholic to Anglican, the formula of purgatory as a cathartic space continues to haunt Shakespeare's stage (Greenblatt). There the Ghost must "fast in fires" till his crimes are "burnt and purged away" (1.5.11–13). The Ghost even speaks in a hellish way, from "under the stage," at the end of his meeting with Hamlet, to make him and others "swear" their loyalty (49).

After hearing the Ghost's story, and its desire to be remembered, Hamlet calls to the "host of heaven," and to "earth," then asks, "shall I couple hell?" (1.5.91–93). The medieval cosmos still pressures Hamlet, through his father's Ghost — whether that figure is seen as a Catholic soul in purgatorial agony, as an ancient fury seeking earthly vengeance, or as a modern manifestation of Hamlet's inner brain theatre.[18] Hamlet vows to obey the Ghost's "commandment all alone," with no mention of God's canon, or how the host of heaven might view this (102). Hamlet does say, "Yes, by heaven," but only as he feels a righteous rage against Gertrude as a "most pernicious woman" (104–05). Here, he is already breaking the Ghost's double commandment: not to let his "soul" contrive against his mother, while avenging the villainous murder of his father (85).

Hamlet's double-bind continues throughout the play. He cannot accuse Claudius of murder, and yet leave his mother's judgment "to heaven" (1.4.46, 86–87). Hamlet does not receive advice from a good or bad angel, or from a saintly Old Man or a scholarly Mephistophilis. He struggles with the double-bind alone and only gets more advice from the Ghost when he misses a chance to kill Claudius and questions Gertrude in her chamber. Even then the Ghost appears merely to whet his "almost blunted purpose" (3.4.112). Hamlet's will is split, not between good and bad angels, or God and Lucifer, but by his father's commandment and the mystery of his mother's desire (Lacan, "Desire").

Ophelia describes Hamlet, to her father, as being "loosèd out of hell" (2.1.83). She then finds her own hell on earth, going mad after Hamlet rejects her and accidentally kills her father. Hamlet's madness is mostly feigned onstage — a trick he uses to carry out the Ghost's command and discern his mother's desire. Yet, he also finds hell on earth.[19] Hamlet moves from God's "canon" against self-slaughter to the changes that Ophelia and other women make in their God-given faces, while nicknaming "God's creatures," to a "realm dismantled ... / Of Jove himself" when Claudius reacts to his mousetrap (1.2.132; 3.1.145–47; 3.2.289). He goes from finding aspects of Jove, Hyperion, Mars, and Mercury in his father's image to the "Rebellious hell" in his mother's bones, to asking for help from "heavenly guards" when the Ghost reappears (3.4.57–59, 83, 104). And he finds a new confidence in the "divinity that shapes our ends," while also finding "providence in the fall of a sparrow," when he returns alive from his execution sentence, as if coming back from the dead for vengeance, like a ghost (5.2.10, 221).

The survival drive to find good or avoid bad territories, evolving over millions of years in our animal ancestors, and eventually becoming the hope of heaven and fear of hell in medieval cosmology, makes a key cultural shift with Shakespeare's Renaissance prince. God may not be present onstage, but His rule against suicide helps the melancholic Hamlet to survive the hell he finds in the rotten state of Denmark — and in his own disturbed spirit. Despite his disgust at female falsity and patriarchal deceit, his inflation of his lost father's image with neoclassical divinities, his horror at his mother's seductiveness, and his vain hope for help from heavenly guards, Hamlet finds a divine providence shaping his tragic end and catastrophic fall. Even if it is just an audience's desire, or the playwright's script, Hamlet discovers a god watching and guiding him — through the twists of fate and soul in this play. Reacting to Denmark's disrespect for its prior king after his death, Hamlet finds a way to fill this lack in his Father's Name, with divine providence, not just the Ghost, as his commander. And he resolves the double-bind that the Ghost gave him by dying with his mother at the play's end.

Hamlet even conceives of heaven and hell within the theatre of his own brain. He tells the king and queen that he has "that within that passes show" — beyond his "trappings" of mournfulness and his "suits of woe" (1.2.85–86). He tells Horatio that there are "more things in heaven and earth" than are dreamed by philosophy (1.4.166–67). And he jokes with Rosencrantz and Guildenstern: "O God, I could be bounded in a nutshell and count myself a king of infinite space, were it not that I have bad dreams" (2.2.258–60). He also speaks to them about mankind being "noble in reason," "infinite in faculties," "in action ... like an angel," and "in apprehension ... like a god" (312–15). Thus, he describes a new Renaissance dream of humans becoming more and more godlike through the infinite spaces of imagination and their transformation of the environment as a superior species — making heaven on earth with their ideal sensors, effectors, and processors: divine apprehension, angelic action, and noble reason. But Hamlet also shows the hellish side of that inner theatre and its external, godlike powers in his tragic perceptions, actions, and flawed reasons for vengeance.

Various classical gods appear in Shakespeare's late romances, and in an earlier play, *As You Like It* (1599–1600), as figures for this Renaissance sense of heaven and hell within the human brain — with its potential for divinity, yet incomplete evolution of limbic Real, right-brain Imaginary, and left-brain Symbolic orders. Hymen, the Greek god of the bridal hymn, brings "mirth" and an exiled duke's daughter back to him "from heaven," after her adventures in the woods away from her uncle's fury (*As* 5.4.108–12). The winged Hymen also acts as a *deus ex machina* bringing "conclusion" to the play "Of these most strange events" (126). A decade later in *Pericles* (1608–09, perhaps co-written with George Wilkins), the lead character miraculously gets his lost daughter back and then is visited in a dream by the ancient mother goddess, Diana of Ephesus (5.1).[20] She tells him to make a sacrifice at her altar — where he also gets back his wife (a priestess of Diana), after believing she died during childbirth at sea. In *The Winter's Tale* (1610–11), Apollo's oracle guides a jealous, vindictive ruler toward the tragicomic recovery of his abandoned daughter and wife, with the latter returning from the dead as a living sculpture.[21] And yet, Leontes initially discredits the oracle, when it states his wife is not guilty of adultery as he believed — until the sudden death of their son, which the king interprets as another sign from the god. At the end of the play, Paulina also admits that she may be accused of witchcraft, of being assisted by "wicked powers" — in bringing the apparently dead Hermione back to life, after Leontes's sixteen years of grief and regret (5.3.91).

Classical gods bring heavenly gifts to disturbed patriarchs as a late plot twist in all three of these plays, moving them from rage and sorrow to tragicomic joy. The gods represent forces of nature and fate, working through the human mind and its social structures to re-imagine a happy ending, via the playwright's and audience's desires. They also show the neuroplastic potential of the human brain and of Shakespeare's culture, looking back to classical deities and sacrificial rites, yet also ahead, in order to stage a better world, beyond repetitive pathways and tragic compulsions.

But the forces of nature and culture, while ultimately beneficial in the pastoral dramas of *As You Like It* and *The Winter's Tale*, and the maritime adventure of *Pericles*, can also twist human fate in hellish directions. In *Cymbeline* (1609–10), Posthumus bears a powerful death-drive, after falsely believing that his wife slept with another man, after sending a servant to kill her, and after regretting this vengeance. He joins a battle between ancient British rebels and the Roman Army, trying to die, through his guilt and despair. He is then captured. While he sleeps in jail, awaiting execution, his family ghosts (his father, mother, and two brothers) appeal to Jupiter for help with the tragic fate of Posthumus. His father asks Jupiter, as divine spectator, why he allowed the villain, Iachimo, to trick Posthumus into believing his wife was unfaithful. The First Brother asks Jupiter why he gave Posthumus sorrow instead of "graces" for his "merits" on the battlefield (5.4.79). The father asks Jupiter to look out his "crystal window" and "No longer exercise / Upon a valiant race thy harsh / And potent injuries" (81–84). The mother also insists that their son is "good" and does not deserve "miseries" (85–86). The father even threatens to take their cause to the assembly of other gods — if Jupiter is too aloof, like a deist clockmaker, as King of the Gods. "Peep through thy marble mansion, help, / Or we poor ghosts will cry / To th' shining synod of the rest / Against thy deity" (87–90).

Jupiter responds with thunder and lightning, appearing on an eagle and throwing thunderbolts. He tells these "petty spirits of region low" that they, as virtual actors, offend his "hearing," as Metamind audience (5.4.93–94). Then he suggests that Posthumus's self-punishments, moving from hellish guilt to purgatorial warfare, may be part of a divine director's plan. "Whom best I love I cross; to make my gift, / The more delay'd, delighted" (101–02). Jupiter reassures them that Posthumus's "trials well are spent"— as he moves through his current purgatory toward catharsis and a tragicomic heaven (104). Jupiter also promises that Posthumus will be reunited with his faithful wife and become "happier much by his affliction made" (108). He gives the ghosts a "tablet" to lay on the sleeping man's breast, "wherein / Our pleasure his full fortune doth confine" (109–10).

When Posthumus wakes, having dreamed at least some of what the theatre audience saw, he finds the tablet. It describes mysterious events involving a "lion's whelp" and "stately cedar" when Posthumus's "miseries" will end (5.4.137–48). Although he cannot decipher this oracle, the rest of the play does it for him (with the help of a Roman Soothsayer at the end), demonstrating how Jupiter — as an ancient mask for the almighty Christian God — was watching and directing the show all along.[22] The Other is present offstage, behind the "marble pavement" of the sky (120), through the desire in Shakespeare's audience to have a Metamind, an ultimate ideal agent, in charge of it all. Shakespeare offers neoclassical deities for this, as his culture moves from a medieval worldview toward a Renaissance sense of heaven and hell within the human brain, with its godlike and devilish forces of apprehension, noble reason, and "infinite ... faculties."

Exposing an internal theatre of gods within the human brain, Diana visits Pericles in a dream and Jupiter visits the ghostly relatives of Posthumus while he is dreaming. But these gods also have holy places outside the brain—

tied to natural forces. Diana, as mother goddess of Ephesus, gives Pericles the miracle of his wife's return, from death in childbirth, only when he visits her temple there. Jupiter returns to his marble mansion in the sky after responding to the ghosts' pleas and threats, not just with a promise to save the life of Posthumus, but also with a fierce display of his lightning and thunder. Shakespeare's use of neoclassical deities in his late plays shows a repeated need for human submission to holy places and natural powers. Yet, in *The Tempest* (1611), he also explores mankind's increasingly godlike powers, to make heaven or hell on earth, through neoclassical spirits and spaces, both patriarchal and maternal.

Here, in his last fully authored play, Shakespeare shows a Renaissance prince trying to control nature, culture, and his enemies, while also confronting his own rage for vengeance. Prospero uses a magic, Mediterranean island for his evolving theatre of justice — having taken it away from a prior colonizer (Sycorax), her god (Setebos), and her bestial son (Caliban), while struggling to master its nature spirits through his "books." It is Prospero, not Jupiter or Neptune, who creates a hellish tempest that humbles his enemies and brings them to his island, separating King Alonso from his son, Ferdinand, and making each believe that the other has died. Yet Prospero also drafts a heavenly vision of his daughter, Miranda, for Ferdinand — and of him for her — manipulating their passions for his own realignment with the king's cultural power back in Italy. First, however, Prospero threatens his servant, Ariel, with a return to hell: to the "cloven pine" where that spirit was trapped for more than a dozen years, venting groans "As fast as mill wheels strike" (1.2.277–81). Ariel was in "torment" with groans that made wolves howl and penetrated "the breasts / Of ever-angry bears" (287–89). Unless Ariel is obedient, Prospero will return him to that place of a "torment / To lay upon the damned" (289–96). But if Ariel does obey, Prospero will "free" him in just two days. Thus, Prospero has a power akin to God the Father in the Wakefield Cycle or *Everyman*, as supernatural judge over this spirit.

With such a threat and yet promise from the human magus, Ariel plays the character of a sea nymph and teases the shipwrecked Ferdinand in a devilish way. The prince hears Ariel's mysterious singing as honoring "Some god o' th' island" (1.2.390). At first, "its sweet air," gives him a taste of heaven, beyond the loss of all his companions in the storm. But then the lyrics of Ariel's song turn toward a tragic magic: the bones and eyes of Ferdinand's father become "coral" and "pearls" at the bottom of the sea (398–99). Ariel's mischief here is like that of Puck and other fairies in *A Midsummer Night's Dream*, toying with the foolish passions of mortals. And yet, Ariel offers a Buddhist-like insight about accepting the transience of life — with the death and decay of a loved one being a "sea-change / Into something rich and strange" (401–02).

4. From Renaissance Rebirths to Postmodern Experiments 143

After this purgatorial song, Prospero grants the prince a glimpse of Miranda as a heavenly "goddess" (1.2.412). She sees Ferdinand, too, as a "thing divine" (418). But Prospero fears the animal lust in the youngsters' idealization of each other — with "both in either's pow'rs" (451). He calls Ferdinand a "traitor" and Miranda a "Foolish wench" (461, 480). He binds up Ferdinand's "spirits, as in a dream" (487). Later, he makes the prince carry logs like a servant, as purgatorial work that purifies his love for Miranda, prior to marriage (3.1). Even then, he warns Ferdinand about "fire i' th' blood" (4.1.53). In a Godlike way, Prospero saves Alonso's life, as well as Ferdinand and the rest of their shipmates, through Ariel's angelic agency. But he makes Alonso believe his son died in the tempest — as purgatorial suffering that readies him to accept Miranda as a daughter-in-law and restore Prospero's dukedom, when Ferdinand returns alive. Prospero also reveals the murderous greed for power of his own brother Antonio and the king's brother Sebastian, creating a purgatory for them, by playing on the repetition compulsion of their sinful ways.

Prospero does not sign a contract with Lucifer. Nor does he need a devil as his servant in order to have magic power.[23] Ariel functions instead like the Good Angel of Faustus, giving Prospero the merciful advice of tender "affections" (5.1.18). And yet, the knowledge in Prospero's books, along with the phallic power of his staff, tempts him like a devil, not only to create a tempest of revenge, but also to lose himself (or his soul) in it. His animal passions for vengeance, alpha-male authority, and territory, after losing his dukedom in Milan and taking control of the island, makes him a brutal master of Caliban and a dictator toward others — until he changes, with Ariel's help, breaking his staff, and drowning his book (54–57). Prior to that, Prospero sees Caliban as evil by nature, "got by the devil himself" upon a "wicked dam" (1.2.319–20). So Prospero, as patriarchal master, becomes evil toward him as well.

When first summoned in the play, Caliban wishes "wicked dew" on Prospero and Miranda, plus a tempestuous wind to "blister" them all over (1.2.321–24). Prospero responds with a promise for him that night of belly cramps, "urchin" (hedgehog) pricks, and bee-sting pinches (325–30). But who started this preexisting fight of wicked wishes and spells? Caliban claims that the island was his, by Sycorax, his mother, but was taken away by Prospero. He was seduced by Prospero — who "made much of" him, gave him water with berries, and taught him how to name the sun and moon (333–36). Initially, Prospero nurtured the young Caliban, replacing his lost mother, though it is not clear from the play whether Sycorax died before Prospero came to the island or was killed by him. Prospero gave food to Caliban and "strok'st" him (333). He offered maternal nurturing in both senses that are vital to primates. For a comforting mother is more crucial to infant development than food, as shown by experiments with infant monkeys, who cling much more to a cloth-covered maternal substitute than to a wire-mother with a bottle (Hrdy 398–99). Prospero also mothered Miranda on the island, from age 3 to 15. But again it is not apparent how her mother in Milan was lost. Prospero claims she was a "piece of virtue," yet does not say what happened to her prior to or during the loss of his dukedom (1.2.56).

There is no mother goddess in this play, like Diana in *Pericles*, helping family members to reunite. But the island itself was a mother to Caliban and becomes a *chora* to Prospero's patriarchal power. Caliban recalls his blissful relation to Mother Nature on the island, where he once was his "own king," but is now "subject to a tyrant" (1.2.342, 3.2.41). There are "sweet airs" on the island, with a "thousand twangling instruments," and voices giving him dreams of clouds that open with riches ready to drop on him, so that when he wakes he cries "to dream again" (3.2.133–40) — like a baby wanting to suckle. Caliban showed Prospero, when he came there, "all the qualities o' th' isle" and learned language from him in return (1.2.337). Thus, in a neuropsychoanalytic sense, Prospero changed from being a nourishing and comforting mother, like the island, shaping Caliban's right-brain ego development as Imaginary mirror, to a harsh superego and phallogocentric patriarch, a No and Name of the Father, distancing Caliban from his island bliss and distorting his natural identity through left-brain Symbolic orders. Or, as Caliban puts it, "you sty me / In this hard rock, whiles you do keep from me / The rest o' th' island" (342–44). Prospero then recalls a trauma in their past relationship, when young Caliban tried "to violate / The honor" of Miranda (347–48). This convinced the patriarch of his adopted boy's "evil nature," a way that he also viewed his brother Antonio, who stole his dukedom (93). Through this view of Caliban, however, Prospero cultivated the boy's evil, giving up on nurture as not sticking to his nature. Thus, the wild child experienced his natural drive for reproduction through Prospero's colonizing mirror and its framework of vengeful restrictions. "Would't had been done! / Thou didst prevent me: I had peopled else / This isle with Calibans" (349–50).

Prospero shaped Caliban's brain and body as sub-human and bestial, viewing him as "Filth" despite the "humane care" he gave him in his own "cell" (1.2.346–47). During the play he tempts Caliban and others, via Ariel, to prove their evil natures, so that he can condemn them. In judging Caliban near the end, Prospero shows this prejudice: "A devil, a born devil, on whose nature / Nurture can never stick" (4.1.188–89). In contrast, he views Ariel as a "Fine apparition" (1.2.317). They express the evil and good within him — as right-brain, Imaginary, holistic sensitivity and limbic, Real, animal passion to his left-brain Symbolic controls. But Caliban also becomes a scapegoat, along with Antonio. Prospero projects upon them his own animal temptations of lust and greed, for reproductive power and greater ego survival.[24] Yet this initial, melodramatic purging of the hero's own potential evil by projecting it onto clear-cut villains, to be battled and beaten by the good, also takes alternative, tragicomic twists in Shakespeare's play, leading to a full catharsis in the end and possible changes in spectators' neural pathways.

Miranda says that she also tried to nurture Caliban with language, to give him "meaning" and "purposes"—when he "gabbled" like a "thing most brutish"—although some editors give this speech to Prospero (1.2.355–57). She (or he) finds that "any print of goodness" will not stick to him, that he is simply from a "vile race," and that he is "Deservedly confined" in a rock prison (352–62). This introduces Caliban as a definite villain, with her as the victim, and her father as the melodramatic avenger. However, all three of them are eventually shown to be tragicomic victims of fate and of their own character flaws, as they struggle heroically with outer and inner evils — not merely projecting them, but also becoming more aware of evil within the good, or helping the audience to do so. Miranda over-idealizes Ferdinand and breaks her father's rule in telling him her name (3.1). But they both submit then to the patriarchal order, changing their views to find joyful play in work and future pleasure through present restraint. Caliban cannot free his identity from the colonizers' gazes, nor use their language for his own welfare, only use it to "curse" and attempt more violence against them (1.2.364). Even when he finds an alternate "god" in the drunken butler Stephano, who seems to have "dropped from heaven," he still worships an alien Symbolic and Imaginary order that represses him as its extimate Real (2.2.134, 145). Yet, in repeating with Stephano and Trinculo his tragic error with Prospero, showing them the best resources of the island (in exchange for the false sustenance of liquor) and attempting to violate Miranda with them, Caliban sees how his desires are defined by the Other — and may even learn to change his fundamental fantasies. For his part, Prospero summons spirits to play ancient goddesses, fantastically celebrating his patriarchal manipulation of his daughter's future, without realizing her need — and Caliban's — for a more nurturing relation to Mother Nature. Yet, despite Prospero's indulgence in theatrical spirits, and his darker ability to open graves and raise the dead through his "potent art," he learns to change his vengeful spells, through Ariel's sympathy with his enemies (5.1.50). Prospero gives up his "rough magic" and, with "nobler reason" against his fury, commits the "rarer action" of mercy rather than vengeance (26–27).

The consensus view of Prospero has changed in recent decades: from a modernist figure of beneficial patriarchal power, as enlightened despot of the island, to a controlling father and repressive colonizer. In various postmodern interpretations of *The Tempest* (feminist, postcolonial, or psychoanalytic),

Prospero may barely realize his sins in the end and still leave a damaging legacy for Miranda, Caliban, and others. However, through a cognitive evolutionary, neuro-psychoanalytic approach, we can see both sides of this godlike thespian, plus other aspects that may have a longer lineage.

Prospero lost his alpha-male authority in Milan and was put on a small boat with his three-year-old daughter, his books, and other provisions. By "providence divine" (and Gonzalo's help), they survived and eventually gained power on the island (1.2.159). Prospero's struggle for survival and dominance thus extends the collective competition of neurons within the brain for which circuits, memes, or egos will be featured in the stage space of consciousness. He demonizes Caliban for attempting to spread his genes, via Miranda, as a new population across the island. Yet Prospero strives for a similar dominance of his race and culture over the island's spirits and resources. Old Gonzalo likewise imagines his own utopian "plantation" when shipwrecked on the island (2.1.139). These are drives that developed in our species over millions of years — to survive in many new environments and alter them, while spreading our biological and cultural progeny, our genes and ideas, in competition with others. That does not mean, however, that the extremes of Social Darwinism are inevitable or desirable. *The Tempest* shows that an increase in human cognitive power, through the left-brain Symbolic order of language, analysis, and books, along with the right-brain Imaginary force of holistic creativity and intuitive perception, may transform limbic, animal passions into the horrors of lust and greed — or the virtues of love and mercy.

Shakespeare's tragicomedy reveals both the good and evil sides of Prospero, with more or less of each being presented and perceived, depending on the brains involved in a particular performance and their evolving cathartic awareness, on both sides of the stage's edge. Toward the end of the play, Prospero conjures his spirit-servants to perform a neoclassical masque in celebration of the "contract of true love" between Miranda and Ferdinand (4.1.84). Spirits play the roles of ancient goddesses Iris (as messenger), Ceres (as harvest, fertility, and mother goddess), and Juno (as queen). This display inspires Ferdinand, as a spectator, to call the island a "Paradise" (124). But Prospero also sends Ariel as an ancient mythic creature to give a tantalizing trick and Tantalus-like hell to Alonso and his courtiers. First, Ariel offers them a heavenly show of natural and cultural bounty — with "sweet music," a banquet, and spirits dancing (3.3.17–20). Then Ariel appears as a "harpy" and makes it all disappear, also making Prospero's enemies, as "men of sin," go "mad" (52–58). Prospero and Ariel create a similar trick for torturing Caliban and his cohorts, with a taste of heavenly apparel becoming a trap for them to be chased by hellish hounds (4.1). Like the ghost of Clytemnestra in *The Oresteia*, Prospero then goads the furious spirits to punish his enemies even further:

"goblins ... grind their joints / With dry convulsions, shorten up their sinews / With agèd cramps...." (257–59).

The Tempest reveals how Prospero, as Renaissance ruler and magician, acts as a *deus ex machina* throughout the play. He uses the machinery of nature and spirit to bring elements of heaven and hell to many on the island. And yet, like Jupiter in *Cymbeline*, Prospero has a providential purpose for the pain and pleasure he causes. He creates various purgatories for his island captives, purifying their animal passions toward higher-order awareness. As he watches over all of this, the tragicomic Prospero also becomes aware of the good and evil within himself, of the temptation for more vengeance, and yet the possibility for a rarer action in virtue. His audience may realize this, too, rationally and intuitively, through their left and right brains — since they have the power, as he tells them in the epilogue, to release him or imprison him on the island with the "spell" of their "good hands," through applause, prayer, mercy, and "indulgence" (8–20).

From Early Modern to Romantic God Ideals: Morality, Kinship, and Dream Plays

Recent studies have found that the ventromedial (lower middle) prefrontal cortex, or VMPFC, is a crucial brain area for developing a "moral sense" (Changeux et al. x). Activity in the left VMPFC "is greatest in people who are very compassionate and caring" (xi).[25] On the other hand, damage to the VMPFC "causes a marked disturbance of social conduct" (Hanna Damasio 43). This is also the area where neuro-psychoanalysts find evidence for a "stimulus barrier" of the superego against the id, protecting the ego from impulsive passions and animal instincts (Kaplan-Solms and Solms 275–76). As I suggested in the introduction, this may locate a key neural connection between Symbolic, Imaginary, and Real processes, which changes through theatrical catharsis (as the purifying of emotions through conscious awareness). For the prefrontal cortex "helps to build mental representations that relate to social interaction" (Changeux 2). It "plays a central role in attributing mental states to others ... and thus contributes to ... the appreciation of aesthetic qualities of artwork and in moral judgment: empathy."

Our sense of good and evil relates to Real, limbic emotions tied to the "homeostatic regulation" of our bodies, regarding what causes health or illness, in individuals and groups (Antonio Damasio, "Neurobiological" 48).[26] Social emotions, such as compassion, embarrassment, shame, guilt, disgust, indignation, contempt, gratitude, awe, and admiration, are "action programs," which produce social feelings and ideas as "scripts" (49–50). Thus, our ancestral "drive for cooperation" and "drive to serve punishments," as basic homeostatic functions, shaped the development of social emotions—and of moral intuition and reasoning, as Imaginary and Symbolic dimensions, from nature to human culture (50–54).

Various "dual process theories" of cognitive psychology use current evidence to verify an "ancient idea" that intuition and reason are two distinct processes in the brain (Kahneman and Sunstein 92). The intuitive process is automatic, effortless, associative, rapid, opaque, and skill-based, in contrast to the rational, reflective process, which is controlled, effortful, deductive, slow, self-aware, and rule-based (92–93). Neuroimaging studies show that moral judgment involves these "two processing streams operating in parallel: a quick and efficient processing stream that provides stereotyped responses based on limited information and a slower, more deliberative processing stream that provides more flexible responses based on a (potentially) much wider range of information" (Greene, "Emotion" 63).[27] In experiments where subjects were given hypothetical moral dilemmas, they followed intuitions in saying they would save a fellow adult or a drowning child, who was "up close and personal"—while sacrificing others at a distance (such as starving children in a foreign country), even though reason and numbers would counsel otherwise (64). Such an intuitive, "personal condition" evolved in us as a priority and "pushes our emotional buttons" today, unlike distanced situations, because our ancestors were never able "to save the lives of anonymous strangers through modest material sacrifices." Even the many donations given after sudden disasters (hurricanes, tsunamis, and earthquakes) soon wane when the suffering of those at a distance is no longer intimate through the news media.

Researchers also theorize that the "mirror neuron system played a fundamental role in the evolution of altruism"—with humans developing a distinct faculty for "imitation" in order to "understand actions done by others" (Rizzolatti and Craighero 109, 119). Mirror neurons in the parietal and frontal cortices enable the "cold" understanding of actions, in monkeys and humans. But humans also have a "mirror mechanism" in the limbic system, including the insula, which enables spectators "to understand the emotions of others"— through an "as-if-loop" involving bodily representations in the somatosensory cortex, as the "neural basis of empathy" (107, 119). Mirror neurons in humans, unlike those in monkeys, fire when a pantomimed gesture is observed or performed (Iacoboni 26–27)—demonstrating a key difference in our ancestors' mimetic communication, as it evolved toward theatre.

Researchers find that there is "a close tie between emotion, self-feeling, and reasoning error inhibition in the human brain," with greater activation in the right VMPFC, as a "paralimbic area," when intuition helps the mind get back on the right logical track (Houdé 139–40). The VMPFC has "strong anatomical reciprocity with the amygdala" in the limbic system, "thereby contributing to the regulation of emotion" (Davidson, "Neural" 73). The orbitofrontal sector of the prefrontal cortex is also key to emotion regulation through the "learning and unlearning of stimulus-incentive associations." And the dorsolateral sector of the PFC is "directly involved in the representation of goal states." One might hypothesize that theatre developed in various human cultures as a further, collective evolution of the mirror neuron system in the primate brain: from cold mirroring, to the mimetic, "as-if" understanding of others' actions, involving empathy and altruism.[28] Theatre also evokes circuits in the somatosensory cortex, limbic system, and PFC — for body representation, emotion regulation, and "reversal learning" of stimulus-incentive behaviors, regarding specific goals and errors in logical judgment. This may produce a catharsis of intuitive feelings in relation to moral reasoning, through quick, associative and slow, reflective pathways.

The Tempest shows Prospero's drive for cooperation on his island, yet also to serve punishments — for the homeostasis of his body, ego, kin (Miranda), and dukedom. For example, he interrupts the spirit masque because he realizes he forgot, and must deal with, the "foul conspiracy" of Caliban and others "against my life" (4.1.139–41). After pondering then how we are "such stuff / As dreams are made on," he admits that he is "vexed" and his "old brain is troubled" (156–59). He has the moral intuition of saving his daughter, as well as himself, from the conspirators. But he does not feel much empathy for Caliban, as a born devil, or for his Italian enemies — until Miranda grieves for the men onboard ship during the tempest and Ariel expresses tender "affections" for them on the island after more such tricks (1.2.5–6; 5.1.18). Prospero uses moral intuition for his stereotyped responses to others, seeing them as good or evil, until the circuitry between his prefrontal lobes, somatosensory cortex, and limbic system changes, through the influence of his allies, Miranda and Ariel, with their intuitive sympathy toward his enemies, as victims of Prospero's wrath. He then uses moral reasoning, fueled by new intuitions, to find the rarer action in virtue than in vengeance: to "drown" his book and bury his staff, avoiding further temptations of power (5.1.54–57). The audience might likewise "unlearn" their circuitry and scripts of tragic errors in judgment, through the related activation of mirror neurons, cooperation and punishment drives, social emotions, altruistic empathy, moral intuitions, and reflective reasoning — while watching this play.

Shakespeare based his tempestuous Mediterranean island and Prospero's colonization of it, at least in part, on the account of an English shipwreck in Bermuda, involving Jamestown colonists in 1609, and the adventures in the prior century of Magellan and company, meeting natives along the coast of South America (Frey 29). Shakespeare's source for the latter, Richard Eden, based his story on Antonio Pigafetta's first-hand memoir about the Spanish encountering natives whom they called Patagonians (big feet). When in distress, these "giants" called out to their "devil," Setebos (15–17).[29] The Spaniards considered, like Trinculo in Shakespeare's play, whether they might take a native back to Europe and make money exhibiting him there (16). In *The Tempest*, Caliban does not have a fully native pedigree. His mother, Sycorax, is from Algiers. Nevertheless, many recent productions have depicted him as a native islander to stress Prospero's colonial imperialism and abuse of his slave.[30] Caliban's name, made up by Shakespeare, seems to be an anagram for "cannibal." Ariel's, on the other hand, is biblical, meaning the lion or altar of God, a poetic name for Jerusalem (Isaiah 29:1–2). It may be easier for Prospero to scapegoat Caliban as foreign and sub–human. Yet, he also threatens Ariel with punishment, and then takes delight in his airy nature-spirit's obedient tricks. Prospero's moral intuition favors Ariel, as well as Miranda, as family — while Caliban becomes the bad seed who can be sacrificed, with subsequent, moral-sounding reasons.

A different approach to the prior residents and gods of a colonized land is shown in the *loa* to *The Divine Narcissus* (1687), the allegorical prelude to a play written, in the medieval tradition, by Sor Juana Inés de la Cruz in New Spain (today's Mexico), about 75 years after Shakespeare's presentation of *The Tempest* in England. Initially, the Spanish characters, Religion (female) and Zeal (male), are shocked at the Aztec characters, America (female) and Occident (male). The Aztecs worship the "God of the Seeds" through Music (another allegorical figure), with song and dance — while also describing their sacrifice of humans, "hearts still beating," and cannibalism of the god's body and blood (1.36, 65). Religion spurs Zeal's moral intuition, "the flames of righteous Christian wrath," against such pagan "idolatry" (2.3–5). Then Zeal bares his sword and threatens the Aztecs with "vengeance," as the agent of a watching God, who grows "weary at the sight" of their religious performances (72–75). Occident, the Aztec warrior, states that Zeal's moral "reasons" make no sense to him (86). Religion threatens the Aztecs further, with being "reduced to ashes" (100). Yet, the Aztecs refuse to end their rites, so the Spanish go to war and beat them.

Once Zeal, as Spanish conquistador, has conquered the Aztecs, Religion changes her tune. Through her moral intuition, she stops Zeal from extending his righteous violence and killing America: "her life is of some worth to us" (3.6). Yet the Aztecs still refuse to capitulate to Religion's "arguments" and give up the worship of their gods (3, 35). Religion then develops her moral reasons, by intuiting certain connections from Christian beliefs and practices to the Aztecs' pagan ideas, sacrificial rites, and theatrical desires — however devilish or horrifying.

Prior to the war, Religion viewed the Aztecs as practicing an "irreverent cult / with which the demon has waylaid you" (2.32–33). After the war, in an aside, she pleads for God's help in understanding "what shadowings / of truths most sacred to our Faith" may be mimicked in the "lies" of the Aztec rites — seeing them as directed by God's rival, the "false, sly, and deceitful snake" (4.16–19). By shifting the blame to the animal-god Satan, she can then empathize altruistically with the Aztecs as striving for the one, true God, albeit through monstrous perversions of the Catholic mass. She uses the biblical parallel of Paul preaching to the Athenians, arguing that they were already worshiping an "Unknown God" that he was clarifying (42). And she tries to do the same with the "God of the Seeds" — Sor Juana's amalgam of numerous Aztec gods, especially Huitzilopoctli, the sun god (Bergland 154; Balsera 296), and Xipe Totec, the god of maize, for whom the Aztecs not only offered heart sacrifices but also flayed and wore the skins of human victims.

Religion first argues theologically that "our true God alone" is the only one through whom the Aztecs receive maize and other crops. While "seeds mature in earth's dark womb," with rain from the sky, Religion says that this is only by "the work of His right hand" (4.65–67). She thus suggests that the Christian God, as patriarchal "Providence" (73), is superior, like the Spanish in war, over the Aztecs' earth and rain gods, Tlaltecuhtli and Tlaloc, who were also fed human sacrifices (though these gods and offerings are not mentioned explicitly). But America asks whether such a God may be "touched" like the image of an Aztec god, "made from seeds of earth / and ... human blood" (81–82). This god-puppet, made of maize dough and blood from human sacrifice, was known as the *ixiptla* and the same term was used for the sacrificial victim as god-actor. While this is not specified in the play, the dialogue suggests that a similar prop would have been used if the play were performed. America refers to it as an "idol, here before you" (80). Religion tells her that Catholic priests may touch God as present in the Host. She also responds to Occident's question about whether her God offers Himself as a sacrifice, via "human blood," by describing the transformation of bread (and wine) into the Body and Blood of Christ in the "Holy Sacrifice of the Mass" (104–15).

Spanish Religion evokes Aztec America's desire to "see this Deity" (4.132) and Religion promises that she will, once she is baptized. Religion also plans to make "a metaphor"— the rest of *The Divine Narcissus*— so that the Aztecs may be converted through a further theatrical allegory about the theology of the Eucharist (156). But Religion wants to take this performance, with its introductory *loa* and Aztec characters, to Madrid for a royal display as well. Zeal questions her plan to take "the whole Indies on / a stage" to the Spanish capital (46–47). However, Religion argues that such an allegorical transfer of culture will not be hindered by "distance" or by "seas" (56–68). And then the "humble Indies" bow to "kiss" the feet of the King and Queen, in the play's ostensible performance in Madrid (64), though this did not actually occur during Sor Juana's lifetime.

Like Caliban, the Aztecs in this play must bow to European power, royalty, and moral reasoning. Sor Juana condenses two centuries of conquest and colonization into a short allegory. Yet, she also shows a change in the Spanish characters: from righteous wrath and brute force to merciful sympathy and theatrical strategies of theological persuasion. The female characters, on both sides of this culture clash, display a better balance than the males of limbic Real, right-brain Imaginary, and left-brain Symbolic orders, using moral intuitions as well as reasoning to discern good and evil in ritual performances — and to alter their alliances. Unfortunately, Sor Juana's own labor with this drama and her other writings did not succeed in pleasing the Church's patriarchy. In 1694, four years after *The Divine Narcissus* was published in Madrid, she was forced to sign a contract in blood (like Marlowe's Faustus) to give up her secular studies and submit to the ordinary performance of her vocation as a nun. She died the following year from a plague in Mexico City.

Sor Juana's extraordinary *loa* stages the early ideological evolution of Euro-American culture as a question of kinship, through conflicting ideals of divinity and theatrical views. Are the Aztecs, in Spanish eyes, fellow human souls yearning for the same God, or subhuman devils and devil-worshipers, like Caliban to Prospero and the Patagonians to Magellan's company? For the natives, are the conquistadors morally persuasive as performers of sacrificial rites or simply a new alpha-male rule, taking over their territory, with a God they must obey? These fundamental questions of kin and god continue to reverberate throughout Western cultural evolution — from the initial, early modern encounters between Old and New World civilizations to our own global identifications today.

In Western culture's many permutations over the past half millennium, a monotheistic God often functioned as the supreme model of good versus evil, for making territories, rules, and rulers of communal belonging, for kinship parameters in reciprocal altruism, and for the sacrificial demands of subordination stress. This monotheistic model continues the medieval consolidation of many gods into one dominant divinity, with His nemesis becoming the prime figure for evil or deception, as seen in the Wakefield Cycle, *Doctor Faustus*, and Sor Juana's *loa*. Yet, the rise of secular humanism from the Renaissance onward offered other godlike ideals for our species' animal passions and divine aspirations — as marked by the Christian, neoclassical, and human powers in Shakespeare's plays.

Hamlet must decide for himself whether his father's apparition is actually from purgatory or a demon from hell deceiving him. He eventually finds, after tragic delays that cause suffering and death for others, that there is a special providence in the sparrow's fall and a "divinity that shapes our ends." In Pericles's case, the pagan mother goddess Diana appears in a dream, combining with the Catholic version of Mary repressed by Protestantism and enabling the miraculous reunion of his family. For Posthumus and his ghostly family, Jupiter becomes a forceful spectator and director, arriving from his marble mansion in the sky with lightning, thunder, and reason, thus answering their ancestral desire for fairness (as shown also in the proto-ethics of capuchin monkeys). And yet, Prospero himself, in Shakespeare's last fully authored play, becomes providential, directing spirits in goddess fertility roles, and bringing both vengeful storms and moral mercy to those on his island.

Other Renaissance playwrights used neoclassical deities and devices, or showed heaven and hell on earth, to represent the culture's move beyond Christian paradigms, toward secular, rational, humanist ideals, though edged with dangerous, tragic passions. In Jean Racine's *Phèdre* (1677), an ancient king rages at his son's apparent betrayal of rational trust. This, along with a queen's suicidal passion for her stepson, leads to the offstage dragging to death of Hippolytus by his own horses, through his father's evocation of Neptune's sea-monster. In Pedro Calderón's *Life Is a Dream* (1636), Basil, the mythical King of Poland, believes a prophecy that his son will be a horrible ruler and so he preemptively intervenes. Like a god, he puts his son in purgatory, heaven, and hell: imprisoning him in a cave, restoring him to the dreamlike pleasures of the palace, and then imprisoning him again — thus producing existential confusion and animal rage in Segismund. With *Tartuffe* (1669), Molière gives a comical, contemporary version of such devilish trickery in the figure of a deceitful holy man. Yet he offers a final vision of godlike justice with the *deus ex machina* of the Messenger, who brings the French king's all-seeing wisdom and rational force for a happy ending.

In these works, the rulers' monstrous vengeance, preemptive strike, and all-knowing justice parallel aspects of Prospero and Jupiter in Shakespeare's late romances. But the Bard's plays offer a wider panoply of explicit gods (or aspects of God), as well as various godlike forces in human reason and emotion, for good or ill. They foreshadow Europe's evolution into an Age of Reason with "enlightened despots," then democratic revolutions with individual rights yet reigns of terror, and then Romantic nostalgia for a lost connection with the "sublime" forces of nature.

Such alternative visions of divinity and morality, in nature and culture, begin to arise in eighteenth century theatre. English sentimental comedy, moving beyond Restoration comedies of manners and intrigue, display the foibles of the upper class, but with more morality. The noble, rational soul of the "man of sentiment," like that of Charles Surface in Richard Brinsley

Sheridan's *The School for Scandal* (1777), eventually surfaces, through various comic twists, and gets rewarded in the end, as if by divine providence. The watchmaker God, in a deist sense, may be watching from farther away than Jupiter's marble mansion in the sky and interacting with humans even less. But the rational sense of honor and justice that God implanted in humans can, according to such plays, create a better, more civilized world on earth. And yet, a darker side to human nature also appears in eighteenth-century theatre, through Carlo Gozzi's proto–Romantic fables and the German Storm and Stress playwrights. For example, Heinrich Wagner's *The Child Murderess* (1776) shows a young mother pushed into insanity by the rational actions of others, without providential remedy — to the point that she punctures the skull of her illegitimate child with a hatpin onstage and licks the blood as her baby dies.

With more commedia-like fun, Gozzi's *The Green Bird* (1765) parodies the "objective philosophy" of the Enlightenment with its foolish twins, Renzo and Barbarina. They tell their parents, Truffaldino and Smeraldina (a.k.a. Arlecchino and Colombina), who adopted them as babies, raised them with love, and were thus driven into poverty, that they did all that out of "self-love," which they should learn to master. The twins leave their parents and become homeless, starving on a beach. They meet the talking statue Calmon, a rationalist like them who was so objective that he turned into stone. When, at his advice, the twins throw a pebble and a mansion suddenly springs up, they become vain with their riches. They allow their parents to join them, but only as their servants. Truffaldino reminds his son that he should know, as an objective philosopher, that the rich help the poor. But Renzo refuses to be that kind of enlightened despot. Barbarina, despite her prior restraint as a rationalist, develops an overwhelming passion to possess a singing apple, dancing water, and a green bird. Renzo also falls passionately in love, but with a female statue, which he struggles to make speak and bring to life. Eventually, Renzo and Barbarina learn to value their passions and self-love, instead of being rationalist hypocrites. As Calmon initially tells them, "Man is only a part of God, and in loving himself, man loves his Creator. Self-love proceeds from heaven, and no one feels it more keenly than the person who acts with compassion, virtue, and pity, assuring himself of an eternal life by loving his divine essence.... Do not follow the lead of those devilish philosophers who reject a supernal and immortal shelter" (255).[31]

The nineteenth-century Romantics sought heaven and divinity through nature and its "sublime" forces, rather than humanist reason, neoclassical ideals, or a Christian sense of God. The noble soul and providential morality of Enlightenment sentimental comedy changed with Romantic tragedy (yet continued into popular melodrama). The soul, or ego, became a perverse hole in human nature, yearning to be filled by a greater, absolute truth — beautiful and grotesque, blissful and horrifying — through the further complexities of fate, heroic error, and Mother Nature. Thus, the rise of an individual self in Western humanism, as guaranteed by the Christian or deist God, acquired a new character. The sense of self shifted from left-brain, rationalist, sequential, pro-social, Symbolic orders toward right-brain intuitive, holistic, antisocial, Imaginary and limbic, emotional, Real powers. Writers such as Goethe, Hugo, and Coleridge defined the opposition between classicism and romanticism in such terms, while working toward "organic" integrations that explored the sublime and grotesque in human nature, with Shakespeare as their model of individual "genius," beyond rational, neoclassical rules (Carlson 181–82, 205–06, 219–21). While all creation participates in a higher truth, human nature, through culture, is divided against itself. Art can help humans become whole again, offering a glimpse of divine truth in material form. Yet it also risks falsifying the sublime, especially with theatre's artificial illusions. So, some Romantic dramatists turned toward pure imagination, writing "closet dramas" not intended for the stage.

The illusion of infinity onstage, through Renaissance and Baroque perspective scenery, with single or multiple vanishing points, was turned outside-in by Romanticism, with a new focus on the sublime within the mind (as foreshadowed by Shakespeare). Tragic heroes displayed good and evil drives, with their minds split between honor and passion, beauty and the grotesque, or heavenly and hellish territories. These properties were produced in and around the human animal by the pressures of society and its warping of nature. Yet there was no clear way to return to the blissful unity of the natural realm — despite the "noble savage" in Rousseau's philosophy and popular American melodramas (Brockett and Hildy 335). Ariel and Caliban could not be merged with Prospero. And dreams, emerging from human "stuff," were still tainted by culture.

In Heinrich von Kleist's *Prince Friedrich of Homburg* (1811), the tainted dream world within the hero's brain is externalized onstage — in a play based on historical characters and warfare of the seventeenth century. In the first scene, the somnambulant prince is found by other nobles in a garden, "half asleep, half awake," weaving a laurel wreath, instead of leading his troops as ordered, after a three hour rest (5). The others play a trick on him, manipulating his inner dream theatre by their external actions. The Elector of Brandenburg winds his chain of office around the wreath and the Princess, his niece, holds it up as the sleepwalking Friedrich follows her and calls out to her as his "bride" (8). They leave him then, retreating up a ramp to the castle and closing the door in his face. He remains standing there with a glove he took from the Princess — finding this in his hand, as objective correlative to the dream, when he wakes (9). Friedrich remembers his dream as reaching "to the gates of heaven," because he tried to follow the others up the ramp and grab the glove (14). Caught up in this mystery of dream and reality intersecting, Prince Friedrich misunderstands his orders later, on the battlefield. Ironically, by disobeying the command he wins victory for his side, though more territory might have been gained if he had obeyed. He is charged with treason by the Elector and given the penalty of death, because this enlightened ruler "demand[s] obedience to the law" (42).

The Prince at first responds nobly, but eventually grovels at the feet of others, begging for his life. Yet, when given the choice of a pardon by the Elector, who is like a father to him, the Prince changes again, preferring to die nobly. As another nobleman reports, Friedrich believed that his dream in the garden was a sign from "heaven" and that "God would grant him" heroic victory on the battlefield, so that he would gain the laurel wreath and the Princess's hand in marriage (83). When the Elector learns of this, he realizes how he and others were godlike manipulators of Friedrich's fate. He pardons the Prince, who then receives what his dream predicted — though he must first suffer a mock execution, before waking to the heaven he had dreamed (like sudden twists of fortune for Segismund in Calderon's play).

The audience sees how the Prince's inner theatre is twisted by other humans, not guided by providence, and how this reveals an underlying existential panic. The Prince, as tragicomic hero, is a noble yet flawed, beautiful and grotesque, romantic dreamer. Like him, Kleist suffered from psychosomatic illnesses: a speech defect, hypochondria, and spasmodic fits of blushing (v). Through this antihero, Kleist both confirms and undermines the godlike, aristocratic ideals of his time as rational and providential, yet dream-based and shaped by human whimsy, or by deeper mental disturbances.[32] When the Prince begs for his life, he goes to the Electress, asking her intercession, like Catholics praying to Mary, as divine mother, to help them reach God. "You seem to me endowed with the powers of Heaven, with the power to save ... and so does Princess Natalia..." (55). Indeed, it is the Princess who goes to the Elector and pleads for a pardon for Friedrich, eventually getting permission to marry him — even though the Prince had previously offended his patriarch by presuming control over that decision when false news arrived that the Elector had died on the battlefield.[33]

This play's many twists of fate point to a particular legacy in humans of mammalian and primate competition for alpha-male leadership, to ensure larger group survival, with females also playing powerful roles in sexual as well as natural selection. In many species, males kill infants who are not their own offspring. But females do this as well, to favor the survival of their own genes. Some mothers even kill and eat the weak infants in their own litter, or abandon an entire litter, to save resources for others.[34] Prince Friedrich, in his initial sleepwalking dream, calls the Elector and Electress his "father" and "mother" (8). They are not his biological parents, but they raised him since he was a boy (xvii). Indeed, the entire play hinges upon the insecurity of the Prince's kinship identity: his special status with or threat to the Elector, as adopted son, impulsive leader, disobedient warrior, and romantic dreamer. As Friedrich says while begging to the Electress for his life, "Of all God's creatures here on earth, I alone am helpless; only I have been abandoned and can do nothing, nothing at all" (55). He appeals to her, and to Princess Natalia, as females who may restore his group identity, personal survival, and reproductive potential (through marriage), if they can persuade the Elector, who also competes with other alpha-males in battle, to pardon and honor him once more. Such heroic insecurity and ironic twists of fate, as dreamed or real, in providential or human hands, reveal the primal, limbic, mammalian passions brewing under the higher cortical rationality of an enlightened, yet romantic society — renegotiating its principles of identity, kinship, hierarchy, sacrifice, survival, and reproduction.

A more radical questioning of civilized rules and human reason, with a poor private instead of a prince as the mentally disturbed everyman, occurs in Georg Büchner's *Woyzeck* (1836).[35] Left as unfinished fragments when he died at 24, with no ordering of scenes, Büchner's play nevertheless inspired later generations of theatre artists, especially in the post–Romantic, naturalist and expressionist movements, by showing certain social pressures producing the violent madness of its tragic hero. In one scene, Private Woyzeck shaves the Captain while a storm rages outside. The Captain calls him "stupid" for not getting his joke about a "North-South" wind and then says he has "no morals" because he sired a child "without the blessings of the Church" (7). But Woyzeck quotes from the Bible about Jesus loving children and this confuses the Captain (8). Woyzeck says he would be virtuous if he had money and "could be a gentleman." In another scene, he tells his friend Andres that an open field where they cut twigs from bushes is "cursed," because it is near "where the toadstools grow ... [and] where the head rolls every night." He stamps on the ground and finds it "hollow." He listens to nature and finds it so quiet that "the world's dead."

Woyzeck suffers from both social and physical afflictions. In addition to the subordination stress of his poverty, of having an illegitimate child to support, and of his low military rank, he eats only peas, as instructed by his Doctor. Woyzeck is paid to do this in an experiment, which produces what the rational scientist calls a "most beautiful *aberratio*"— in the patient's symptoms of madness (11). The Doctor is annoyed, however, when he sees "subject Woyzeck" pissing on a wall instead of saving his urine to be tested. Woyzeck attributes this behavior to his "Nature," but the Doctor tells him that man has a free will to control nature. "In Mankind alone we see glorified the individual's will to freedom!" But he demands that Woyzeck produce more urine, as in their "contract," showing there are limits to this poor man's freedom. Woyzeck cannot give urine, yet offers his Doctor something even better. He asks about a "double nature" that he seems to have (perhaps due to the peas), with "a terrible voice saying things" to him. He also sees the "shapes the toadstools make," which he wants to "read." This delights the Doctor, who then increases Woyzeck's salary because his "*Idée fixe*" accompanies an otherwise "rational state."

Such ravings reflect the pressures of society upon this poor private — and the lure of the Romantic sublime, evoking scientific wonder (as with Darwin and Freud). Woyzeck's schizoid perceptions show the "double nature" of the human animal, sculpted by his cultural environment and inner neurotheatre. They also indicate the hollowness of our nature, its lack of being, with primal drives radically transformed by language and culture. At times, Woyzeck seems rational. At other times, he stamps on the ground and finds it hollow, or wants to read the shapes of toadstools that "grow up out of the earth" (11). In the place where toadstools grow, he hears a deadly quiet in nature, but also knows that a head rolls there every night. Woyzeck demonstrates a new awareness in modern culture that nature's sublime power within human beings lies not only in the higher orders of reason and science, reflecting an orderly watchmaker God. It also appears at the grotesque edges of madness: with psychotic voices and hallucinatory fantasies revealing the evolutionary heritage of right-brain Imaginary and subcortical Real orders as well.

In Woyzeck's psyche, a foreclosed Symbolic order returns in the Real, producing Imaginary visions that seem so real to his double nature that he wants to read them — and voices that pressure him so much that he eventually kills Marie, the mother of his child. He lacks significance (or a "master signifier" to tie his psyche in place),[36] except as a scapegoat under class, military, and scientific subordination stress, with authorities directing his foolishness, immorality, and *aberratio*. Yet, other fragments of the play show aberrations in the society around Woyzeck, as rational and irrational theatre. He visits a fairground with Marie and they see the Charlatan, who displays a costumed monkey. The Charlatan calls it "nothing" as a creature that "God created," but something else through human "Art" (9). The performing monkey reflects the play's tragic hero, who bears the weight of being nothing, in nature and God's creation, while becoming perverse as a soldier and scientific subject. "It walks upright. Wears coat and pants. And even carries a saber! This monkey here is a regular soldier.... So what if he *is* still on the bottom rung of the human ladder!" Marie and Woyzeck also go inside a fairground booth, where they see a talented horse that has "brute reason," according to the Proprietor (10). The horse puts "human society to shame," as a "professor at our university." It nods when asked if there is a "jackass" in the audience, thus showing that it is a "person," yet also a "beast." And it offers a lesson to the audience: "Man must be natural!" Meanwhile, the Drum-Major gazes at Marie, remarking that she is "a piece," with eyes that are like "looking down a well ... or up a chimney," and that she could "breed Drum-Majors" (9).

Here, the competitive, theatrical, evolutionary drives of other human animals around Woyzeck become apparent. They seem more rational than he does with his diet of peas. But the Drum-Major has a "brute reason" like the horse, or like the monkey strutting in the soldier's suit: to steal Marie and breed with her. Indeed, Woyzeck questions Marie about earrings she says she "found," wondering how she could have found two of them (10). And she replies, "Am I not human?" He sees their baby sweating in its sleep, commenting that everything "under the sun works," and gives Marie money. When he leaves, her guilt gets the best of her and she says she could run herself through with a knife — which is a desire Woyzeck eventually fulfills for her. She also says: "We'll all end up in hell ... man, woman, and child." But in another scene in her room (which may be earlier or later), Marie asks the Drum-Major to "march" for her, admiring his chest "as broad as a bull's" and his beard "like a lion" (11). When she demands that he touch her, he sees "devils" in her eyes. Thus, the animal poses and passions of apparently rational humans around Woyzeck help to drive him mad, as they compete beyond moral restraints, through beauty and theatrics, to breed stronger offspring. Yet, such scenes also show the female's evolutionary drive to gain survival resources for herself and her progeny, plus the best possible seed, even if those come from different men — while this conflicts with the male's drive to provide only for his own kin (Buss; Hrdy).[37]

Likewise, in another scene the Captain teases Woyzeck about finding hairs from a Drum-Major's "beard" in his soup (12). This joke may refer to the double nature (of two beards) in the Drum-Major's leonine anatomy. The Captain also teases Woyzeck about finding hairs on "a certain pair of lips" if he hurries home. Yet he claims, with such pressure of shame, that he is "only trying to help" Woyzeck. Meanwhile, the Doctor observes with pleasure his patient's irregular pulse, tight facial muscles, and occasional twitches. Woyzeck raves about the Captain "just playing" with him and about the earth being hot as hell, but himself cold as ice, and thinks therefore that hell is cold — suggesting that society's hot pressures create his cold hellish alienation. He also talks about the "beautiful, hard, grey sky" on which he could pound a nail and hang himself. In a related scene, Woyzeck rages at Marie for being a whore, "beautiful as sin" (13). But she repeats her desire of being stabbed instead of loved by him. "I'd rather have a knife in my body than your hands touch me." In a different scene, Marie calls her child "a sword" in her heart — after reading from the Bible about Jesus freeing a woman caught in adultery by the Pharisees, though she is not able to pray for such forgiveness herself (16).

All of this taunting, testing, raving, and deadly desiring, shows that the gods creating Woyzeck's hell (and Marie's)[38] are in the people and social structures around him — and in his own psyche — even if he sees the sky as his gallows. Thus, he says to Marie: "Every man's a chasm. It makes you dizzy when you look down in" (13). But Woyzeck also projects a perverse vision of providence upon the sky, calling the sun coming through the clouds: "God emptying His bedpan on the world" (15). In other, fragmentary scenes, Woyzeck watches Marie and the Drum-Major, dancing together. He stretches his body on the ground, in an open field, and listens to the voices telling him to "Stab the goat-bitch dead" (14). And he tells Andres that he hears voices in violins and in the walls, telling him again and again to "stab." In another scene, he does stab Marie, after saying her lips are as hot as coals, yet he would give up "heaven" to kiss them again (17). This shows the perverse cultural shaping of Woyzeck's neurotheatre, as a fallen angel, in cosmic madness, stabbing Marie through vengeful love.

After acting out against Marie (in a Lacanian *passage à l'acte*),[39] fulfilling her desire for death and the Doctor's for aberration, Woyzeck flees. But he returns to his social mirror and natural crime scene — according to the play's apparent order. He goes to an inn and asks others there, "Do I look like I murdered somebody? Do I look like a murderer?" (18). He then goes back to the pond where he killed Marie and finds her pale corpse with "red beads" of blood around her neck — asking her how she earned that necklace with her sins. He finds the knife he used on her, and he throws it out farther in the water, finding it again and throwing it again. Such repetition compulsions show the synaptic rutting of Woyzeck's inner theatre, despite its paradoxical shifts between rational and irrational visions. Social and psychotic pressures push Woyzeck to express his biological drives of survival, reproduction, and kinship in a twisted way, by killing Marie, the apparent mother of his child. He finds a hell on earth, through its hollowness and his own, through the alpha-male Drum-Major usurping his place with Marie (and the Captain shaming him), through the Doctor demanding his urine and madness, through God emptying His urine as sunshine, and through vengeance against Marie for taking another mate when he was providing for her and their child.

In the fragmented scenes of this play, Büchner hollows out the Cartesian *cogito* (guaranteed by reason or God) and the Enlightenment man of sentiment, while revealing grotesque Romantic shadows of nature, society, God, and the individual mind. There is also, somewhere in the play's unclear order of love, jealousy, madness, and guilt, a scene where Marie and several children listen to a Grandmother telling an almost postmodern, existentialist story about the non-existence of "heaven" (17). In this "Once upon a time" tale, a little girl finds herself extremely alienated like Woyzeck, with everyone around her "dead." She looks for someone, "night and day," but meets no one on earth. So, she travels toward heaven. But when she arrives at the "friendly" moon, it is just "a piece of rotten wood." When she gets to the sun, it is just "a faded sunflower." When she reaches the stars, they are "little golden flies ... caught in a spider's web." She wants to go back to earth but it is "an upside down pot." So she stays in the heavens "all alone" and cries. Where can Western theatre go, with its cultural permutations of human evolution and projections of a divine cosmos, after this?

Melodramatic, Realist, Anti-realist, Metatheatrical, and Postmodern GIPPs

Romantic tragedies investigated the sublime extremes of nature's drives within the human brain, through right-cortical and limbic rebellions against left-cortical rationality. Romantic antiheroes, such as Friedrich and Woyzeck, displayed paranoid egos, schizoid world views, and perversely violent or suicidal behaviors — with God being distant, whimsical, malicious, or non-existent. But these plays were not a dead end. They influenced later theatrical movements, which evoked new paradigms of kinship selection, through sympathy or fear, self-sacrifice or scapegoating, with godlike ideals as guiding identity and purpose principles (GIPPs). Melodramas transformed the rebellious, socially outcast heroes of Romantic tragedy (with their beautiful and grotesque passions, with conflicts of reason, honor, and madness, with displays of violence and supernatural elements) into simpler characters, with good triumphing over evil through violent, spiritual struggles, to confirm popular values. Melodrama thus became the dominant genre in American theatre, cinema, and television, from the eighteenth to twenty-first centuries. But extreme Romantic works such as *Woyzeck* also inspired the radical experiments of naturalist and expressionist plays, which questioned conventional social goods as Western culture headed into two world wars.

A prime example of the conservative moral values, and yet potentially progressive politics of melodrama occurs with the most popular play on the nineteenth-century American stage: *Uncle Tom's Cabin*. The various adaptations of Harriet Beecher Stowe's abolitionist novel were performed by up to 500 touring companies at a time, from the mid 1800s to the early 1900s. In George L. Aiken's version, published in 1858, which served as the script for most productions, the slave Tom becomes a mythical hero to the little white girl, Eva, after he saves her life (when she falls off a steamboat) and she then persuades her father, St. Clare, to buy him (2.2). For Eva, according to her father, this slave's "Methodist hymns are better than an opera, and the traps and little bits of trash in his pockets a mine of jewels, and he the most wonderful Tom that ever wore a black skin." The play argues for a profound shift in kinship selection, from the white ideology of superiority over blacks to fellowship with them — or even a childlike idolization. But it is not only a white girl and her father in New Orleans who make this shift. Ophelia, St. Clare's visiting cousin from Vermont, admits that she despises blacks so much that she cannot bear to be touched by the orphan, Topsy. Yet she eventually adopts this "wicked" black girl as her own daughter, converting her from a "shiftless heathen" slave to a moral citizen, who becomes good because she feels loved for the first time in her life. This is predicted by the angelic Eva, who not only tells the devilish Topsy that she loves her, but also that she may become one of the "spirits bright," whom Uncle Tom sings about in his Christian hymns, and go to heaven — as Eva herself does in the play's final tableau (2.4, 6.6). Topsy hopes that she will turn into "a little brack angel" when she dies, because she has become less wicked, after Eva's death, in thinking of her whenever she is tempted (4.3).

Although the name, "Uncle Tom," eventually became pejorative in the Civil Rights Era a century later, in this play the tragic Tom demonstrates exemplary Christian piety. He refuses to give up his faith in God, despite the despair of his fellow slave, Cassy, under the brutality of their master, Simon Legree. Tom prays "Heaven help us" when sold to Legree at auction (5.1). Later, Cassy tells Tom that they are in "the Devil's hands" as slaves of Legree (6.1). She exclaims to God, regarding their white slave-master Devil: "Father in Heaven! what was he and is he!" And Tom asks: "Oh heaven! have you quite forgot us poor critters?" Yet then, Tom tries to lead Cassy back toward belief in "Our Heavenly Father." She replies that she used to see a picture of Him over the altar as a girl, but cannot believe that He is with her anymore: "there's nothing here but sin, and long, long despair!"

However, the villain Legree is also afraid of devils. Cassy tells him that he fears "the Devil" in her and he agrees (6.3). Another slave, Sambo, shows Legree a "witch thing" he found tied to Tom's neck, to protect him from feeling pain, Sambo thinks. It was actually a lock of hair and a silver dollar that Tom wore, in a packet tied to his neck, as a relic of the angelic Eva after her death — though she failed to obtain his freedom from her father. The lock of hair reminds Legree of his own dead mother. He tells Cassy how he became more and more "brutal" by rejecting his mother and her pleas that he give up his life of sin: "I set all the force of my rough nature against the conviction of my conscience." He then shows that he is "bewitched sure enough" when he sees the image of his dead mother as a ghost — and mixes her power to haunt his conscience with Cassy's and Tom's. Legree repeats the evil drive of his rough nature, against the goodness he fears — as a devil opposed to his own. He beats Tom to death, when the noble slave refuses his command to beat a female slave, Emmeline, who did not want to become an object of her master's pleasure. And yet, the martyred Tom still finds a melodramatic triumph at the end, seeing the gates of paradise as he dies: *"Heaven has come! I've got the victory, the Lord has given it to me!"* (6.5).

Uncle Tom's Cabin uses a conventional cosmological framework, along with its melodramatic formula of good getting victory over evil, to demonstrate a progressive shift in moral kinship. Topsy changes from a wicked heathen to black angel, through the love of Eva and Ophelia. Tom, with his Christian piety and self-sacrifice, is idolized by Eva and then included, along with her father, in the final tableau of her ascent heavenward.[40] And yet, other slaves are shown rebelling against slavery in another way — by running away from their masters. Eliza carries her son, Harry, on a block of ice across the Ohio River to reach the North. Her husband, George Harris, who has a different master, escapes separately from her. The play eventually shows George, Eliza, and Harry triumphing over the slave catchers in two further, spectacular scenes — as they are reunited and make their way to freedom in Canada with the help of Quakers in Ohio and Pennsylvania.

Such action-packed spectacles, plot twists, rhetorical turns, and heavenly tableaux were used in this extremely popular play as political devices to evoke audience fear and sympathy for the tragic sacrifice of Tom and the melodramatic survival of the Harris family. Also, George and Eliza are light skinned, aiding their escape to the North and perhaps the moral kinship felt by many in the white audience. They and other black characters were performed by white actors, wearing some degree of blackface makeup, enabling white spectators to extend their compassion across racial differences, despite the minstrelsy caricature of Topsy and her "breakdown" dances.[41] White characters (even the comical Phineas), siding with the black family in its escape to the North and helping Topsy to reform, also helped the white audience to connect kinship emotions with the struggling blacks and good whites (the slave-owners Shelby and St. Clare) against the evil white villains (the lawyer Marks and brutal slave-owner Legree). This is exemplified, too, when George Harris, disguised as a white man, is recognized by a former white boss, Mr. Wilson, who sympathizes with him, but tells him his fate as a slave might be an "indication of providence" to which he should return (2.3). George asks in response whether a white man captured by Indians might also see an "indication of providence" if a stray horse helped him to escape — rather than accepting his servitude as ordained by God.

Neuroscientific research on the communication of emotions demonstrates how performers' facial expressions, poses, gestures, and vocal prosody may have "automatic" effects on spectators (Niedenthal et al. 25). Various studies show the embodiment of corresponding emotions in the spectator through mimetic responses to smiling faces spliced into a comic film, to gaze direction and bodily posture in political leaders, and to the happy or sad prosody of a recorded voice (24–25). The gesture of arm flexion (movement toward the body) facilitates the identification of positive information, whereas arm extension encourages a negative perception (32). Facial mimicry, which changes emotional states, even occurs "in the absence of conscious recognition of the stimulus" (Atkinson and Adolphs 170). Such unconscious, bodily and emotional mimicry, which is "somewhat ... lacking in detail" compared to conscious experience, uses a "subcortical route" and is followed by conscious processing that uses both higher, cortical and lower, subcortical regions (173). The initial, unconscious processing of emotional mimicry provides "a crude assessment of the value of the stimulus (good/bad, harmful/pleasant)," motivating approach or avoidance behaviors (151). The subsequent, slower, less automatic processing includes "more complex attributions and rationalizations of the causes of an emotion...." Thus, the original, white spectators of *Uncle Tom's Cabin* may have had initial emotional reactions that were stereotypical, through melodramatic identifications of good or bad, pleasant or harmful characters, as well as automatic emotional mimicry involving attraction to or fear of the black-faced and other white actors.[42] Yet, the plot twists revealing complex, sympathetic struggles for the black heroes (Eliza, George, and Tom) and tragic sources of the white villain's violence (in Legree) may also move spectators toward new political views — of kinship with black slaves, against brutality in white slave-masters, who may also be victims of their own evil.[43]

While popular melodramas like *Uncle Tom's Cabin* maintained a moral Christian framework, the new experiments of naturalist and realist dramas in the late nineteenth century challenged that worldview. Influenced by Darwin, naturalists tried to replicate onstage a particular urban environment and its effects on abject, immoral characters — like a scientist studying a natural biosphere as shaping animal instincts. The upper-class humans and other forces that constructed society's "lower depths" (as Maxim Gorki entitled one of his plays) became godlike or devilish powers — in the audience as well as behind the scenes — which might worsen that hell or redeem it somehow. This is also suggested in George Bernard Shaw's *Major Barbara* (1905), a realistic comedy of manners about a Dionysian munitions maker who changes the beliefs of his Salvation Army daughter and her Greek professor fiancé, when he shows them how money and weaponry might be needed, even if morally impure, for social welfare and cultural evolution.

In a similar vein, psychological realism challenged its audiences with characters' sympathetic, yet immoral acts, pointing to a new awareness of evolutionary, social, and metaphysical problems. This is exemplified in Henrik Ibsen's *A Doll's House* (1879) and *Hedda Gabler* (1891), which expose a modern question about Mother Nature — and the nature of mothers — that still haunts us today. Ophelia in *Uncle Tom's Cabin* reformed the orphaned slave Topsy, from "wicked" to "brack angel," giving her a nurturing mother, freedom in the North, and Christian morals. But Nora in *A Doll's House* leaves her three children, committing social suicide (after she considers drowning herself) when her husband cannot value how she broke the law to save his life. Likewise, the pregnant Hedda feels so trapped by her marriage that she pressures a former lover to shoot himself and then she also commits suicide. This reiterates the Romantic impulse toward a tragic, grotesque, yet foolish escape from social moors and thus shows the uncertainty of natural, nurturing laws.

Woyzeck was influential on the naturalists in revealing the grotesque suffering of poor characters onstage, with humans dependent upon the social environment for their sense of providence. Yet, various anti-realist movements in the late nineteenth and early twentieth centuries went further in staging the mythic dimensions of good and evil, of sacrificial beauty and the grotesque, as potential powers of the human animal. The German expressionists in the 1900s-20s (also influenced by *Woyzeck*) showed the hellish providence of modern social pressures, not through a realistic or naturalistic replication of the urban environment, but by expressing the madness of characters through subjective distortions of the scenery onstage. Their "station plays" mimicked the Stations of the Cross in Roman Catholic churches, but with an ordinary everyman as the suffering Christ figure and no divine plan of redemption — except through the audience, watching like God the Father, with the power to change human fate, yet also desiring sacrificial entertainment.

Bertolt Brecht's epic theatre, with its alienation effects, emerged out of the expressionist movement and influenced later, postmodern theatre artists, challenging spectators, in their godlike position over the characters, to move from right-brain compassion to left-brain analysis, in order to change the fate of fellow humans in their own social world. Brecht mocks the insufficient interest of divine spectators in his *Good Person of Szechwan* (1938–40), when three oriental gods, disguised as humans, visit an immoral town and promise to save it if they can find just one good person. When they find her, barely surviving as a poor prostitute, they offer seed money, like investment bankers, but otherwise refuse to help her. So she creates an evil capitalist double of herself to keep her good character alive.[44]

At the turn of the century, the French symbolists tried to give audiences a divine experience in the theatre, through transcendent words and mystical images, with dreamlike scenery, mood-making lights, chanting actors, and other evocative shapes and sound effects. Alfred Jarry went in another neo-romantic direction with his *Ubu* plays (1896), presenting a grotesque tyrant as Anti-Christ, pushed by his wife to take and keep power through murder and torture, with a toilet brush as his scepter. Unfortunately, this monstrous parody of Macbeth also foreshadowed many brutal dictators in the twentieth century, performing as demonic powers in real life. During World War I, the dadaists reflected modern chaos in the forms of their work, prophesying the eruption of many future devils in the global conflicts and technologies of a post–Enlightenment era.

And yet, in the early twentieth century, certain expressionist, surrealist, and symbolist playwrights expressed the hope of finding a new sense of spiritual forces within the mind — offering positive possibilities for a collective future. August Strindberg's *A Dream Play* (1902) mixes mythic images from East and West into the dreamlike experiences of a compassionate, Christ-like deity, the Daughter of Indra, journeying through the modern world, observing and participating in the suffering of humans — while the audience watches at a distance like the god Indra, yet also interacts imaginatively like his Daughter. In the preface and prologue to his play, *The Breasts of Tiresias* (1917), Guillaume Apollinaire compares the surrealist style to the technology of the wheel in relation to the human leg, an extension that involves a radical change in form. The surrealist playwright thus becomes "the creating god / Directing at his will / Sounds gestures movements masses colors / ... to bring forth life itself in all its truth" (qtd. in Carlson 344). The symbolist dramatist, W. B. Yeats, created poetic plays drawing on Christian themes and Irish myths, while advocating the use of ritualistic masks, dances, chants, and a chorus, like he found in Japanese Noh. With a similar nostalgia for transcendent meaning, despite modern alienation, T. S. Eliot wrote poetic dramas that returned to prior rituals through choral figures, embodied furies, and sacrificial heroes, basing his *Murder in the Cathedral* (1935) on the medieval martyrdom of Thomas à Becket and *The Family Reunion* (1939) on Aeschylus's *Oresteia*. Luigi Pirandello also showed the continued yearning for metaphysical ideals, in a modern context, with his nearly absurdist *Six Characters in Search of an Author* (1921). In it, fictional characters meet an acting troop rehearsing another Pirandello play, seeking from them (and thus the unacknowledged audience) a godlike Author to complete their existence.

Influenced by the surrealist movement and Dalcroze's eurhythmics, Antonin Artaud advocated a ritual theatre of cruelty, which would show the spiritual forces within the actor's body, like a victim "burnt at the stake, signaling through the flames" (*Theater* 13). In his short play, *The Spurt of Blood* (1924), Artaud depicts God as a huge hand grabbing a whore by her hair, while a Gigantic Voice commands: "Look at your body" (*Collected* 64). Then, according to the stage directions, the Whore's clothes become transparent and her body is "naked and hideous" (65). She bites the wrist of God and blood shoots across the stage. Here, as in his writings about theatre, which greatly influenced later, postmodern experiments, Artaud combines Jarry's rebellious and grotesque absurdity with a deadly earnest, symbolist, and surreal striving for a better sense of God, as bloody, not just metaphysical, through the body and brain of the human animal.[45]

In American theatre, Eugene O'Neill explored the vexed relations between mankind and God, drawing on various realist and anti-realist devices, while often using ancient plots as prototypes for his plays. Thornton Wilder also investigated the modern nostalgia for divine providence and presence. *The Skin of Our Teeth* (1942) moves from Ice Age prehistory to modern warfare, with an everyman's family battling endlessly for survival. In *Our Town* (1938), Wilder focuses more hopefully on small-town America as having an address in "the Mind of God," while the audience knows in their "bones," according to the metatheatrical Stage Manager, that something is "eternal about every human being" (28, 52).[46]

After the traumas of the Second World War in Europe, existentialist and absurdist playwrights created allegorical settings showing the absence of God. Hell is other people, in Jean-Paul Sartre's *No Exit* (1944), or being stuck in a room with them, for who knows how long, with no recourse to any metaphysical authority, except perhaps the unseen theatre spectators. In Samuel Beckett's *Waiting for Godot* (1953) and *Endgame* (1957), the search for an Author persists, or for others beyond the stage as audience, but the only thing found is more waiting, or "zero" outside the window. In Jean Genet's *The Screens* (1961), victims of the Algerian War meet in an afterlife underworld above the stage, yet with no God or Allah as judge.

And yet, prominent playwrights continued to depict God, the devil, or various spirits in metatheatrical dramas, despite the apparent absence of divine providence in the twentieth century's war and holocaust theatres. In Arthur Miller's *The Crucible* (1953), seventeenth-century Salem girls demonstrate their possession by devilish spirits, sent by others they accuse as witches — in a parallel to the modern scapegoating of communists during the McCarthy era. In Archibald MacLeish's Pulitzer-winning *J.B.* (1956), two circus vendors put on God and Satan masks to act out the cosmic framework for a modern Job and his family, tested by adversity toward nihilistic despair. MacLeish offers a "Distant Voice" from offstage, suggesting that a metaphysical spectator may still be acting at the edges of our world. In Peter Shaffer's *Equus* (1973), a modern British boy, under the care of an envious psychiatrist, experiences violent, ecstatic passions, while possessed by an equine god, after his atheistic father replaced a suffering Christ poster in the boy's bedroom with a horse's head and its staring eyes (44–45). Actors in horse masks onstage make present the boy's cathartic rite of sacrifice, under the voyeuristic gaze of the therapist, theatre spectators, and animal deity. In Shaffer's historical drama, *Amadeus* (1980), Antonio Salieri speaks to the "Ghosts of the Future" in his theatre audience, asking if they sympathize with his campaign to test God's sacrificial reasons for making Mozart, instead of himself, the musical prodigy (5). "I felt the pity God can never feel," says Salieri. "I weakened God's flute to thinness. God blew — as He must — without cease. The flute split in the mouth of His insatiable need" (91). Thus, questions of divine spectatorship, plotting, and purpose lingered after the godless horrors of World War II, in relation to the Bible, unconscious forces, and history — while the theatre audience also became a prime focus for metaphysical inquiry.

This exploration of the metaphysical through the physical presence of actors and audience also occurred, with a more progressive political edge, during the Vietnam era in America, when the theories of Artaud and Brecht, plus the ritualistic "poor theatre" of Jerzy Grotowski, triggered new performance experiments. Actors' choral bodies gained godlike authority, regarding current social issues and Euripides's ancient text, in the Performance Group's *Dionysus in 69*, which was directed by Richard Schechner (1969) and loosely based on *The Bacchae*. As with the prior work of the Living Theatre, audiences of the Performance Group participated in shaping spaces and rubbing flesh. This created an "environmental theatre," as Schechner termed it, involving not only the stage world, but also the world as a stage, making both into ritual territories. Likewise, the Open Theatre, led by Joseph Chaikin, demonstrated the power of ensemble acting to create hieroglyphic forms in an Artaudian sense and provoke the audience toward political thought in a Brechtian way. Their production of Jean-Claude van Itallie's *The Serpent* (1969) moved from a description of brain surgery, combined with poetry from T. S. Eliot, to iconic reenactments of the John Kennedy, Robert Kennedy, and Martin Luther King assassinations, and then to Eve, Adam, and the Serpent in the Garden of Eden. This offered both clinical and biblical views of recent, collective traumas, with the audience placed in a godlike position of comparing, judging, and perhaps changing their social world. Other companies during this era also used happenings and guerrilla theatre to transform ordinary, public places into ritual, politicized spaces, where the theatre god, Dionysus, became a postmodern presence.

Since the 1960s, Judeo-Christian, perverse, and multicultural gods have emerged in various postmodern plays. In Aimé Césaire's *A Tempest* (1969), a Caribbean recasting of Shakespeare's play, the African god Eshu disrupts Prospero's masque in honor of his daughter's betrothal, mingling with the island spirits who are playing ancient goddesses and shocking the aristocratic Italian spectators, as well as the postmodern audience. This is perhaps due to a subversive trick by the rebellious Ariel, or his "mistake" as Prospero says (47). Elsewhere in the play, Ariel and Caliban debate the best way to convert or overthrow Prospero's rule, as different types of rebels, akin to Martin Luther King and Malcolm X. Likewise, Eshu brings a trickster spirit from West Africa, via the slave trade of later centuries, to Prospero's Renaissance colony. This god of the crossroads is still invoked today in the syncretic orisha religions of Latin America, as the first god summoned in spirit possession rites. In Césaire's scene, Miranda sees him as a "devil." But Eshu replies: "God to my friends, the Devil to my enemies! And laughs for all!" He partakes in the party's drink, saying he prefers "dogs" as an offering. Then he sings about himself as "a merry elf ... [who] can whip you with his dick," before leaving the party and causing Prospero's "confused" brain (48–49).

Like Césaire, Nigerian playwright Wole Soyinka mixed African, Christian, and ancient Greek deities in his play, *The Bacchae of Euripides: A Communion Rite*, which was commissioned and first performed by London's National Theatre in 1973. (Soyinka was then in exile in England after being imprisoned twice in Nigeria, in 1965 and 1967–69, for his political activities.) The orgiastic theatre god, Dionysos, even becomes associated with Jesus Christ. Soyinka adds two pantomimed wedding scenes, shown to Pentheus by the theatre god, as displaying "the past and future legends of Dionysos" (*Collected* 285). The latter mime involves "a traditional Christ-figure," wearing the "thorn-ivory-crown of Dionysos," who changes water into wine (286). Thus, the subsequent dismembering of Pentheus by the bacchae offstage, as described by one of his officers, indicates not only a bloody African revenge of the slave chorus against Christian colonial oppression, but also a comingling of Dionysian and Christian sacrifice. In the end, Soyinka transforms the grotesque tragedy of Agave murdering her son, and her deluded ecstasy with his decapitated head, into a tragicomic communion. Jets of blood spurt from the head, mounted on a thyrsus, but it turns into "wine" when others drink it (307). In his contemporary writings about African ritual theatre, Soyinka relates the orishas Obatala and Ogun to Nietzsche's Apollonian and Dionysian orders of ancient Greek culture (*Myth* 140–41). One can find these spirits, too, in Soyinka's play about Dionysos.⁴⁷ For Soyinka calls Ogun "the elder brother to Dionysos" in a note at the start of his *Bacchae* (*Collected* 234).

In the same decade, the Christian God-man appeared directly onstage in the rock musicals *Jesus Christ Superstar* (1971, Andrew Lloyd Webber and Tim Rice) and *Godspell* (1971, Stephen Schwartz and John-Michael Tebelak), which were also made into movies. Causing controversy at the time, these musicals stressed the human drama of biblical figures, using modern lyrics and rhythms from a rebellious and clownish youth culture.

Through rebels and clowns, Luis Valdez (like Césaire and Soyinka) presented colonized gods persisting in a bilingual Mexican-American culture. In the stage musical and later film, *Zoot Suit* (1978 and 1981), El Pachuco, a 1940s-style Chicano gangster, becomes a primal, ambiguous alter-ego to the main character, with the colors and smoking mirror (or marijuana) of the Aztec sorcery god, Tezcatlipoca.⁴⁸ In an earlier play, *Dark Root of a Scream* (1967), a dead Vietnam soldier named Quetzalcóatl embodies the return of the Aztec feathered-serpent god of that name (Huerta, "Representation"). In *Bernabé* (1970), a "cosmic idiot" lays in and with the earth, La Tierra, who then appears as a character onstage, dressed like a "soldier woman of the Mexican Revolution, 1910" (135, 158). She eventually turns out to be the Aztec earth goddess, Coatlicue, uniting with Bernabé after his death (167).⁴⁹ With the news of his death at the end of the play, Bernabé's uncle calls it "God's will," suggesting a Christian sense of providence. Yet, the audience has seen Bernabé's earlier, visionary interaction with the figures of La Tierra and her brother, La Luna (the moon), who appears as a "1942 style," zoot-suited Pachuco, as well as their father, El Sol, as "Tonatiuh, the Aztec Sun God" (157, 162). The same actors who play these gods also play mortal characters in Bernabé's life. This indicates how his daily life of poverty and subordination stress as a mentally retarded farm worker is transformed, through a dreamlike metaphysics, into the sacrificial heroism of an "ancient" lineage going back to the Aztecs (164). He is thus "the first of a new raza cósmica [cosmic race] that shall inherit the earth," according to El Sol, who then demands Bernabé's death, through the Aztec rite of heart sacrifice, before he can unite with La Tierra (165). But in mockery of such fantasies, Torres (played by the same actor as La Luna) tells Bernabé that La Tierra is already possessed by others today, who buy, sell, or rent her (148). Valdez leaves it to the audience to decide whether Bernabé is an idiot, a crazy dreamer, or a prophet who might herald the return of the Real—from the material poverty of his farm workers' theatre (El Teatro Campesino) to sacrificial gods, or other GIPPs in their lives.

Such ambiguous, indigenous gods, from the repressed Real of a postcolonial community, also appear in the plays of Tomson Highway, a Native American, Canadian playwright. In *The Rez Sisters* (1986) and *Dry Lips Oughta*

Move to Kapuskasing (1989), Highway presents the trickster god Nanabush, as male in the former and female in the latter — the opposite of the lead characters in each of these dramas about reservation life. Nanabush takes the form of various human figures in the dream visions of others, returning with scenes from the past, while challenging their memories and American Indian identities. With similar ironic twists, yet with an Anglo-feminist longing for Mediterranean fertility and mother goddesses, characters in Caryl Churchill's *Cloud 9* (1979) — the postcolonial Victoria, her girlfriend Lin, and her gay brother Edward — chant together, in drunken half-seriousness: "Innin, Innana, Nana, Nut, Anat, Anahita, Istar, Isis" (94). They call them "back through time, before Jehovah, before Christ, before men drove you out and burnt your temples," to give back "what we were ... the women we can't be." Instead, they get Victoria's estranged husband Martin, and then the ghost of Lin's brother, a British soldier killed in Northern Ireland, who now just wants a "fuck" (98). Churchill also invokes the ancient fertility god Dionysus in *A Mouthful of Birds* (1987), her modernized, feminist version of *The Bacchae*, co-written with David Lan.

The repressed desires of former gods and ancestral spirits percolate to the surface in various postmodern plays, as alternatives to Judeo-Christian norms.[50] These figures express Real, subcortical drives and limbic passions — of sacrifice, rebellion, or celebration — erupting through right-brain, Imaginary, conflicting identities and left-brain, Symbolic, ruling orders. Given the secular humanism and monotheism that dominates current Western culture, the immortality of such minority gods depends on spectators' beliefs and interests, as well as those of artists, through the creative, consumer networks of their interacting brain theatres.

Angels also appear as guiding identity and purpose principles (GIPPs) in some postmodern dramas, becoming heralds of millennial change in the late twentieth century. In José Rivera's *Marisol* (1993), the title character is a yuppie Puerto Rican in New York, whose Guardian Angel appears as "a young black woman in ripped jeans, sneakers, and black T-shirt" with crude silver wings hanging limply from her "diamond-studded black leather jacket" (5). As a "suffering, burnt-out soldier of some lost cause," the Angel represents Marisol's family beliefs, in her continued Catholic prayers, yet also the changing environment around her (6). The Angel apparently saves Marisol from a crazy man on the subway and from her violent neighbors, and then appears in her sleep. She tells her that the universe is sick "with amnesia, boredom, and neurotic obsessions ... [because] God is old and dying and taking the rest of us with Him" (15). So the angels have decided to rebel: to kill "our senile God ... and restore the vitality of the universe with His blood" (16). Marisol resists this message, repeating childhood phrases about God's goodness and greatness. But the Angel tells her she is leaving as her guardian and Marisol must fight on her own, without trusting "luck or prayer or mercy or other people" (17). Later, Marisol becomes like the homeless man with a golf club who assaulted her on the subway, fighting other crazy people she meets and struggling to hold onto her sanity. Her Angel appears again, at the end of Act One, in military fatigues, with face camouflage, medals, a machine gun, and her bloody wings torn off.

Midway through Act Two, Marisol finds herself "at the rim of the apocalypse" and prays once more to her lost Guardian Angel for help because "God has stopped looking" (55). But then she doubts that and prays instead to God for a deal: to spy for Him against the angels in order to save herself. Through such cosmic confusion and various crazies she meets (including her friend, June, as a Nazi skinhead lighting people on fire, and June's brother, Lenny, who believes Marisol made him pregnant), she attains a final, postmortem vision of the new universe, after being shot to death by the "Woman With Furs" (66–68). She sees homeless people joining the heavenly host in their revolution against God—while stage lights shine into audience's eyes, so that spectators also see "the wild light of the new millennium," involving new ideas, powers, miracles, and "hope" (68).

The old God, in a Judeo-Christian sense, is not the enemy, but the Absent One in Tony Kushner's *Angels in America* (1993), also set in New York. Prior Walter, a gay man suffering from AIDS (and abandoned by his boyfriend, Louis), is visited first by his ghostly ancestors and then by an Angel who calls him "Prophet," crashing through his bedroom ceiling at the end of Part One of the two-play series (124–25). This reflects other plots in *Part One: Millennium Approaches*. A Mormon couple, Joe and Harper, lose faith in their marriage and prophetic tradition, as Joe realizes and admits to her that he is homosexual. Roy Cohn, a famous, ruthless lawyer, refuses to acknowledge that he has AIDS and is visited by the historical ghost of Ethel Rosenberg, whom he helped to sentence to death as a spy. In *Part Two: Perestroika*, these plots continue into Joe's full experience of his sexuality, Harper's separation from him and move to San Francisco despite her agoraphobia, and Roy Cohn's death after Ethel's ghost comforts him with a ritual Jewish song. Prior is told by his Angel that God left the heavenly host on the day of the San Francisco earthquake of 1906—because humans are so migratory and inventive, driving Him away (196–97).[51] In a vision of heaven as San Francisco's City Hall, Prior learns that the bureaucratic angels (America, Europa, Africanii, Oceania, Asiatica, Australia, and Antarctica) want humans to stop changing so much, as they "maintain surveillance over Human Mischief" through an old radio (281). But he fights against their view, wrestling with his personal Angel of America and refusing to become their prophet.

4. From Renaissance Rebirths to Postmodern Experiments 173

In 1912, near the start of the twentieth century, Rainer Marie Rilke asked, "Who, if I cried, would hear me then in the angelic orders?"[52] Near the end of that century, the deadliest and most destructive in human history, *Angels in America* and *Marisol* gave opposite visions of the angelic orders, as supernatural audience and actors. In Kushner's play, aloof yet sexually powerful angels try to contain the hyper-evolution of the human species and its diverse cultures, in order to bring God back. In Rivera's drama, scrappy, wingless, urban angels lead street people in a revolution against God, to change the world even more. Yet in both plays, a lead character wrestles with an angel, like Jacob in the Bible. In each case, a postmodern angel inspires sexual ecstasy, prophetic insights, and extreme visions that push the human character toward insanity. Both dramas display the tragicomic terrors of urban life, of minority alienation, of disease and madness — extended into a fanciful cosmic reordering, with the old, patriarchal Other as absent or a senile enemy.

These plays express particular developments in collective limbic emotions fueling right-brain Imaginary flights of rebellious fantasy figures against left-brain Symbolic executive powers. Rivera presents, through the sacrificial vision of Marisol, the Real (dis)order of suffering street people and violent madmen — rather than devils — aligning with fallen angels against a negligent God to make change. Kushner offers, through the AIDS-related hallucinations of Prior, the abject pain of the gay community, in bodies and politics, rebelling against heaven's traditional executives and their repressive, fateful cosmology. But whether angels are placed in the left-brain Symbolic order, through God's absence, with an anti-prophet perversely challenging their authority (in Kushner's play), or in the right-brain Imaginary as perverse freedom fighters alongside humans, seeking a more diverse universe (in Rivera's), both plays show fundamental anxieties and hopes for change through the brain's collective, theatrical orders, at the brink of a new millennium.

In these 1990s plays, there is a mix of earlier melodramatic figures of good and evil, with tragic twists in perspectives. They also combine earlier realist and anti-realist agendas, through psychological explorations and flourishes of symbolist, surrealist, Artaudian, expressionist, Brechtian, and futurist styles. And yet, such aspects of postmodern divinities might also be found in the plays mentioned above that draw on ethnic minority spirits to question the dominant God and great works of the Western tradition. The plays considered throughout this chapter express a continued evolutionary struggle in our species, articulated by specific cultures and performers, to redefine human kinship ties, altruistic ideals, sacrificial demands, and survival threats, regarding the good or evil of supernatural figures, as guiding identity and purpose principles (or their negatives), which reproduce through us mortals.

This has not been a complete survey. Alternative mixtures of godlike

GIPPs appear in other plays, especially from parts of the world — Asia, the Middle East, Australia, Oceania, and most of Africa and Latin America — beyond the limits of this book. It is worth noting, too, that *Angels in America* was a highly controversial play in the early 1990s and was almost banned from performance in Charlotte, North Carolina, in 1996. But it became an award-winning HBO movie in 2003, directed by Mike Nichols with major Hollywood stars: Al Pacino as Roy Cohn, Meryl Streep as Ethyl Rosenberg (and Joe's mother), and Emma Thompson as the Angel "America" who visits Prior. Thus, at the turn of the millennium, the American mass media extended the power of supernatural figures in Western theatre, through new permutations of traditional ideals, as the following chapters consider.

5

Cosmic Forces on the Movie Screen

One can glimpse — through scripts and other evidence of theatrical performances — specific figures of gods, angels, and devils circulating in the networks of human brains that formed particular cultures of the past. Those prior layers of supernatural figures in Western culture continue to influence its global, stage and screen media. Yet one can also find distinctive shifts between the leading theatre deities of earlier eras: from prehistoric cave spirits to ancient gods onstage, to the medieval Judeo-Christian God, and then to Renaissance humanist ideals, which still involve neoclassical figures. Such experiments with revised divinities and transcendent perspectives continue in the Enlightenment sense of a distant deist God and with Romantic passions for the sublime powers of nature as beautiful or grotesque.

The aspiration for a divine viewpoint also arose during the modernist era, at the turn from the nineteenth to twentieth century, despite the scientific attitude of the naturalist and realist movements. Spectators were given a godlike view, through the "fourth wall" at the edge of the stage, with naturalism's stress on the human cultural environment as a determining habitat and realism's exploration of psychological truths, ego illusions, and unconscious conflicts. In the same period, the spectator became godlike through various anti–realist styles: transcending ordinary reality by viewing mythic, dreamlike forces (surrealism), paranoid, schizoid pressures (expressionism), and ecstatic, poetic experiences (symbolism). There were also Artaudian rites of primal sacrifice and Brechtian devices for a more alienated, socially critical awareness, which led, beyond absurdist questions about God and reason, in the 1950s and 1960s, to the postmodern return of diverse, minority deities in our current era. But a significant change occurred in the medium of theatre's supernatural characters and spectators as popular melodrama, romantic comedy, political satire, mystery plays, and classical tragedy gained new, technological powers through cinema.

With the rise of realism in Euro-American culture, from the early modern to postmodern periods, supernatural figures did not appear prominently onstage, except as ethnic and queer alternatives. Yet their Judeo-Christian forms arose often on the silver screen, especially in biblical epics, comedies, and horror films. This persistence of cosmic beings onscreen may reveal an uncanny kinship with the God Experiences of our prehistoric ancestors. Their communion with supernatural visions, through the extended-mind[1] of cave-wall art, may share a kinship with cinema audiences today. We are still entranced by dreamlike images projected onto a shared surface—with associations from our neural theatres moving beyond the edges of the screen. Of course, there are great differences in these visionary technologies. Paleolithic humans went deep into the bowels of the earth, developing ritual performances through hallucinatory visions, expressed as paintings, etchings, and sculptures on specific rock walls, which seemed to move by firelight, probably with echoing music and sound effects.[2] Today's filmmakers use high-tech devices that capture moving images from life, or combine them with animations, and project them in a light and sound show, for the mass audience to experience in a multitude of cinematic and home theatre chambers. Yet the convention of spectators sitting in the dark to watch a movie on a wall, with voices and music forming around them, may create a ritual akin to the mystical trances of our cave artist ancestors, shared through prehistoric media.

The inner visions of geometric designs, dreamlike figures, and hybrid forms that were painted on cave walls (according to Lewis-Williams's theory) were probably alienating when experienced alone, like mystical or temporal-lobe experiences can be today. But by interacting with these visions, through the cave-wall membrane, even painting onto their hands pressed against the walls, our ancestors created a shared performance space inside the earth.[3] They thus materialized the supernatural not only in cave art, but also through interconnected neural networks, like those of theatre and cinema. Unlike the figures in current films, few of the cave images are human. Most are of powerful animals, some of human and animal hybrid figures. Perhaps this shows the initial emergence of a distinct ego awareness (and Theory of Metamind) in our ancestors—showing the power of Other animals and yet the human potential to use nature and master it, through artistic tools. If so, such biblical "dominion" over animals and the environment also relates to the many ego-mirrors of today's screen media, with powerful stars, avatars, and fictional worlds extending the godlike power of human minds, from agricultural and industrial to virtual revolutions.

In order to consider this lineage more fully, I will go back even further in time—using evolutionary psychology and primatology. How might the social networks, performance concerns, and theories of mind in other primates reflect the proto-theatrical drives that evolved 30 million years ago, when we shared a common ancestor with such animals? (This assumes that the behaviors and cognitive abilities of other primates today are similar to those of our primal ancestors.) What distinctive aspects emerged in the human Theory of Mind, which were eventually projected as godlike figures on historical stages and screens?

Cultural Selection: From Baboon to Human Metaphysics

Evolutionary psychologist Geoffrey Miller has argued that sexual selection may have played an even more significant role than "natural" (environmental) selection in the big brains and distinctive talents of our species. The human brain is costly, taking 25 percent of metabolic energy, while being 2 percent of total body weight (134). "Sexual selection ... transformed a small, efficient ape-style brain into a huge, energy-hungry handicap spewing out luxury behaviors like conversation, music, and art." Yet, such wastefulness works as a fitness indicator, like the peacock's tail, showing opulence in order to attract mates and increase one's genetic offspring. Moreover, mate choice could be "the single most powerful moral filter from one generation to the next," favoring altruism or heroism, even at the risk of survival according to environmental selection (307). "Status based on moral leadership is a legacy of the great apes," who have structured their groups as dominance hierarchies for "at least five million years ... striving to attain status through their moral leaderships, rather than just through their physical strength" (319).

Both mechanisms of evolution, sexual and environmental, can be seen in the cultural selection powers of movies, especially in relation to the prior legacy of historical theatre stages and Paleolithic cave performances. The "extended phenotypes"[4] of cultural ideals onscreen affect not just the male gaze but also female mate choices in real life—through beautiful bodies, fetishized actors, compelling characters, and fantasy scenarios. The "stars" onscreen, playing dreamlike movie figures, exercise spectators' emotions, intellects, and bodies through the mirror neuron systems in their brains. The neural circuitry of the mass audience becomes pruned through cultural values, in romantic partnership or violent survival, displayed as fiction onscreen. Yet, cinema idols are also shaped in the other direction, across the screen membrane, by mass audience desires, critics' choices, and ticket sales.

This is somewhat akin to theatre's function in earlier periods of Western history, displaying reproductive and survival ideals, as perverse pleasures or moral truths. Such ideals onstage were shaped by audience demands and dominant (state, church, or individual) sponsors, while affecting them also in return, via artistic choices. Today, theatre actors share their mortality with the audience, in a collective space, like prehistoric shamans with their cave rites. But movie viewers see "immortal" performances onscreen, which far outlast the actors who have aged or died since the film was made.[5] This is also like the visionary animals and hybrid figures of prehistoric cave art, lasting long beyond the human actors and artists who created them.

The movie audience gains an illusory, divine power to jump between experiences and time-points, while moving in a ghostly way through various spaces, unseen by characters, or across vast landscapes. Cinema spectators may cut instantly between different viewpoints within a scene or perceive multiple layers of meaning through montages or look down from above like angels. More devilishly, movie voyeurs may enter intimate spaces, where bodies and emotions are unveiled. Film viewers sometimes take sadistic pleasure in the suffering of characters onscreen or identify with them masochistically. Through these and other cinematic devices, mimicking the magic of the brain's inner theatre of perceptions and dreams, audiences get both realistic and fantastic, objective and subjective views of film characters and environments, with collective perspectives beyond any individual, mortal experience of life.

Theatre and cinema, though related, have become distinctive extensions of our ancestors' cave performances and painted projections—showing the ape legacy of our excessive brain powers and yet morally constraining cultures. Non-human primates do not paint animals on cave walls.[6] Nor do they create stages and audience areas for displaying natural and supernatural figures. Nor do they capture scenes on film and transform them in a godlike way, creating new realities and dreams for others. But baboons do have a social "metaphysics," according to biologist Dorothy Cheney and psychologist Robert Seyfarth, who studied these primates for six years in the wild. They argue that baboons provide a key to understanding the human sense of metaphysics, as a "product of evolution" (4). As with our ancestors, the large size of baboon groups meant that natural selection favored "rule-governed classes" as a way for individuals to keep track of complex social relationships.[7] These and other monkeys not only have mirror neurons for goal-driven gestures, but also "facial identity" and "facial expression" neurons in the temporal lobes that process those types of social communication (126). They attend to others' gaze directions and use gazes "to target opponents and to recruit support in aggressive alliances," or avoid eye contact when being threatened (166). They also use directed vocalizations to understand the intent of others (175).

And yet humans, from the age of four onwards, have a full Theory of Mind (ToM) about others' beliefs and knowledge, which baboons lack, even when enacting deceptive behaviors (Cheney and Seyfarth 152–57).[8] The baboon metaphysics of social rank relations, with awareness of others' intentions and emotions, but not of their distinct perspectives, show the foundation in our ancestors of a further metaphysics, with theories about others' beliefs and viewpoints, using distinctive brain areas that evolved later (157–59). This eventually developed into theories about higher alpha minds, as shown in theatrical representations of Egyptian hybrid gods, Greek Olympian deities, and the medieval Christian God at the top of a Great Chain of Being with rule-governed classes: supernatural, human, and animal.

Baboons give contact calls repeatedly throughout the day. As observed, too, with birds and other mammals, "signalers do not recognize whether or not their audience is ignorant or already knowledgeable about the information being conveyed" (Cheney and Seyfarth 162). Baboons and other monkeys use "social cues, gaze direction, and call type" to announce their intentions and assess those of others (183). But they "probably do not recognize when someone is attempting to manipulate their beliefs...." Baboons also "do not act deliberately to inform ignorant individuals, nor do they attempt to correct or rectify false beliefs in others or instruct others in correct usage or response to calls" (263).

Like humans, some species of monkeys and apes use tools, even selecting and modifying them for use (Cheney and Seyfarth 187). Yet they rarely teach their young how to use them; "learning is passive and involves little active intervention" (190). Experiments with chimpanzees show that they are often "indifferent to the welfare" of others, whereas humans as young as 18 months old recognize the needs of a companion and help that person to achieve a goal (193). Between human brains there is a mirroring of emotions, as well as gestures and facial expressions. Representations of "pain, disgust, and shame in others activate many of the same areas of the brain as those activated when we experience or imagine ourselves" with those emotions (192). Monkeys and apes may have some ability to recognize the emotions of others,[9] but they lack the human "motivation to collaborate and to share goals, emotions, and knowledge with others" (193). For example, male chimpanzees cooperate in hunting, but "they do not consistently coordinate their respective roles...." Unlike other primates, humans recognize "an obligation to help the group" (Tomasello, "How" 15). Human infants, but not chimpanzees, collaborate in group pretense, as collective play activity. Even earlier, at one year of age, human pointing and gesturing shows a cooperative communication and shared intentionality distinct from other primates (*Origins* 333). These are initial steps toward participating in social institutions with certain roles and powers — or in the meta-role-playing of theatre and cinema. Such a difference may also exemplify how human self–awareness evolved, in the brain's theatre of consciousness,[10] allowing our ancestors "to predict the behavior of others on the basis of introspection about our own motives, thoughts, and beliefs" (Cheney and Seyfarth 201).

A full Theory of Mind, through empathy about others' beliefs, knowledge, motivation, and experiences — with attempts to rectify false beliefs, to teach those who are ignorant, and to predict behavior based on self and other awareness — appears distinctive to humans. It is also a primary aspect of theatre, from prehistoric cave performances (with shamanic ritual ordeals and visionary art) to the many stages of Western culture, and to current screen media. So is human collaboration through role-playing, from childhood games to adult performances. Yet humans share with baboons a social metaphysics of communal rank and relationship knowledge, which needs continual updating as power alignments shift in the networks between brains. In primate evolution, this became the basis for our bigger brains' departure from the ancestor we share with baboons 30 million years ago. Humans developed a full ToM, as the inner theatre of self and Other awareness about beliefs, knowledge, and predictive behaviors. Eventually, this inner theatre was also shared through external theatres displaying the superhuman minds of gods and devils, in a projected, cosmic hierarchy of alpha leaders and seductive rebels.

Today, television shows and videogames bear various visions of angels, devils, witches, and vampires, as good and evil forces that the popular audience believes, to some degree, nostalgically, fervently, or playfully. Also, "Google Earth" gives Web users a godlike view — as if watching the human world in a cosmic theatre, with the power to fly down and examine specific details. It reminds us that we are being watched from above, in this way, by satellites in space. Likewise, cell phones, email, chat rooms, blogs, YouTube, Facebook, and Twitter offer new media for postmodern humans to engage in a social metaphysics far beyond, yet akin to that of our primate relatives, performing for and watching as the collective Other. Such social media involve persistent contact calls, facial expressions (or emoticons), and ritual gestures to figure one's current role in the intentions and emotions of a shifting, rule-governed, hierarchical network. This shows our continued primate drive to belong in specific communities, even with others whom we never meet in person: to know and mimic the desires of the Other, against the alienating effects of language, culture, and technology. Yet, the hundred-year-old technology of the movie theatre still bears the biggest cave wall for communal visions, gathering spectators in the dark for ritual belonging and egoistic projections. How do supernatural characters and settings in cinema relate back to our ancestors' spiritual and social metaphysics in Paleòlithic caves, or in various stages of Euro-American theatre? Do they also extend the environmental and sexual selection pressures that produced our wastefully big brains, with cultural fitness indicators and moral filters? Do they point toward new theories of the Other's metamind in our current millennium, through changing rules of human hierarchies, as culture wars with glimpses of transcendent truths?

God Comedies

Shamanic animal spirits, ancient Egyptian hybrid gods, and Greek deities rarely appear in Hollywood cinema, at least until recently.[11] Yet aspects of these beings sometimes appear through Judeo-Christian figures of God, angels, and devils onscreen. The *Oh, God!* comedies show a post–Vietnam-War nostalgia for a divine imperial ruler to return to earth, as benevolent cosmic spectator — during the Carter and Reagan eras. The short, elderly, and bespectacled George Burns presents a comical form of this kindly grandfather God, as Metamind viewer who wants humans to remember and return to His goodness, like God in the Wakefield Cycle and *Everyman*. He gives targeted attention and help, yet also applies pressure to certain humans, with His demonic twin shown as well, in the third film of the series. The *Oh, God!* God is like Zeus in his distant watching and judging of humans, like Prometheus in his helpful, yet ambiguous gifts, and like Dionysus or shamanic spirit-guides in His trickster performances and transcendent, yet bestial temptations. He is akin to Apollo (or shamanic spirits), providing dreamlike visions with symbolic meanings. And He is like Athena outwitting the Furies (or Horus and Osiris subduing Set), as he offers an ultimate moral filter against his counterpart, but with dangerous sacrificial risks.

In the first film of the series, *Oh, God!* (1977), the ultimate Being appears onscreen as a frail looking, witty, yet calmly demanding old man — a comical counterpart to the white-bearded superpower image, or medieval king and judge.[12] This movie increased George Burns's significance to popular culture, reviving his persona for a new generation near the end of his life, after his long career in vaudeville, radio, and television. As God, he first leaves notes for Jerry Landers (John Denver, another media icon), who is the assistant manager of "Food World" in Los Angeles. These clues lead Jerry to the 27th floor of a 17-floor building, where he meets God as a little old man in a fishing cap with thick glasses. God puts the blame on mankind, instead of chaotic, natural forces (such as the desert, raging sea, and storm threats symbolized by Set) for the sky turning into "mud" and for the "trash" given to kids in their cereal. He picks Jerry as an Everyman prophet, like the millionth man over a bridge, to spread the word that God "lives," that He is "watching," and that the world "can work" if people nourish each other instead of killing and if they delight in the world, as made by God, instead of ruining it. Jerry asks why God does not fix everything Himself. God replies that He gave humans "free will" to improve the world or not. Yet, God admits that He made mistakes, comically offering the ostrich and avocado (with its pit too big) as examples. God even confesses that He is not omniscient: He does not know the future. This allows for more tension as the movie continues its comical twists with God plaguing Jerry as his antagonist, while Jerry appears crazy to others, like prophets of the past.

God's ambiguous gift to Jerry (and the film audience) of dreamlike glimpses of the Supreme Being in manageable form eventually leads to fame for the grocery store manager, who gets in the newspaper, on the local television news, and then on the national *Dinah Shore Show*. There, a police-artist's sketch helps Jerry share his unique vision of God, on paper and the television screen — like the prehistoric art on cave walls, extending an inner theatre toward others, but in a modern, secular rite. This sketch also suggests a criminal element to God's visit, disturbing the conventional moral and social order. Jerry's wife and son are embarrassed by how he is ridiculed in the media — as it also plays the role of a Dionysian trickster. Yet God again gives targeted help to Jerry, when he must answer 50 theological questions, some in Aramaic, while alone in his hotel room, as a test of his visions (somewhat like a shamanic ritual ordeal). God visits as a busboy and gives Jerry the answers. But when Jerry relays God's message that He is unhappy with preachers who make money from people's beliefs, he is accused of libel. In court, people laugh at him as crazy. He calls on God as his only witness. Although at first God does not appear, Jerry suggests to his audience that, for a moment, they hoped He might. This extends the comedy to the movie audience as well, who might realize a similar trick played on them, in that moment, by the Dionysian Jerry.

When God does visit the courtroom, a bit late, others see Him like Jerry. He reminds the court that the mass audience of *The Exorcist* movie was easily convinced that the devil exists. He also does a card trick and a disappearing act like a stage magician, leaving no trace of His visit on the courtroom tape recorder. Jerry is acquitted, but still loses his job. God visits him alone once more to console him with the idea that he planted good "seeds" that may someday bear fruit. Like Apollo to Orestes, God gives unique, divine counsel to the alienated Jerry, helping him to subdue a chorus of furies mocking him, through the mass (or dream) media. Like Athena with Orestes, God helps Jerry in court, altering the environmental and sexual selection paradigms.

He provides a new moral filter for Jerry to continue reproducing his genes and memes, through family and culture.

Like animal spirits in prehistoric art, projected from the artist's inner theatre through the cave wall to a larger audience, God appears to the world, according to this comical film, through one person's hallucinations, which are shared with the movie audience. This plot device is repeated in the sequel, *Oh, God! Book II* (1980). George Burns again plays God, but this time visits a little girl named Tracy and pressures her to convey His divine point of view, as cosmic spectator and Metamind. Tracy first hears His voice in a restaurant, then gets His message in a fortune cookie, and then sees Him as an old man in the back seat of her father's car, while he leaves her there briefly. Instead of being concerned for souls, like God in medieval cycle and morality plays, this God is an ecologist, even more than in the first film, worried about pollution and the scarcity of eagles. He has been watching "millions" of planets and finds that earth is "always a problem" because He "put people on it." When Tracy tells Him she cannot do anything about that as a child, God consoles her that she is in "good company" as His prophet. He takes her father's place in the driver's seat and laments that He is not in people's thoughts anymore, though He is still "owner of the store." This suggests the fading of traditional monotheism, but also implies the comical theme of the movie: God is a capitalist and needs better advertising.

Tracy's father is an ad-man, so she comes up with potential slogans. But they all seem derivative. "You're in good hands with God" (like with Allstate Insurance). "God is bullish on humanity" (mimicking the Merrill Lynch brokerage cliché, though also suggesting Dionysian hybridity). Such failures to create a good God slogan also relate to Tracy's family life: her parents are divorced and her father is too busy to help her because he has a girlfriend with "big boobs" (Tracy says). Earth is suffering because humans have forgotten God, as a fading father figure in their neural theatres, and Tracy suffers the loss of her father to a woman who is not her mom. God could be merely an imaginary friend for her — with her inner theatre projecting an ideal Father as her private hallucination.[13] At times, the film shifts to that adult point of view, with Tracy apparently talking to no one. Her parents' concern about this brings them temporarily together. A psychologist tells them that Tracy produced this God-friend in order to have that effect, while also emulating her father's career. Yet Tracy finds help from an Asian American boy, Shingo, and other friends at school, who believe her vision of God. They join her in creating a "Think God" slogan and writing it all over town.

Tracy also sees God appear on television, instead of Johnny Carson on the Tonight Show. But when she tries to contact Him again, as ultra–alpha male, through that mass medium as her mystical cave wall, it fails her. Tracy is alienated in her prophet role, despite her slogan spreading across the world. She runs away from home, fearing she will be put in an institution, because adults still view her as "bonkers." (God warned her of this earlier — if she told them about seeing Him.) A homeless man scares her at the train station, like a devilish figure, but then God appears to her at last, explaining that He had been "busy." Tracy asks Him a profound theological question: why does He let bad things happen? God replies by insisting on the binary nature of reality. There is nothing He can do about pain or suffering because it is "built into the system." There cannot be a top without a bottom, or pleasure without pain. Here, God shows His left-cortex bias, appealing to the girl's binary operator (in her left inferior parietal lobe) to rationalize her limbic and right-brain suffering of panic and fear, due to social alienation. He also turns this comical film into a morality play, telling her not to give up because that will not solve the problem.

God drives Tracy home in the sidecar of His white motorcycle, driving in a crazy way, because He never drove such a vehicle before. This provides a comic spectacle, as cops see Tracy in the sidecar with no one driving the motorcycle. Yet it also points to the postmodern problem of moving beyond the modern dethroning of God as cosmic judge. If God is not in the driver's seat (as Nietzsche prophesied in the late 1800s), how will Western culture and other patriarchies continue — with new ideals of divine order and morality?

A panel of psychologists meets with Tracy and her parents. They decide that if she is not crazy, they must be, so she will have to go to an institution. But then God enters the room as "Dr. Stevens," apologizing for being "late" due to "traffic." He causes the miracle of a chandelier disappearing and reappearing, and of day outside suddenly turning into night and back to day. This proves to the shrinks that reality is not as certain as they think. They let Tracy go home with her parents, who are now reunited after seven months of separation. God appears to her once more, while the family celebrates in a Chinese restaurant. She credits God with bringing her parents back together. But He says they did it on their own. Then He promises to call on Tracy again if He has a "real big problem" and disappears. This cute, happy ending nevertheless suggests that God is lacking in a postmodern world, sometimes needing help from humans for His goals, just as they are lacking their full being in nature and may need to believe in supernatural powers to fill life with meaning.

Here and in the first film of the series, the nostalgic notion of God as cosmic spectator, judge, and helper (yet wanting help from humans) is mixed with the comical figure of a short old man in thick glasses, appearing to His alienated prophets as an imaginary friend and then to the others, too, as a distinguished psychiatrist or courtroom witness, and as miracle-making trickster. He does not throw thunderbolts like Jove in Shakespeare's *Cymbeline*.

He does rule from above, although without omniscience and omnipotence. He does not suffer, like Prometheus, for a rebellion against Zeus. Yet, He overturns the moral orders of cops, doctors, theologians, and lawyers, with His dangerous, powerful gifts and hallucinatory visions. He also affects the (im)moral filters of the cultural environment and its sexy pressures — through advertising and the mass media — comically suggesting a critique of the capitalist modes of survival and reproduction. Like Osiris/Horus and Apollo/Athena, as well as the medieval God, He judges and subdues those who are against Him. But He may be absent at times, or too busy to attend to all areas of His store as "owner." In comical ways, He also becomes an alternative threat to the social order, like Set, the Furies, Dionysus, and Lucifer/Satan. This double sense of an executive, yet trickster God reveals how the left and right brain battles of postmodern metaphysics (even with secular skepticism) may be based in social rank knowledge and alpha competition, as an evolving Theory of Mind related to that of other primates, magnified by human reason and remnant animal drives.

In the third film of the series, the rational and animal dimensions of God are developed further, not as a hybrid creature on cave walls, but as good and evil twins onscreen, expressing conflicts in the human neurotheatre. *Oh God, You Devil* (1984) begins with an ominous scene, as a father in New York in 1960 prays at the bedside of his sick son. God (George Burns), standing in the street outside, in a jacket and white cap, suggests that He helped the boy survive and hopes that he won't be "another disappointment." The boy grows up to become Bobby Sheldon, a songwriter trying to get his big break. He eventually meets the agent, Harry Tophet (George Burns as the Devil), who wears red-tinted glasses and lights his cigar with his thumb. Harry also drives a car with "HOT" on the license plate and a computer built into its dashboard, on which he watches evil leaders in the world. When Bobby, located elsewhere, says to himself that he would sell his soul to the devil to be a success in the music business, Harry's computer shows his face and name, but also that he is watched over by God (because of the father's prayers). But Harry says that Bobby asked for him, so the "rule" is that he can try for his soul now. The Devil then quips about God using "tricks." Ironically, God is a trickster to this Devil, and the Devil insists on a rule in offering Bobby a Faustian bargain. The social metaphysics of human rule-governed classes and rank knowledge are shown here as a cosmic theatre of both God and Devil watching over or possessing (and directing) certain souls.

Harry meets Bobby while he is working as a singer at a neurologist's outdoor wedding reception. Harry says he will make Bobby a star, with plenty of money for the baby he is hoping to have with his wife. Bobby wonders how Harry knew that and sees further evidence of his trickster powers when

The Devil (George Burns) lights a cigar with his thumb (*Oh God, You Devil*).

a waiter's pants fall down, causing a series of farcical mishaps. The film also shows rock star "Billy Wayne" at the end of his Faustian contract of seven years, getting a phone call from the Devil and then disappearing in a flame. Likewise, the Devil, as agent Harry Tophet, gives Bobby a contract offer. But Bobby is unsure, so Harry scratches out a line and says it will be an "open" contract, for a "trial period." Bobby signs and sends away his former agent, Charlie (who is no angel, just cheerful and incompetent). Harry changes Bobby's name to "Billy Wayne," suddenly getting him a big record deal, with producers who were dismissing him before. Bobby, now Billy, calls home and learns that his former wife lives with another man as "Bobby" (the man the audience saw as "Billy Wayne"). Thus, the Devil is more of a trickster here than God was in the prior two films. Wanting to bring souls to him, like the *Everyman* God, the Devil promises Bobby: goods, fame, and success in propagating his music, which would also help in his animal drive to produce offspring. Like God in the second film, Harry the Devil works within the capitalist system, but he then tricks Bobby by taking away his wife and potential family. This presents a more distinct, morality-play message: the choice between good and evil means, for Bobby/Billy, having an ordinary family, focused on biological reproduction and gradual labor toward cultural success, or getting sudden stardom with many sexual pleasures and the seeding of the mass audience with his artistic, egoistic memes.

Going the latter route, Bobby (as Billy) gets a mansion to live in and

Harry as his Mephistophilis in a red sweater, answering his Faustian, theological questions. The Devil admits he is God's rival, but says that God rarely makes personal appearances on earth. When Bobby asks if God "quit," the Devil answers with a question that resonates across the horrors of the twentieth century (especially for Jews and other victimized believers): "When was the last time He did something? You tell me."

Despite the Devil's warning that he will just get upset, Bobby (now Billy) goes to a restaurant where he used to eat with his wife, Wendy. He finds her there with the former Billy (now Bobby). She is three months pregnant and they say they have been married six years, but the film's protagonist believes the baby is from his seed, not the imposter's. He tries to get out of his contract with Harry, who devilishly tells him to use his sex appeal as a rock star to retake Wendy. Instead, Bobby/Billy looks for God through a Catholic priest, who thinks he is crazy, and a Jewish rabbi, who tells him that God will find him, not the other way around. Then he hears a black preacher in the street who tells him to find God in the desert. So, Bobby as Billy Wayne tries to page God from his hotel room in Las Vegas, where he is scheduled to give a concert. God calls him on the phone, convincing him that He is the one who made the beautiful rainbow outside the window. Bobby wants his wife and his former life back; God tells him to praise Him "in the highest" and hangs up. Bobby does, but nothing happens. God seems absent again in this postmodern Everyman's desert ordeal of psychological dismemberment in a hotel-casino, his alienation from family through pop-stardom, and his capitalist deal with the Devil, becoming alpha in the current social metaphysics, yet losing his "soul."

Bobby despairs during his concert and takes pills to overdose while in his dressing room. Meanwhile, God meets the Devil at a poker table. The twin powers barter for "the kid"—discussing the Pope's soul, a new Black Death, and a small pox epidemic as tradable options. God claims that the Devil does not have "clear title" to Bobby, because he lied about the contract's "trial period." So, they play five card draw for Bobby's soul. God raises the stakes, offering millions of souls on His list. The Devil folds at this bid and God reveals that He was bluffing. He had "nothing" in his cards, but then tears up the Devil's contract. This gambling allegory shows a postmodern God who is still in a binary rivalry with the Devil, over the souls that each one wants. Together they form a supernatural, desiring Metamind, as extension of the human neurotheatre with its social metaphysics: an executive, rational, left brain and a rebellious, intuitive, right brain, each drawing in different ways on limbic emotions and brainstem drives.

The Devil draws the cards of Bobby's lust for personal success and his initial fear of not having enough money to start a family, involving

The Devil (George Burns) deals cards to God (George Burns) as they gamble in Las Vegas over Bobby's soul and contract (*Oh God, You Devil*).

ego-memetic and bio-genetic drives for reproduction, in record albums and children. The Devil plays with those cards during the film, extending Bobby's limbic drives, like higher neural pathways in his brain, toward new social networks of his altered identity and fate, which the film represents in his choice to sign a metaphysical contract, change his name, and become a "star." God has nothing to offer Bobby in return, except his old life. Bobby doubts God's power to make such a miracle and commits suicide, giving into the Devil. God shows, however, that he is a trickster like his twin, at least at five card draw. He intervenes to give targeted help to Bobby — though not, the Devil suggests, to many others. The movie audience may laugh and applaud at this, identifying with Bobby's temptations toward evil, despair, and death, yet relieved when God saves him. But the capuchin sense of fairness in our primate brains might also rebel here, like Cain in the Wakefield Cycle, against a trickster God who enjoys gambling and sacrificial risks.

The film plays further with the metaphysics of social rank, through the human desire to know the Metamind of God, as divine spectator, and thus glimpse an extended, left-brain reason for evolutionary suffering. After the other Billy Wayne dies as an overdosing rock star, God at last appears to Bobby. He explains that His evil twin always wanted to be like Him. He says His interest in Bobby is due to his father, who was an honorable man and gave a "beautiful little prayer" when Bobby was sick with scarlet fever as a child, singing a song to cheer him. God insists that He watches over people, telling Bobby to go back to his wife and become a firm but loving father, while resisting temptation, because the next time God will not bail him out. Five years later, when Bobby's daughter has a fever and he prays to God, he hears God's voice joining his in singing the same song that his father did, when he was sick.

The three *Oh, God!* films extend ancient god aspects and medieval allegorical concerns, through modernist existential questions, into a postmodern era of consumerist, mass-media drives, involving patriarchal and Western imperial crises. In the first two films, God alone watches over the world, concerned about humans forgetting Him and polluting the world. He visits an everyman (or everychild) prophet who spreads that message through television and advertising. Thus, George Burns gives a comical portrayal of God. Yet he represents immortal power over adversity like Osiris/Horus and Apollo/Athena, while offering help to mankind like Prometheus and playful tricks like Dionysus. He reflects the postmodern nostalgia for one ruling God as cosmic (if not comic) Spectator, who could still visit earth, embodying the best aspects of our evolution, from animal emotions, including Promethean compassion, to right-brain playfulness and left-brain authority. But the projection of such elements from our big brains toward an ultimate theatrical Metamind raises questions about the evil aspects of human nature in its cultural evolution — extrapolated as a darker side of God, or his twin, the Devil.

The *Oh, God!* films show the "owner" of the universe choosing to work through the neurotheatre of human brains (in Jerry's and Tracy's visions of Him) and through the evolution of cultural good and evil (His image on television, made by a police artist, His courtroom trickery, His use of children and advertizing, His crazy driving, His challenging of reality principles, and His gambling). The first two comedies focus on whether people are mad to see God and believe in His miracles or magic tricks. The third film stresses a binary metaphysical battle, with God and the Devil competing for souls, like distinct evolutionary forces drawing humans (or the everyman Bobby and his twin Billy) into different fates. But the performance by George Burns of both the Devil and God, as trickster ally to the film's everyman and as mostly absent, yet helpful gambler, shows such evil and good forces as linked in the evolution of neural and cultural networks.

God and the Devil compete for souls as their currency of power. They seem to need each other — as the Devil says twice in the film, once to Bobby and once to God. God jokes in return that He could get along fine without his twin. Yet that joke opens a question as to whether God is right or merely bluffing. Does God need the Devil so that humans have a "free will" to choose against Him? Or because there is a Gnostic symmetry in the universe, like pleasure and pain on earth? Or because the binary operator in the human brain projects a prime mover for cultural evil as well as good? Do humans need a Devil persona for the animal terrors and temptations that twist in our enormous brains, from the balanced (though often cruel) survival, status, and reproduction drives in nature to the various desires of imperialist cultures and consumer media?

God also appeared as a character in the recent comedies, *Bruce Almighty* (2003) and *Evan Almighty* (2006), both directed by Tom Shadyac. Performed by the black actor, Morgan Freeman (whose name recalls the slave ancestry of many African Americans), this depiction of God primed the neurotheatre of the mass audience for Barack Obama's election as the first black President of the United States in 2008. Yet, Freeman plays not only the supreme Spectator, with specific desires targeting the human lead, but also a Mephistophilian trickster — even more than George Burns as God or Devil.

The first film casts Jim Carrey as the divine contact point, recalling his earlier movie, *The Truman Show* (1998, Peter Weir), in which his character's entire life is constructed as a reality show, unbeknownst to him initially, with a godlike director watching from above, along with millions of others at their television screens. In *Bruce Almighty*, Carrey plays the disgruntled newscaster, Bruce Nolan, who complains to God about the world's unfairness, like Cain in the Wakefield Cycle. Bruce is furious about not getting the anchor position at the local news station in Buffalo, New York. He challenges God to "smite" him and criticizes the divine Alpha for not doing His job. Like the

God (Morgan Freeman), dressed as an electrician, claps to turn off a heavenly light from above, while Bruce (Jim Carrey) watches (*Bruce Almighty*).

Devil in the final *Oh, God!* film, tuning into Bobby's despair and giving him the dream of stardom, God gives Bruce supernatural power, which also becomes his nightmare. In this case, the Faustian trick does not involve signing a contract or selling one's soul. Yet Bruce learns that the temptations and complexities of being "almighty" make the job excruciatingly difficult. They tear at his heart, eventually bringing him to his knees in a rainstorm. Despite the comical premise and farcical scenes, Morgan Freeman's God is akin to Dionysus in *The Bacchae*, when he lures Pentheus to his tragic dismemberment with the illusion of godlike power in spying on the possessed women and their rite of sacrifice. Here, however, in a postmodern twist, the divine Metamind watches Bruce get torn apart psychologically, not by pretending to be a female worshipper of a possessive deity, but by playing God Himself.

Freeman's God is not short and stooped like Burns's. He has white hair and a trim beard, yet stands tall and strong, without the need for glasses. He first appears to Bruce in a janitor's uniform, then as an electrician, with a bright light gleaming behind him. He claps to turn it off, and repeats the gesture with the "Clapper" jingle, showing that He also watches television and finds commercials infectious. Then He shows His formal attire, appearing in an all white suit and black shoes. He lets Bruce look in the file drawer of his life, which shoots out at great length, almost knocking Bruce out of the window, and then drags him along the floor at high speed as it closes. God also changes Bruce's hand to have seven fingers, when the skeptical, postmodern

Bruce (Jim Carrey) is dragged by a file drawer of his life's accounts while God (Morgan Freeman) observes at a distance (***Bruce Almighty***).

God (Morgan Freeman) reads Bruce's life-file while Bruce (Jim Carrey) hits the file cabinet, after being pulled back in by it.

everyman challenges the alpha Metamind to tell him how many fingers he is holding out behind his back.

After His jokes and tricks, God tells Bruce His plan. Bruce does not believe at first that he has God's power — until he successfully parts the red soup in his lunch bowl. God then explains the "rules" of playing God: that he should not tell anyone he is God and that he cannot interfere with human "free will." Here again God, as ultimate Metamind, provides a guarantee against free will being illusory, which current neuroscience and much of early modern to postmodern theatre explores. Yet, taking on God's job, with a supernatural will over others, also proves tricky for Bruce.

God allows Bruce to cause suffering to others, although the film does not show a Cain-like character complaining about such unfairness. Bruce pulls the moon closer to earth as a romantic gift for his girlfriend, Grace (Jennifer Aniston), but that causes severe flooding in Asia. (This suggests the social metaphysics of First World pleasures causing environmental disasters in the Third World.) He then overuses his power to make fun of his rival and replace him in the alpha news job. Bruce also hears more and more voices in his head, as if he is going mad. God takes him instantly to the top of Mt. Everest, where they enjoy the view and do not feel the cold. He explains that the voices are people's prayers (from just a small part of Buffalo). It is Bruce's job, as God, to help them — instead of using his power for ego pleasure and social rank.

Bruce answers prayers by email at high speed, making many people happy (and suggesting how mass-media power becomes godlike today). But he loses "Grace" when she sees him being kissed by a beautiful woman, attracted to his alpha status. Indeed, he yells at Grace that she cannot leave him, because he is "alpha and omega." The film shows that she still has the "free will" to do so, providing tension in this romantic comedy. It also demonstrates the power of sexual selection in human evolution, even beyond social rank and super-natural selection, determining reproductive immortality.

Like Bobby in *Oh God, You Devil*, who gains stardom while losing his wife, Bruce has the power to do almost anything as a man-God, but not bring back the woman he loves. God visits and commiserates with Bruce about not being able to "make somebody love you without affecting free will." And then, Bruce's choice to answer everyone's prayer also backfires. People riot and destroy property, through anger and joy, due to a lottery scandal and their hockey team winning the championship. Bruce despairs and has trouble finding God again, like Tracy in the second *Oh, God!* film. But eventually he gets the divine Spectator's advice that people do not know what they really want. God assures him, however, that humans have the power to make miracles themselves in ordinary ways, which is better than getting divine gifts.

This romantic comedy and supernatural fantasy is also a morality play, with its everyman learning a postmodern lesson. He does not purge his body to free his soul as in *Everyman*, or choose the advice of a good or bad angel like Faustus. *Bruce Almighty* presents a parable of humans becoming more godlike in power, through the miracles of cinema and other technologies, while needing more wisdom about desire, about serving others, and about the complexity of good actions having evil consequences. Yet this film also exemplifies the premodern belief in a divine Metamind, watching and keeping order from above, like God in the Wakefield Cycle or *Everyman*. Freeman's God reminds us of this, after His visits with Bruce on earth (unlike His medieval counterpart, who usually sends angels or Death as messengers). God leaves in His janitor's uniform, walking up a stepladder into a brilliant white light, and tells Bruce to "keep looking up"—where God is also looking down on us.

Bruce receives Grace's prayer to God to make her stop loving him, because that causes her to suffer. Torn by this, Bruce asks God to take away his power. He then gets hit by a truck and goes to heaven. Both God and Bruce are dressed in white there, with a pure white background — reconfirming a Western binary associating white with good and black with evil, despite the casting of Morgan Freeman as God. Bruce decides to answer Grace's prayer and let her be happy without him. God rewards him for this, returning him to earth, a survivor and reproducer with Grace once again. In the end, the Metamind's emotions seem to be love and forgiveness, along with a trickster's wit, while watching and sometimes intervening in the human realm. Yet God's initial response to Bruce's complaints, giving him absolute power, turned out to be a targeted revenge, not just a helpful gift and lesson. Bruce gives into that temptation, too, when he takes supernatural vengeance on his rival, Evan (Steve Carell), making him speak in an extremely tongue-twisted stutter while anchoring the television news.

God shows more of His trickster pleasure and biblical wrath in the sequel, *Evan Almighty* (2006). Like Burns in the *Oh, God!* trilogy, Freeman as God calls upon a human prophet and pressures him with a message for the world, turning him into a social scapegoat. Evan, Bruce's rival, becomes the lead in the second *Almighty* film. Elected to a higher status, as U.S. Congressman, he moves with his wife, Joan (Lauren Graham), and three sons into a huge, new home. With his wife's encouragement, the egoistic Evan prays for God's help to "change the world." Old-fashioned tools and lumber mysteriously arrive and God visits him with a command to build an ark, which also signifies "Acts of Random Kindness" that will gradually change the world. Evan resists his divine duty, until animals follow him everywhere and his hair and beard grow unavoidably long, making him look biblical. His repression of nature, in deferring a hike with his sons and refusing to let them adopt a pet dog, comes back to haunt him — through the spirited animals, around and in his body, as sublime tricksters.

Evan's sons join him in the building project, using antique technology, as dictated by the Bible and a manual God gave him, *Ark Building for Dummies*. Yet Joan and others do not believe the flood will come, as Evan prophesied, on September 22 (a date doubling the fateful 9–11). Joan takes his sons away. But Evan continues his work on the ark, getting help from the animals and obeying a higher survival call, beyond his wife's selection power.

God appears to Joan then and convinces her to go back to Evan, undoing the initial comic friction between them over the ark-building, which was used in the Wakefield Cycle for a similar story, a half millennium before. God reminds her of the moral in the biblical flood: the human family and animal pairs must survive together. But this morality movie also shows God's ecological bent, like Burns expressed in the first two *Oh, God!* films. Here, God causes or allows a natural disaster, apparently to expose the evil of Evan's rival, the alpha-male Congressman Long (John Goodman). Long ignored code violations to build a dam, creating Long Lake and the neighborhood where Evan lives. He also pressures Evan to cosponsor a bill to allow housing developments in the National Parks. When Evan defies him, Long brings the police and a wrecking ball to destroy the ark. But the dam breaks and Evan, with his animals, family, and neighbors, ride the flood waters to safety on the steps of Capitol Hill. Good triumphs and evil is exposed, through the hero's trust in divine hallucinations and biblical technology, through a father, sons, and animals working together, and through God's wrathful judgment as manifest in the destructive force of nature.

The homes destroyed in the flood, and possible lives lost, are not shown at the film's end. Instead, the audience enjoys a thrilling ride on the ark, speeding through America's capitol, via computer generated imagery. This movie was billed as the first "green" film: minimizing carbon emissions, recycling the ark's wood, and planting trees. Thus, Shadyac made God in his own image, with a concern about nature related to the production of this movie's miracles — and its fantasy destruction of Washington as the site of governmental evil, yet also of potential rebirth.

But the wasteful danger of human nature is shown as well. At the end of the flood scene, Evan warns Congressman Long about the human "wolves" behind him, waiting to take him down, as alpha politician. Ironically, this suggests that the evil of men governing against nature emerged within nature itself, as an aspect of God's creation. Wolves and primates vie for status in their groups, with big-brained humans evolving alpha leaders of tribes, religions, and nations, whose godlike powers are magnified by modern technologies and postmodern media. The *Almighty* films show not only a kindly God as divine Spectator, helping humans be good, but also a Dionysian trickster. As comedies, they reveal a tragic flaw in God's targeted Promethean gifts. Evil alpha temptations arise, with complex consequences, when media celebrities and politicians become man-gods, with animal passions and drives.

Despite the final image in *Evan Almighty* of God dancing with Evan's family, during their nature hike, and giving the audience a new commandment, "Thou shalt dance," the movie illustrates a dangerous crisis in the Western tradition of patriarchal rule. Evan needs a bestial, biblical madness to make him into a good father and social leader who might change the world for the better. Such Dionysian passion, revealed also by Jim Carey's Bruce, usually appears in the form of Lucifer, Satan, and other devils, tempting or possessing humans, from medieval and Renaissance dramas to recent horror movies. But it is sent by a trickster God in the *Almighty* comedies. It also arises in spirited animals, such as Bruce's dog, who will not learn to urinate properly, and Evan's ark friends.

The cross between spirit and animal in human nature, extended to the cosmic realm, with God as an aloof yet concerned Spectator, appears, too, through various angel characters in recent movies. They are sometimes messengers of God, even as the figure of Death, or warriors against evil, as in medieval drama, or helpers to humans (like Knowledge in *Everyman*).[14] But they are also depicted as invisible watchers, hearing the inner thoughts of humans and taking notes. Some are tempted to fall to earth, to participate in our sexual and cultural selection, or to express the primal wrath of God and potential destructiveness of nature, outside and within mankind. Such fallen angels are thus related to the many devils populating movies in the last several decades. Why do these manifestations of a divine viewpoint, with human desires and animal emotions, continue to appear in various forms, from prehistoric caves to theatrical stages and cinema screens? Which structures in our evolving brains project them — or sense their existence beyond us — as aspects of God in popular belief?

Godlike yet Animal Angels

A recent study on the neuroanatomy of religious belief investigated three dimensions: "God's perceived level of involvement, God's perceived emotion, and doctrinal/experiential religious knowledge" (Kapogiannis et al. 4876). It found that these dimensions involve neural networks used for a Theory of Mind (ToM) regarding the other's intent, emotion, imagery, and language. This demonstrates that religiosity is integrated with "networks used in social cognition" (4879). More specifically, the study explored how beliefs about "God's involvement and God's anger" engage ToM-related prefrontal and posterior brain areas (4876).[15] Various statements were used to quantify beliefs and their effects on the brain. A low dimension of supernatural involvement was expressed as "'God is removed from the world' or 'Life has no higher purpose,'" a high dimension as "'God's will guides my acts,' 'God protects one's life,' or 'God is punishing'" (4877). God's perceived emotions, from love to anger, included statements about forgiveness, protection, and an afterlife reward or, on the other side, wrath and punishment. The gradient of doctrinal to experiential religiosity was measured with abstract concepts ("A source of creation exists") or moral, practical ideas. (This did not include mystical experiences as considered in chapter 1 here.)

Brains were mapped in 40 subjects using functional Magnetic Resonance Imaging (fMRI). Statements about God's high involvement with humans "did not produce a reliable pattern of activity" in the brain (4877). But statements reflecting God's non–involvement "engaged a lateral network concerned with understanding agents' actions" (4878). This suggests an attempt by subjects' brains to understand the Metamind's actions in the world — even more so with statements about a distant God or about life having no purpose. "ToM processes were engaged to understand God's intent and resolve the negative emotional significance of his lack of involvement." Likewise, statements regarding God's anger involved "emotional ToM" in brain areas for detecting the "facial expression and linguistic content" of other humans' emotions. Statements involving God's love stimulated brain areas for positive emotions and the "suppression of sadness."

From prehistoric caves and ancient and medieval theatres to current cinemas, various gods have been shown, with animal or human faces, often speaking a human language. To perceive and revise them in their own neurotheatres, audience brains used Theory of Mind structures, which evolved from our primate ancestors for social rank knowledge and eventually enabled humans to imagine not only the intentions and emotions of others, but also their distinct perspectives (beyond baboon metaphysics). As shown by recent God films, and demonstrated by the fMRI study, even ideas of a distant God, or doubts about life's purpose without God, activate brain areas for understanding God's viewpoint. A sense of God's love, protection, forgiveness, and afterlife rewards suppresses existential sadness and activates positive emotions in the audience (as with statements in the belief experiment). Yet, ideas of God's anger, wrath, and punishments — merely suggested by the *Oh, God* films, but shown more in the *Almighty* movies with a farcical file cabinet and destructive flood — involve neural networks for interpreting facial and linguistic expression in other humans, which also developed as key elements of primate rank knowledge and human social cognition.

The wings on conventional Christian angels, along with their human bodies and facial expressions, indicate their hybridity as both animal and transcendent figures, watching from above like God, with a Metamind viewpoint, yet sensing our primal social passions. They represent the Mother Nature and alpha-Father aspects of God: divine love, protective care, and judgment about afterlife reward or punishment.[16] In movies, angels may visit earth as divine messengers, with or without wings, expressing the Other's mind and knowledge of destiny, or demonstrating God's primal wrath or love.[17] Sometimes, Hollywood angels fall from paradise, departing from the divine like Lucifer (or Adam and Eve), though not due to a direct conflict with God. Instead, they desire the experience of human bodily sensations, even at the price of mortality — perhaps reflecting the movie audience on the other side of the screen, tempted to cross.

This is the case with *Der Himmel über Berlin* (*Wings of Desire*, 1987), which inspired a similar depiction in the Hollywood movie, *City of Angels* (1998). In the former, angels wear dark overcoats, instead of wings, and hang out in the library or elsewhere in the city, invisibly eavesdropping on the inner thoughts of humans and sometimes putting a hand on them to shift their rutting neural circuits. Or they fail to help and feel the anguish of the person's despair. Such guardian angels also take notes and discuss their findings with each other like anthropologists, although in childlike wonder. Director Wim Wenders uses black and white film to show the angels' perspectives, viewing the essence of things. He also uses overlapping voiceovers as they muse about mortals' thoughts. The angels, Damiel and Cassiel, marvel at the tragic energy

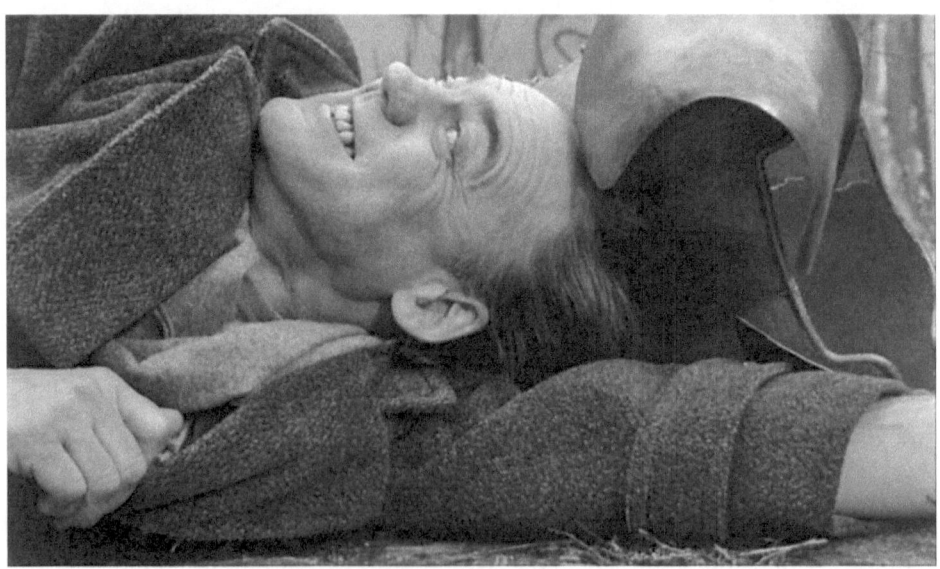

Top: The angels, Damiel (Bruno Ganz, left) and Cassiel (Otto Sander), share notes on mortals while sitting in a new car in a BMW showroom. *Bottom:* Damiel (Bruno Ganz) smiles when hit by his falling armor and feeling pain for the first time, after becoming mortal at the Berlin Wall (both photographs from *Wings of Desire*).

Damiel (Bruno Ganz) appears in the dream of a circus aerialist (Solveig Dommartin) in *Wings of Desire*.

of war, despite the success of human evolution.[18] Homer, an old man they follow, also recalls the Berlin sites that used to be around him but were destroyed during World War II. Historical footage of such destruction is shown (and of dead bodies, even children's corpses), as if also recalled by the angels or the movie viewer. When Damiel (Bruno Ganz) falls in love with an aerialist and falls to earth to join her, he lands next to the Berlin Wall and experiences the full array of human senses — as the film turns to color. He is hit by his falling armor, yet enjoys a first-time taste of his own blood, plus other ordinary sensations, as a mortal actor in time, instead of angelic spectator. American actor Peter Falk, who is in Berlin shooting a war movie, is also a former angel (as well as a television star), so he helps with advice. The human Damiel eventually finds his beloved, who seems to recognize her soul mate, after seeing him (earlier in the film) as an angel with wings in her dream. But their soulful romance is undercut by Homer's continued mourning for a lost, pre–war Berlin and the movie viewer's perception of Damiel's loss, as he joins in the folly of human sensations and passions.

The connection between movie stars, cinema spectators, and angels is rehearsed once more in the Hollywood film, *City of Angels*—although without Peter Handke's poetic dialogue, the documentary war footage, or *Trauerspiel* (mourning play) sensibility. Neither film shows God in person, yet both suggest that angels are extensions of God's nurturing concern, which might become a loving incarnation. Angels are ideal spectators and interlocutors, like God is to Jerry, Tracy, Bruce, and Evan in their films, though they are judged as "bonkers" by their peers. The angels here also function like God in the Wakefield Cycle, wanting to be close to humans, yet allowing them free will — or like ancient Greek gods sometimes allowing choice in their mating with mortals. Damiel in the first film and Seth in the second fall to earth through their envy of mortal sensations. But Seth, even more than Damiel, finds both heaven and hell on earth,[19] as a lacking, sensing, and mortal human in love with a doctor, who dies suddenly at the film's end, just after he sacrifices his immortality to be with her fully.

The pseudo-angelic power of movie viewers to fly through space, to take leaps in time, to be an invisible voyeur (as with the fourth-wall tradition of modern stage realism), and to travel inside characters' minds with voiceovers

Top: The angels, Seth (Nicolas Cage, left) and Cassiel (Andre Braugher), share notes while sitting on an L.A. freeway sign. *Bottom:* Angels stand on an L.A. beach at dawn (both photographs from *City of Angels*).

and subjective point of view (POV) shots is reflected in the nurturing yet perverse talents of angels in these films. In *City of Angels*, set in Los Angeles, the heavenly beings, with their dark overcoats and no wings, watch mortal life from a freeway sign, or stand together on the beach at dawn, showing their tie to nature, or in the airport control tower, with a connection to the current technology of flight. Seth (Nicolas Cage) is introduced by the film as a guardian angel, helping a little girl at death. He reassures her spirit that she is going "home" and that her mournful mother will someday understand. Thus, the movie viewer is given a cosmic perspective, transcending life and death, like at the start of the Wakefield Cycle and *Everyman*, but without the character of God and with a romantic, postmodern angel.

Heart surgeon Maggie Rice (Meg Ryan) wonders "who" they are fighting as doctors and nurses trying to save lives. Meanwhile, Seth watches her work and becomes enamored of her, like the combination of a lusty Greek god, the medieval figure of Death,[20] and a shamanic spirit guide — as psychopomp for those she fails to revive. Unlike the angelic Damiel, who only visits his beloved in a dream, Seth, while still an angel, appears to Maggie in an empty hospital corridor, looking human. His presence fills her lacking being and answers her existential, career-oriented despair, especially when she looks into his dreamy eyes. They meet elsewhere, too, and eventually kiss. Like the movie audience, he can only sense her through sight and sound, not touch. Yet he and the spectators can gaze at her, invisibly, while she takes a bath.

Seth's perverse angelic power becomes apparent to Maggie when she invites him into her home. While helping her to make dinner, he cuts his thumb, but the knife causes no wound or blood. This horrifies Maggie, as if a virtual creature or videogame avatar had crossed into her reality. She calls

Maggie (Meg Ryan) is watched invisibly by Seth (Nicolas Cage) in her bathroom (*City of Angels*).

Top: The archangel Michael (John Travolta) appears initially in his underwear, smoking and scratching. *Bottom:* Michael (John Travolta) brings the dog Sparky back to life (*Michael*).

him a "freak" and a "liar." Later, he falls from heaven to become human and stop her from marrying another man. But he loses her to death, when a logging truck hits her while she is biking. This final twist evokes in Seth, and perhaps the movie viewer, a primate (or capuchin) rage against providential unfairness, especially after his (rhesus-like) sacrifice of heavenly bliss. And yet, it makes his memories of being with Maggie — and the film's romantic ending — all the more valuable, like mortal life.

Fallen angels show further human perversities in the romantic comedy *Michael* (1996), the satire *Dogma* (1999), and the noir, graphic novel inspired *Constantine* (2005). In *Michael*, the winged archangel (played by John Travolta) becomes a childlike trickster indulging in human pleasures: smoking, drinking, piling sugar on his cereal, seducing a series of women, finding obscure tourist sites in Middle America, and knocking heads with a bull, instead of remaining an aloof warrior in the spirit realm. Yet he also brings a pet dog back to life and plays cupid to two tabloid news reporters, who fall in love and commit to marriage.

In *Dogma*, two outcast angels, Bartleby (Ben Affleck) and Loki (Matt Damon), a former Angel of Death, want to return to heaven through a Roman Catholic loophole, a church archway in New Jersey that forgives all sins. As they travel from Wisconsin to New Jersey, they go on a killing spree against sinners, like in Old Testament times when God was more wrathful (or like serial killers and vigilantes today). Even worse, if these angels, banished forever from heaven, become human, go through the archway, and die, they will go to heaven, proving that God is fallible and ending all existence. Hence, much is at stake in this spoof of the Church's supernatural orthodoxy.

In order to stop them, Metatron (Alan Rickman), the Voice of God, appears as a flaming angel to abortion clinic worker Bethany (Laura Fiorentino). He shows his wings and lack of genitalia to prove his angelic identity. Then he sends Bethany on a mission to stop the bad angels. She is helped by two joint-smoking prophets, Jay and Silent Bob (Jason Mewes and director Kevin Smith), the reincarnated thirteenth apostle, Rufus (Chris Rock), and a muse turned stripper, Serendipity (Salma Hayek). They are opposed by three teen devils on rollerblades with hockey sticks and a "shit demon" from Golgotha — under the command of the horned devil, Azrael (Jason Lee). Azrael, who is also a former muse, bears resentment against God because he was punished with Lucifer and other devils, though he remained neutral in that battle. He kidnaps Bethany's team because he wants existence to end, in revenge against God's unfairness — again showing monkey morality, extrapolated from social metaphysics to the cosmic order.

Bartleby had been a more cautious angel, holding Loki back in his lust for guns and bloodshed. But, like the Wakefield Satan, he has an "epiphany"

The former Angel of Death, Loki (Matt Damon), mortal and drunk without wings, points up at the wrathful angel Bartleby (*Dogma*).

about God being unfair in giving humans free choice and forgiveness, instead of making them servants like the angels. Loki accuses Bartleby of sounding like Lucifer and cautions him about making "war on God." Then, with a capuchin-like rage at cosmic unfairness, Bartleby massacres a crowd of people gathered outside the special church archway. He also kills his friend, the mortal and drunken Loki, who tries to stop him from going through the arch. However, Bethany saves the day — and the universe. After learning she is a blood relative of Jesus, she escapes from Azrael and frees God from an old man's body. (He was hiding inside the old man to play a videogame at a beach arcade, but was beaten unconscious by the teen hockey demons.) God appears, with Megatron, from inside the church, stopping Bartleby's revolution, hugging him compassionately, and then making him explode in punishment. Yet God turns out to be a Goddess (pop singer Alanis Morissette), who also does a playful handstand and presses Bethany's nose in answer to her question of why humans exist. This depiction of God, unlike the medieval Father and Son, offers a new sense of the divine, even if only half believed, in a more secular world, as feminine and "funny."[21] Akin to tricksters of the past, but more tidy, She clears the streets of the massacre, by making the dead disappear.

Such crossings between angelic and devilish natures, with the divine Spectator as a playful yet neat, pop star (or sneaky, videogaming, old man), not a super–parent, reflect a postmodern sense of the brain's complexity, as well as the media's power to shape collective dreams of the cosmos. The human brain is not just a binary system of good and evil spirits, like the voices Faustus hears, but an intersection of many goals, emotions, and instinctual values. Today's mass media, although initially Euro-American, are mixing

Top: Bartleby (Ben Affleck) descends in front of the church's "Catholicism Wow" sign after his massacre of the human crowd. *Bottom:* God (Alanis Morissette) surveys Bartleby's massacre before cleaning it up (both photographs from *Dogma*).

cultures from across the globe, extending the brain's networked plasticity to create new versions of angels, devils, and God as fictional possibilities, not dogmatic molds. And yet, there is often resistance to such satirical, pastiche gods, and to the various human demons they reflect. *Dogma* evoked strong complaints about its parody of Catholic ideals and supernatural figures, even before the movie was released. When it was, the angels' killing sprees were not shown as fully as scripted, due to budget limits and the coinciding Columbine High School massacre. Trimming the violence of fallen angels in such a movie may, however, make it easier to emulate — in the neural circuitry of some spectators' brains — especially if a goddess comes, as *deus ex machina*, to clean the streets, like the replay button in a videogame.

Top: Bethany (Laura Fiorentino) has her nose pressed by God (Alanis Morissette) after her question about the meaning of existence (*Dogma*). *Middle:* The archangel Gabriel (Tilda Swinton) looks surprised when stopped by Lucifer from bringing his son, Mammon, to the earth's surface (*Constantine*). *Bottom:* Lucifer (Peter Storemare) holds Constantine (Keanu Reeves) back from ascending into divine light (*Constantine*).

A similar, rebellious angel premise occurs in *Constantine*, though in a serious, noir tone. The androgynous archangel Gabriel (Tilda Swinton) wants to create havoc on earth by bringing Mammon, the son of Lucifer, to the surface, through a woman's womb.[22] Like *Dogma*'s Bartleby and the Wakefield's Satan, Gabriel expresses envy about God's favoritism with humans. Yet, John Constantine (Keanu Reeves), the exorcist-detective, says that God has "no plan" and is like "a kid with an ant farm," suggesting that He helps no one, just toys with and watches His creations.[23] God only appears in the film toward the end, as a bright light from heaven, not as a personified character. But He does act at that point, stopping Lucifer (Peter Storemare) from dragging Constantine to hell, after the hero sacrifices himself to save a victim of suicidal despair. Then Constantine rises toward heaven and the light of God, but Lucifer holds him back.

Though ultimately foiled by Constantine and Lucifer, during God's apparent absence, Gabriel almost becomes godlike, deceiving others to bring chaos to earth and accelerate human evolution through adversity: "to inspire mankind to be all that was intended." This shows a projection into the supernatural realm of Machiavellian Intelligence, as well as capuchin monkey resentment, in emulation of and rebellion against the alpha Spectator. Gabriel claims it is "unfair" that humans are redeemed by simply repenting. "I'll make you worthy of His love.... It's only in the face of horror that you truly find your nobler selves."

Cathartic Devils

While fueled by a money-making, entertainment drive, Hollywood may act in a similar way, influencing the neural circuitry of the mass audience through the dream machine of cinema. Horror onscreen may hone the "nobler selves" of movie viewers, as they identify with the victims struggling to survive against supernatural memes reproducing through and around them. Cinema thus extends the "threat rehearsal" mechanism of mammalian dreams (Revonsuo) toward a collective experience and potential catharsis. However, most horror films, with their melodramatic plot structure, reinforce good versus evil stereotypes, increasing a postmodern paranoia that God is not involved in the balancing act and that only a few humans can survive the infectious nature of evil and its supernatural violence, through wit or luck (or by being the sole virgin in a slasher movie). For example, *Rosemary's Baby* (1968) cultivates the fear of evil in one's neighbors and of the monstrous reproductive drive within a woman's own body. *The Exorcist* (1973), so popular that it spawned sequels and prequels for decades (1977, 1989, and 2004/2005),[24] evokes fears of bodily possession by demons, of a girl's rebellious sexual power,[25] of occult game-playing (with the Ouija Board), and of failing patriarchal rites. Yet, in the latter vein, the tragic conflicts of Father Damien Karras (Jason Miller), haunted by his mother's death in poverty, his own crisis of faith as a Jesuit psychologist, and his struggle with the devil in himself, may go further, for some spectators, toward a complex catharsis of fear and desire.[26] This encourages a fuller Theory of Mind about the perspectives (beliefs, thoughts, and motives) of others or the Other, beyond clear-cut good and evil.

In such movies, including the more historically accurate *Exorcism of Emily Rose* (2005), the victim becomes bestial in being possessed by the devil, which increases during the rite, as the demon within fights for control of its bodily territory. This triggers fear and panic in movie viewers, not only those who believe in spirit possession, but also many who do not. Devil possession illustrates the fierce potential of a subcortical animal within each human being: limbic mammalian emotions and brainstem instincts competing with left-cortical Symbolic (patriarchal or priestly) rules. Such scenes also show the "devil's advocate" of the right cortex presenting uncanny bodily signs, primal sounds, and strange languages — as Imaginary, rebellious alternatives to the left hemisphere's prosocial, rational, verbal order. Possessed victims demonstrate a violent regression to the polymorphously perverse, infantile body, which has a different mimetic discourse, on the other side of nostalgic fantasies about blissful communion with the (m)Other's voice and body parts.

Movie devils often appear as distinct beings with bestial attributes, showing a continued desire, even in our secular humanist culture, to understand the animal heritage of our interconnected brains. As indicated by the research mentioned above, mapping believers' brains, movie viewers interpret various passions in the Other onscreen, using ToM circuitry. Mirror and intuition neurons in audience brains fire with the facial and linguistic expressions, as well as perverse mimetic gestures and primal sounds, of devil characters — as if spectators were doing or feeling the same themselves. Thus, movie devils may possess not only the humans onscreen, but also those in the audience, as cinematic spirits.

In *Jacob's Ladder* (1990), animal-devil signs are shown with flash frames of a penis-like tail on a homeless man in a New York subway car, with horns exposed on a nurse's head when her hat falls off, and with the wings, claws, and horn-like tongue of Jacob's sexy Latina girlfriend, Jezzie (Elizabeth Peña), dancing at a party. Yet Jacob (Tim Robbins) eventually learns — as stated by his chiropractor Louie (Danny Aiello) — that such demonic apparitions may actually be angels, trying to free him, as he clings to his bodily illusions at the end of his life. With this horror film, the scene by scene triggering of fear, panic, rage, and lust systems in audience brains is not just melodramatic, good

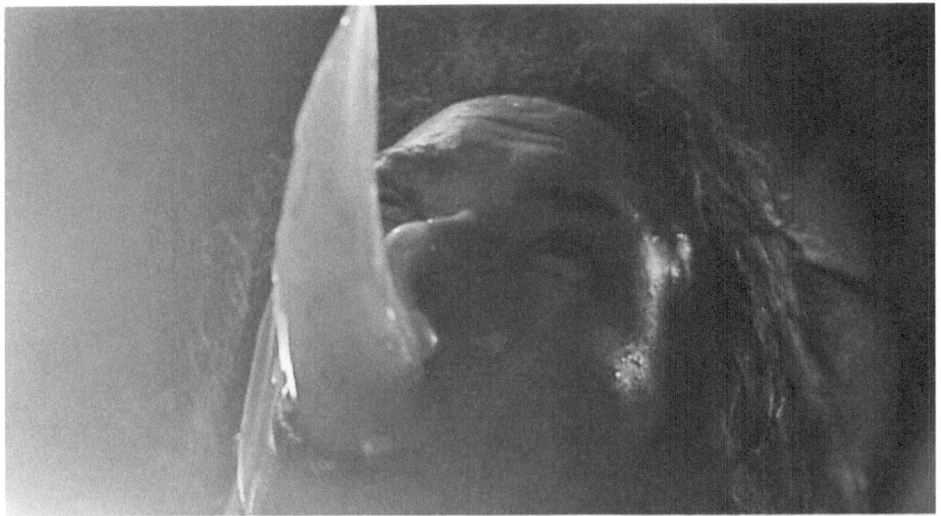

Top: While dancing at a party, Jacob's girlfriend, Jezzie (Elizabeth Peña), turns into a demon with a horn emerging from her mouth, in Jacob's view (*Jacob's Ladder*). *Bottom:* Apelike, devilish corpses chase Constantine (Keanu Reeves) in hell (*Constantine*).

versus evil entertainment. It is also focused toward a tragic awareness of one's own connection to evil in perceiving the animal in others as demonic. Such a catharsis may alter the ventromedial filters of prefrontal-limbic networks, between angelic superego aspirations and devilish libidinal passions, in the social metaphysics of spectators' inner and collective theatres.[27]

Crouching, apelike, devilish corpse-figures chase the hero of *Constantine* during his visit to hell. On earth, he battles a ghoul formed of many swarming insects flying together as one organism — thus playing upon various

Insects form a collective ghoul figure, attacking Constantine on a city street (*Constantine*).

fears in movie viewers, not only of infectious bugs, but also of their own multiple selves, formed by billions of brain and body cells. Yet, a devil character is depicted more sympathetically, as part human and underworld beast, in *Hellboy* (2004), which was also made from a comic book series. The titular character, adopted by humans during World War II as a baby devil, works for the FBI as an adult, battling his evil kin, while alienated from humans as a "freak." He files down his goat-like Mithras/Pan (or director "del Toro") horns, but still has fiery red skin, a long tail, and a crablike oversized right fist. He fights mechanized human zombies and beastly demons that have fangs, tentacles, claws, and boar-like bodies.[28] At the climax, Hellboy's horns sprout again, with his temptation to side with evil and open a cosmic channel for chaos to reign on earth. Yet he pulls off his horns and uses one to stab his nemesis, expressing a tragic conflict within himself, as well as giving the audience a melodramatic victory over evil. More of that comes in the second half of the climax as Hellboy destroys a giant octopus-like "god," born from the villain's belly, and revives his human girlfriend from death, with a magic spell he knows. This animal-human devil-warrior reveals how a stereotype can be misleading, with social alienation pushing him toward evil, though he was trained to fight for the good. He then wins a tragicomic, melodramatic redemption through love and trickery.

Other graphic novel–inspired films, with a less explicit sense of hell, display hybrid, animal-human figures that reflect the Western tradition of angels and devils. The *Batman* series (1989, 1992, 1995, 1997, and 2005) has led most recently to *The Dark Knight* (2008), with the masked and caped crusader protecting Gotham and its citizens like a guardian and warrior angel, especially when they project his search-light logo into the sky, as if praying for

Top: Hellboy (Ron Perlman) files down his horns while looking in a mirror. *Bottom:* A boar-like demon prepares to fight Hellboy while attacking humans in a subway station (both photographs from *Hellboy*).

help. But Batman (Christian Bale) becomes suspect to civil authorities as a vampire–like threat to their corrupt order, especially through the Joker's trickster plots, which tempt the "dark knight" to show his animal and devil as well as angel aspects.

Batman's melodramatic battle against evil takes various tragic twists — like that of his friend, the District Attorney (Aaron Eckhart), who crosses

Hellboy (Ron Perlman) pulls off his horns, during a fight with the villain and the evil within himself (*Hellboy*).

over to the dark side, joining the devilish Joker (Heath Ledger) in becoming a complex, sympathetic villain.[29] Both of them are "agents of chaos," as the Joker says of himself. Thus, chance is the "only morality in a cruel world," according to the D.A., suggesting that nature's amoral evolutionary drive, cruel battles for survival, and random mutations are more powerful than human laws. During the film, Batman becomes like Oedipus, revealing a tragic flaw in his righteous, Apollonian hubris, as he fights to save the city, yet violates its laws. The Joker and D.A. become like Dionysus and Set, trickster figures twisting logic and events to show sublime forces beyond human control, even in the twenty-first century. The Joker also gives a face to the mysterious threat of America's enemies in the age of terrorism, teasing evil out of the good, and rage out of the rational, like a post–Enlightenment Mephistophilis.

The Faust paradigm has been used, too, in various comedies onscreen. In *The Devil and Max Devlin* (1981), Bill Cosby plays the devil, Barney Satin, who pressures Max (Elliott Gould), after he dies and goes to hell, and again when he revives on earth, to get innocent people to sign a Faustian contract. But Max eventually defies him and the black Mephistophilis threatens him again in a fiery, cave-like hell, this time appearing in a hybrid form, with donkey ears, small horns, red Afro and skin, blackened face, and pitchfork (perhaps satirizing traditional, racist allusions to the devil as "black man" or "red demon"). But a female Mephistophilis may also appear onscreen as a sexy

Bill Cosby appears as the devil in a cave-like hell (*The Devil and Max Devlin*).

seductress. Elizabeth Hurly plays such a trickster in *Bedazzled* (2000),[31] luring the nerdy Elliot (Brendan Fraser) into signing a comically thick contract and twisting his fantasies of a more popular self into various chambers of hell on earth.[32] She even conspires with Elliot's guardian angel (a black man he meets while in prison near the end of the film),[30] as they teach him to value his soul, to let go of his dream girl, and to find her look-alike with a personality more akin to his own. Like the Bad and Good Angels onscreen, movie viewers take pleasure in watching Elliot's transformations and tribulations. But they may also get a warning from this morality play, which reflects the scapegoating sacrifices and impossible desires produced by our mass media's social metaphysics, refashioning the sexual selection drive.

A view of Good and Bad Angels working together in a topsy-turvy, tragicomic cosmos is shown in more complicated ways by a European film in several languages, *Sin Noticias de Dios* (2001), which has the English title, *Don't Tempt Me*.[33] This parody of noir, spy, gangster, and buddy film genres reflects a sense of social metaphysics by showing the business orders of angels and devils without a sign of God (as in *Angels in America*). Heaven is a French cabaret in nostalgic black and white. Hell is a rowdy prison with a junk-food cafeteria. And earth has corrupt cops. Yet, a sexy devil and angel (Penélope Cruz and Victoria Abril), who initially compete for the soul of Manny, an ex-boxer, decide then to work together to keep him alive until he can go to heaven. They collaborate because the "Operations Manager" of hell says that a coup against him will succeed if Manny goes there, tipping the balance of souls.

Top: Daryl Van Horne (Jack Nicholson) gets angry at his three girlfriends, after they use magic against him. *Bottom:* Van Horne turns monstrous when banished from his mansion by three witches, his former girlfriends, in *The Witches of Eastwick.*

Top: Van Horne (Jack Nicholson) returns to his progeny through video screens (*The Witches of Eastwick*). *Bottom:* The fertile Christabella (Connie Nielsen), with her father, the lawyer-devil John Milton (Al Pacino), and his penthouse wall sculpture (*The Devil's Advocate*).

The film begins with the devil and angel in a theological debate, just before they rob a grocery store where they both had worked (which occurs near the end of the film). The devil says the angel is relying on God, who cannot help her, because He is either too weak to stop evil or does not want to, or both. The angel replies that God is "nothing" or exists in the future, and retranslates Exodus 3:14 as "I shall be who I shall be." Later in the film, when they get shot during the robbery, they are surprised to find that they bleed. But they realize that the old order has changed and they are being punished for "fighting evil with evil." This may provoke a tragicomic awareness in the movie audience also, beyond binaries and stereotypes of the past.

Violent, sexy devils sometimes appear onscreen to embody the evil Other's passion for reproducing with human females. This represents not just an uncanny metaphysical threat, as with lusty Greek gods. It relates, too, to Mother Nature's monstrous evolutionary drives in mammalian infanticide and ape rape. But rather than showing infanticide, the child of the human womb appears demonic, inspiring desires, even in the audience, to destroy such a kid as anti–kin. Or, instead of showing a predetermined rape by Satan,

Top: Wall figures move erotically as Christabella (Connie Nielsen) seduces Kevin (Keanu Reeves) in *The Devil's Advocate*. *Bottom:* Erotic wall figures become tormented souls in hellfire at Kevin's foiling of Satan's plot.

the devil is a sophisticated charmer who becomes bestial when denied. The former can be seen in *Rosemary's Baby* (1968) and the *Omen* series (1976, 1978, 1981, 1991 for television, and 2006). The latter appears in *The Witches of Eastwick* (1987, George Miller) with Jack Nicholson as a magical playboy, yet procreating devil, Daryl Van Horne. Murderous and mating aspects of the devil also arise in *Angel Heart* (1987, Alan Parker) with Robert DeNiro as the supernatural accountant Louis Cyphre or Lucifer—and the shining eyes of a demonic child. Likewise, in *The Devil's Advocate* (1997, Taylor Hackford), Al Pacino plays the Nietzschean demon, John Milton, bearing a lawyer's will to power.[34] Milton tells his protégé Kevin (Keanu Reeves) that he is "not a puppeteer," because he allows free will. Unlike God, he is a "fan of man." This Miltonic Lucifer calls God a "sadist" for giving mankind pleasures that are then forbidden, as He watches and laughs from above. Humans, he says, are God's "gag reel." Then the bas-relief of erotic bodies on the wall of Milton's penthouse office swirl in movement, like baroque angels or prehistoric cave art in firelight, almost seducing Kevin to mate with his half-sister, Satan's lawyer-daughter, Christabella, to create the Anti-Christ. But Kevin foils his father's plan.

Such Lucifer and Satan figures in modern dress reflect the persistent traces of ancient gods, medieval Christian figures, Renaissance aspirations, Enlightenment ideals, and Romantic rebels—as persistent structures in Western culture's pruning of neural pathways and evolutionary drives in recent generations. But Hollywood has also presented dream/horror images of fallen angels and rising devils, of the divine and animal in human characters, interacting onscreen and perhaps possessing the audience, through biblical epics that are even more akin to medieval plays. One of these was a surprise hit in the new millennium, showing the power of the multiplex wall, as a cave-like membrane, for millions to experience the Son's masochistic suffering and the Father's mournful spectatorship, as if in a transcendent, yet historical trance, possessed by Him.

A Postmodern Passion

There have been many movies about Jesus. Examples include *The Gospel According to Matthew* (1964, Pier Paolo Pasolini), *The Greatest Story Ever Told* (1965, George Stevens), *Jesus Christ Superstar* (1973, Norman Jewison, based on the stage musical), *Godspell* (1973, David Greene, also a stage musical), *Jesus of Nazareth* (a 1977 television miniseries), *The Life of Brian* (1979, Terry Jones), *The Last Temptation of Christ* (1988, Martin Scorsese), *The Book of Life* (1998, Hal Hartley), *The Life of Jesus: The Revolutionary* (1999, Robert Marcarelli), and *The Nativity Story* (2006, Catherine Hardwicke). But only one such film became a blockbuster hit, drawing on renewed religious sentiments in the mass audience, after apocalyptic fears from Y2K and 9–11: *The Passion of the Christ* (2004, Mel Gibson). Its success surprised many in Hollywood, not only because the era of biblical epics was thought to be over, but also because this film was presented in Latin, Hebrew, and Aramaic, with vernacular subtitles.

According to the Internet Movie Database, the movie grossed over 370 million dollars in the U.S. alone, more than 10 times its estimated production budget, becoming the highest grossing R-rated or subtitled film in U.S. history. Catholic and Protestant church groups made pilgrimages to movie theatres. One woman in Wichita, Kansas (Peggy Scott) died from a heart attack she suffered while watching the sacrifice of Christ onscreen. As with the passion plays of medieval cycle dramas, such as *The Scourging* and *The Crucifixion* in the Wakefield Cycle, many Christian movie viewers experienced ecstatic identifications with the sufferings of Jesus, showing God's tragic, yet redemptive love for man. But other viewers criticized its explicit violence. Journalists called it "one of the cruelest movies in the history of the cinema" (Denby, *New Yorker*); "a biblical night of the living dead" and "a cinematic torture chamber" (Wilmington, *Chicago Tribune*); a "snuff movie — *The Jesus Chainsaw Massacre*" (Edelstein, slate.com); "the Gospel according to the Marquis de Sade" (Ansen, *Newsweek*), "an incredibly obtuse piece of macho-masochism" (Bradshaw, *London Guardian*); and an "obscene movie" through Gibson's "perverse imagination," which shows "not the Passion of the Christ, but the sick love of physical abuse," with its scourging scene "approaching the pornographic" (Carroll, *Boston Globe*).

The movie was also criticized for its apparent anti–Semitism. Gibson shows the Jews beating Jesus when they arrest him, the Jewish priests demanding that Jesus be executed for blasphemy, and the Jewish rabble choosing to free Barabbas instead of their "King."[35] Yet, the film focuses, too, on Veronica wiping the bloody face of Jesus and offering him water, Simon of Cyrene refusing at first to help carry his cross but then challenging the sadism of the Roman soldiers, Mary as suffering mother, Mary Magdalene helping her to mop up her son's blood, Judas tormented by his guilt, and Jesus himself in great sympathetic pain. All these characters are Jews as well.[36] The Romans emphasize this with Simon, calling him "Jew" when he rebukes them. The film also shows the Jewish high priest, Caiaphas, about to release Jesus with no evidence against him — until Jesus insists that he is the "Son of God."

Like the overall plot of the Wakefield Cycle, Gibson's film offers its mass audience a good versus evil melodrama of innocent victims threatened and tortured by mostly one-dimensional villains, with Jesus emerging from his victimization to claim victory over evil and death in the end. However, with our greater sensitivity (or guilt) today about the horrors of anti–Semitism, the Roman torturers become easier scapegoats for audience fear and hatred, as representatives of human villainy, rather than the Jews, who were the typical villains in medieval drama. Mocking and yet materializing the social metaphysics of Christ's passion, as messianic King of the Jews, the sadistic Romans turn Jesus into an object of ridicule and sacrifice. This enables today's viewers to identify with Christ's suffering as God and man onscreen, giving cosmic meaning to their own. It shows God in a human body enduring tremendous pain, thus elevating the good souls in the movie audience above the villainous Romans and Jews. But does the film also bear tragic edges with its villains, victims, and cosmic hero, through specific plot twists, recognition scenes, and presentations of sacrificial violence?

Roman Catholics who have meditated on the fourteen Stations of the Cross in church and prayed the five Sorrowful Mysteries on their rosary beads will recognize many parts of Gibson's station play: the agony in the garden, the scourging at the pillar, the crowning of thorns, the carrying of the cross, the wiping of Jesus's face by Veronica, Simon helping with the cross, the nailing of Jesus's body to the cross, and the crucifixion scene. But the way Gibson presents the scourging, midway through the film, almost upstages the rest of its redemptive identifications. His wrists chained to a post in an open courtyard, Jesus (played by Jim Caviezel) is beaten again and again by two smirking and laughing soldiers, who also make animal snarls in jest. When the canes they use do not bring enough blood, they turn to whips, including one with metal hooks that pulls the flesh off Jesus's back. Spurting blood pools on the ground. Reaction shots show the agony of Jesus's mother and Mary Magdalene watching, unable to stop the monstrosity. Here the voluptuous

The flesh of Jesus (Jim Caviezel) is pulled by a hooked whip, used by a Roman soldier to scourge him in *The Passion of the Christ*.

violence nears slasher-film territory, with the audience contemplating not only God's willingness to be a human sacrifice, but also the filmmaker's will to show so many bloody wounds. And yet, Jesus is abused again in later scenes: with a crown of thorns pressed into his head and further beatings while he is mocked as king, with beatings on the road to Calvary, and with his arm wrenched out of its socket, then nails hammered into his hands and feet as his body is affixed to the cross. These scenes are intercut with flashbacks of the ritual offerings of bread and wine during the Last Supper to connect Jesus's symbolic and physical sacrifices of body and blood. Mary's memories of happier times with Jesus are also shown as she watches his suffering.

The extreme violence of this mainstream film may distance some spectators into thinking about how the special effects could look so brutal without actually harming the actor. (Caviezel did suffer a 14-inch scar on his back from a mistake made in filming the scourging and a shoulder separation from the 150-pound cross.) But others in the audience might experience what theatre artist Antonin Artaud advocated half a century ago: a cruelty in the performance that evokes a ritual transformation of spirit and body in those watching, as they identify with the suffering of the tragic character and actor, as scapegoat and martyr. Even as fiction, the "violence of the thought" of what was done to Jesus, displayed onscreen, might produce a real-life catharsis in the audience (*Theater* 82) — through mirror and intuition neurons.[37]

There is a "risk" to theatrical cruelty, as Artaud admitted (*Theater* 82). Various viewers of Gibson's *Passion*, depending on their precinematic sense of Jesus, might distance themselves from the blood rite as obscene or objectify themselves with the character and actor onscreen for the sake of the Other's painful pleasure (and Lacanian *jouissance*). The threat of meaninglessness in spectators' ordinary lives could be psychologically overcome by a willingness to suffer more, vicariously and in real life, identifying with the masochistic fetish of their martyred savior in the movie. (This logic of transcendent sadomasochism may fuel terrorist activities as well.) The danger of sacrificial identification in Aristotelian tragedy was criticized by Bertolt Brecht a half century ago. But it becomes exponentially greater through cinema's mass-audience powers. Many spectators are moved to embrace their fates and accept more suffering, as an audience of "little Oedipuses," identifying with that tragic hero, or in this case little Jesuses, imitating Christ by submitting to injustice (Brecht 87).

And yet, *pace* Brecht, that may be what is best for certain viewers, if they choose to have their faith in God's will strengthened through this film's gut-wrenching violence. For them, the passion shows an ultimate moral filter in Jesus, as the culturally selected image of an alpha-male Metamind, with God lowering Himself to become a fully human actor, suffering terrible cruelties through compassion for others. Instead of evoking sympathy and fear as a watching and judging patriarch, or a gambling and wrathful trickster (like in the *Oh, God!* and *Almighty* comedies), this deity may inspire changes in spectators' ventromedial filters, between frontal and limbic lobes, through His submission to a tragic fate, despite the movie's overall melodrama. Christ's passion onscreen might purify audience emotions by evoking complex identifications, as God experiences the *hamartia* (tragic flaw or original sin) of being a human animal.

At the start of Gibson's *Passion*, Jesus prays in the garden to the Other part of his divine being, asking the Father to "defend" him and "save" him from "traps they set"—as shown to the movie viewer with previous intercuts of the Jewish High Priests paying Judas for information on where they can arrest their enemy. Already, the film shifts toward melodrama, from the cruelty of the alpha–Spectator's demand that Jesus suffer and die, to a villainous conspiracy against him. The melodramatic conflict of good versus evil is increased then with a figure added to this nighttime garden scene. Unlike in the Wakefield Cycle or the Bible, Satan appears as a female face in the shadows (Rosalinda Celentano), tempting Jesus with despair at the sacrifice demanded by God, as heavenly accountant. "No one man can carry this burden. I tell you. It is far too heavy. Saving their souls is too costly." Then Jesus prays to his Father to let the fateful "chalice" pass, yet quickly accepts His divine will, while looking up at a full moon being covered by clouds. After questioning the identity of the Father and Jesus, Satan lifts his/her robe and sends a snake toward the praying messiah. Yet, Jesus defies his melodramatic opposite, standing and stomping near the snake. The externalizing of Christ's human, existential doubts in a Satanic face and phallic snake ties this scene to the prophecy of Genesis 3:15 in an earlier garden, about enmity between Eve's offspring and the serpent: "He will strike at your head, while you strike at his heal." It also creates a cosmic villain for Jesus to battle throughout the film, making the savior into a warrior hero, even in his suffering (like Gibson himself as an action star, especially in *Brave Heart*).[38] Yet, showing Satan with a feminine face and phallic snake suggests the power of sexual selection in human evolution, as the animal opposite to divine, cultural aspirations.

The snake in the garden signifies the animal fears and desires of Jesus's mortal nature, tempting him to flee from the cup of suffering that his divine Father wants him to drink,[39] while evoking the startle response of sympathetic panic in many spectators. Jesus's stomping on the ground near the snake may show the rise of his human higher-order consciousness and divine spirit over his animal drives and emotions. Yet, the continued appearance of horror-film demons throughout Gibson's movie plays upon the "paranoic" structure of

Satan (Rosalinda Celentano) with a strange child (Davide Marotta) watches Christ's scourging (*The Passion of the Christ*).

the ego (Lacan, *Écrits* 5). The fear, panic, and rage systems in spectator's animal brains become reflected by these screen monsters, while Jesus exemplifies the threatened yet heroic, human and divine ego. Many spectators may find hope in Jesus's sacrifice onscreen — with his struggle against, victimization by, and ultimate triumph over such animal drives in himself and others. But the movie's melodramatic demons might also evoke a greater fear of evil, panic at loss, and rage against perceived villains in real life.

Gibson's film offers various sacrificial identifications, with many devils involved. It shows Peter, cowering at the bottom of a bridge, after failing to stop the Jewish soldiers from arresting Jesus in the garden. Coincidentally, Jesus falls from the road at the top of the same bridge and hangs by chains that bind him, staring at his betrayer, until he is hauled back up for his journey toward further suffering. Peter then hears and glimpses a roaring demonic face rush past him. He turns his face toward the cinema audience in a similar expression of terror, not just guilt, at his complicity in Jesus's torture. Likewise, Judas is chased by demonic children and then finds himself alone with the decaying corpse of a donkey, its teeth smiling grotesquely as Judas ties a rope to a tree branch and hangs himself, tormented to death by his role in God's plan. Satan and other devils, as well as Jesus's mother and Mary Magdalene, watch from the crowd as he is scourged, carries his cross, is nailed to it, flipped over while on it, and raised upon it to die in agony. Satan even appears, during the scourging, as a perverse Madonna figure holding a strange, elderly faced child — suggesting an alternative Anti-Christ evolution of human nature through sexual and cultural selection. And yet, at the climax of Jesus's suffering during the crucifixion, he calls to his Father to forgive mankind, despite the animal cruelties of the human brain, which Satan represents. In

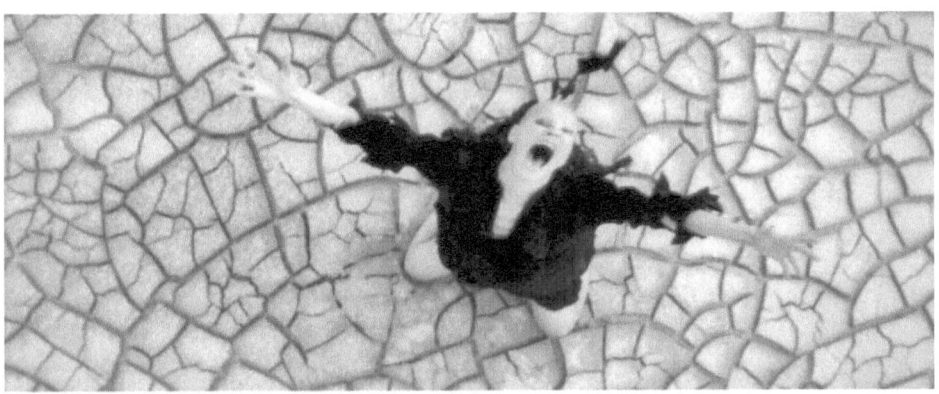

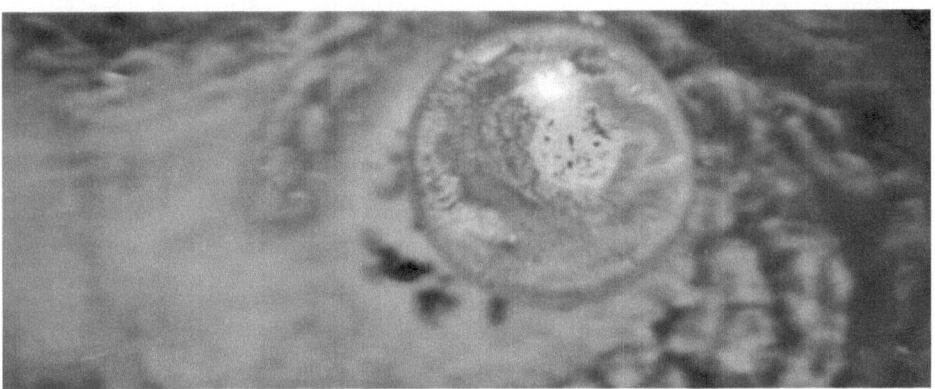

Satan (Rosalinda Celentano) screams in the pit after Christ's crucifixion. God's tear falls toward the crucifixion scene (both photographs from *The Passion of the Christ*).

contrast to Christ's victorious sacrifice on the Golgotha hilltop, Satan is shown twisting and screaming in defeat, standing on cracked clay at the bottom of a pit. This inverts the prior social metaphysics of the Jewish High Priests, Pilate, Herod, and various Roman soldiers, condemning Jesus and beating his body into a bloody ruin, with Satan and her Anti-Christ child as gloating spectators over him.

In Gibson's passion play, unlike the Wakefield Cycle, God the Father is never shown. The divine source of Jesus's agonizing love is merely suggested by the moon over him in the initial garden scene, with aerial shots over the crucifixion climax, a tear drop falling from far above it (containing the scene in it and causing an earthquake with high winds when it hits the ground), and with a bright light offscreen that Jesus walks toward as he leaves the tomb at the film's end. The ostensible cruelty of God the Father, in demanding his

Son's agony, is turned into a divine tear drop (though with a wrathful nature) and happy ending sunlight, along with the nurturing love and compassionate pain of Jesus's mother and Mary Magdalene. They are often shown as witnesses to Christ's passion. They mop up the blood after the scourging (with towels provided by Pilate's wife), peer under the cross while Jesus is nailed to it, and cradle his corpse after it is taken down. At these points, as with God's teardrop as an overhead lens, the film stresses sympathetic spectatorship, like recent angel films, rather than a strict Father judging mankind, as in medieval drama, or trickster and bestial deities like Dionysus, the Furies, and various movie devils. God, Mary, and Mary Magdalene watch with the mass audience, suffering in identification with the movie's hero. Will this help the Other within the audience to achieve a complex tragic catharsis, purifying compassion circuits, rather than evoking melodramatic fear and righteousness, with a passion for divine vengeance?

Despite its dangers of sacrificial repetition through stereotyped villains and its sadomasochistic temptations of violent entertainment, *The Passion of the Christ* serves to remind us of the tragic cruelties and costs in nature's demands on humans, as godlike creatures. Like God the Creator in the Wakefield Cycle, demanding of Jesus that he suffer and die for the sins of mankind, and Jesus at the end of that epic assigning souls to heaven or hell based on their fulfillment of his command to love one another, Gibson's film suggests a tragic flaw in God's (or Mother Nature's) gift of human life, of an evolving brain, and of transformative cultures. The expansive knowledge of good and evil has given us godlike powers, yet also transformed our animal brains into monstrously cruel organs — competing for transcendence as vengeful, sacrificial, divinely inflated, and hollow egos. Fallen from paradise, we are aware that we must die someday. Through specific experiences of family, society, and nature, nurturing or threatening our existence, we seek a greater cosmic meaning through our evolving brains. Gibson shows Jesus tempted in the garden to despair, to avoid his tragic fate of loving mankind because it is "too costly." Instead, Jesus accepts the Father's will and emerges in the end as a melodramatic victor over evil and death. But Gibson also shows the tragic cost of this divine love in the victims and villains around Jesus. Peter, the Jewish high priests, Mary Magdalene, Mary, Veronica, Simon of Cyrene, Pilate and his wife, the cruel Roman soldiers, the foolish King Herod, and the suicidal Judas react in various ways to Jesus as the embodiment of God's demand for sacrifice. This reveals the tragic edges in different audience identifications: fearing, sympathizing with, imitating, and to some degree causing the bloody displays onscreen. God is not shown in multiple ways, with an epic scope, as in the Wakefield Cycle. Yet, with the overhead cinematic shots of Gibson's passion play, the mass-media audience becomes the supernatural Other (not just observers at the foot of the cross), watching and crying through God's tragic point of view. Yet, they also look down in melodramatic triumph at Satan trapped in the pit.

Theatre and cinema, along with other mass media technologies, express sacrificial devotions, though often in the guise of selfish entertainment. Whereas ritual sacrifice in prior eras, or in religious contexts today, might emphasize an oblatory hope and divine demand (giving up something valuable as the only way to get something else from the gods), mass-media sacrifice stresses free choice and enjoyment. For a small fee, the spectator experiences vicarious conflicts, dangers, sufferings, and triumphs of characters onstage or onscreen — with illusions of violence that make the sacrifice apparently painless in real life. Yet, the pleasure of such godlike spectatorship (with immortal cinematography and editing magic, flying through space and jumping in time) also makes a sacrificial demand on the viewer. Like God's gift of "free will," and nature's gift of higher-order awareness, theatre and cinema offer supernatural power at a price. Not only the artists and producers behind the scenes, but also the spectators watching the show submit to the sacrifice, which involves their lifetimes, emotions, and thoughts, thus shaping their neural pathways and their culture's evolution. A medieval religious play like the Wakefield Cycle or a current movie like *The Passion of the Christ* may promote the melodramatic shaping of sacrifice in reality, with the passions of repeated vengeance (and Hollywood money-making). Or it may advocate the tragic awareness of such sacrificial urges. This depends on how the violence is displayed and how it is perceived, interpreted, and used by each spectator. For it was not just the lady in Wichita, suffering a heart attack during Gibson's passion play, who gave her life for that experience. We also choose how our brains and culture will evolve when we watch, like gods, the sufferings of others, onstage or onscreen, and participate in that sacrificial offering by shaping how it plays out, inside our minds and outside the theatre.

God, angels, and devils have appeared in recent movies as a judgmental sky Father, compassionate (m)Other, dutiful messenger, righteous or rebellious warrior, transformative trickster, and possessive, procreating, or murderous beast — and in combinations of such aspects. This may evoke melodramatic identifications with clear-cut, supernatural, good and evil characters, raising the human drama to cosmic significance, yet justifying a violent vengeance on stereotyped scapegoats in real life. Or it may, with complex twists in characters and plot, convey tragic, Artaudian cruelties, purifying compassionate neural networks as moral filters (even at the risk of masochistic abjection), and tragicomic, Brechtian parodies, clarifying problematic social metaphysics (as with *Dogma*). These supernatural types and audience effects, as symptoms of cultural selection,[40] express and continue to shape the evolving theatres within and between our brains, like in Paleolithic caves and various historical stages of live performances. Such effects are magnified, however, even beyond cinema, through the vast broadcast reach of television and its intimate, weekly ritual of a "home theatre," as considered in the next chapter.

6

Millennium in the Home Theatre

After the shocking news of September 11, 2001, and the subsequent wars against international terrorists, it is easy for Americans to forget the homebred paranoia of the 1990s, with serial killers, suicide cults (Heaven's Gate and Waco), the Oklahoma City and Atlanta bombings, the Unabomber, and a looming computer crisis for Y2K. Yet, the Manichean (good versus evil) worldview of the Bush administration, paralleling that of Islamic terrorists, continued many of the desires and fears of the late 1990s, refocused by the 2001 disaster. Television's *Millennium* (1996–99)[1] foreshadowed America's post–9/11 melodrama of avenging innocent victims by punishing "evildoers" in Afghanistan and Iraq. The series focused on homebred, supernatural terrorism, yet presented a similar sacrificial context: the lure of partaking in a cosmic battle as a prophet or martyr, instead of being "left behind" in apocalyptic times.

Such millennialism, both secular and religious, reveals a perverse danger in current American idealism. Despite the lack of success in recent American-led wars, an imperial hubris persists in Western culture's "Manifest Destiny," as it spreads mass-media consumerism, democracy, and humanist rights, in competition with other godlike ideals and perspectives. Millennial crusading persists into the twenty-first century (with a revised deadline of 2012 or later). This organizes our biologically inherited survival, status, and reproduction anxieties into focused fears and desires like the social metaphysics of other primates for millions of years, inside a certain kin group and against those threatening it.

Television, with its angels and devils, as well as witches, vampires, and aliens, plays a crucial role in America's millennial paranoia and sacrificial illusions. Television creates a sacred space of dramatic magic within the home, as the hearth of the living room or bedroom, as the shrine of mass culture where Americans spend hours of worship each day, despite the recent competition from computer screens and handheld devices. Americans still devote themselves to their favorite television shows with more ritual loyalty than they pay to cinema or live theatre.[2] Television bears a vast potential, in millions of home theatres, to inspire cathartic insights about the flaws in American hubris, like ancient tragic rites, or to evoke paranoid fears and copycat violence against perceived threats in real life.[3] This chapter explores such positive and negative possibilities in *Millennium* through neuro-psychoanalytic and theatrical theories of catharsis.[4]

Cultural theorists writing about film and television, such as Slavoj Zizek and Jean Baudrillard, sometimes draw upon the work of Lacan, Brecht, and Artaud.[5] But I will use those three in relation to neuroscience while exploring the power of television as an insightful cathartic yet dangerously mimetic medium, with *Millennium* as a case study. Certain shows in that series exemplify Oedipal paradigms, revealing the structures of the brain's inner theatre as it develops in childhood, through specific family and cultural experiences. Such Oedipal structures are expressed in *Millennium* through parental figures, devilish criminals, angelic forces, communal sacrifices, and mysterious signs of God's apocalyptic desires (or His absence during a final cosmic battle). These also reflect the legacy of our ancestor's supernatural and heroic projections, from cave walls to ancient, medieval, and modern stages, as well as recent screen media.

Joining the Group through Television

The missionary zeal, melodramatic heroism, and cult ideology of twenty-first-century American millennialism, including that of President Bush as an evangelical Christian, were prefigured in the television series *Millennium*.[6] The series involved a covert, mystical cult called the "Millennium Group," fighting to interpret apocalyptic events and act righteously against evil forces, but with no depiction of God's role, control, or existence.[7] The fictional Millennium Group worked with law enforcement, but also moved beyond conventional human rights when a divine mission demanded it, even at the cost of innocent victims as collateral damage.

In the series plot, the Group enrolls and then alienates *Millennium*'s tragic hero, Frank Black, employing his "gift" of profiling serial killers and sensing demonic presences (akin to Constantine in his apocalyptic film). Initially, Frank leaves the FBI and becomes a consultant with the Group. But eventually he distrusts it and tries to escape, because it ruthlessly demands the allegiance of its members against rival cults and, in Frank's case, his own family. He even suspects that the Group may have caused the apocalyptic spread of a certain biological weapon, after he is inoculated by a vaccine that the Group controls. At the end of the second season, an outbreak of the Marburg virus kills Frank's wife — making her and other victims bleed to death through the pores of their skin.[8]

Specific episodes involve various anxieties that persisted across the millennial year 2000 (or 2001): biological weapons, viral plagues, terrorist bombers, serial killers, pornography, marital infidelity, family violence, kidnapped hostages, geopolitical apocalypse, media fetishism, cyberspace perversities, and the supernatural. Although Islamic mass-media terrorists do not appear, *Millennium* prophetically addresses how television itself becomes a "cursed gift" like Frank's psychic flashes, repeating the violent scenes experienced by others. (Similar flashes of violence were used in the subsequent television series *CSI* and its various clones, to quickly show forensic evidence.) *Millennium* exemplifies how television may offer tragic insights into evil terrors and vengeful desires, in heroes and victims as well as villains. Yet it also reflects how television news magnifies and replays mass-media events like 9/11, producing the expansive, repetitive, posttraumatic paranoia that increases terrorism's effects.[9] Television has a tremendous potential in the twenty-first century to express unconscious social and personal fantasies — in ways that may evoke copycat violence or positive, cathartic insights.[10] It can propagate melodramatic stereotypes of good versus evil that justify vindictive vengeance and preemptive strikes. Or it can show the complex causes and tragic effects of apocalyptic terror, moral polarities, and sacrificial violence.

The pages ahead examine various episodes of *Millennium* to consider how spectators might be drawn into identifying more melodramatically or tragically: toward "catharsis" as a simple release of emotion or as a more complex distillation and full awareness of unconscious fear and aggression. As effects researchers argue,[11] the repeated watching of television, from childhood on, tends to evoke a prevalent belief in its melodramatic scripts, where "aggression is almost always justified" through stereotypes of good and evil characters, whether in fictional or "reality based" programs (Shrum 258–59).[12] Melodrama may build a sense of communal connection, with spectacular conflicts, extreme emotions, and collective cheers or jeers, providing a simple moral world that reassured nineteenth-century American stage audiences (even beyond immigrants' language barriers) and continues to console millions of screen spectators today. However, melodramatic violence is typically shown without painful consequences for clear-cut heroes and villains, instead of evoking a more tragic awareness of the complex sources and effects of violence for all involved.[13] Melodramatic television violence stimulates the fear and aggression of viewers with vicarious dangers,[14] desensitizes them to the complexities of perpetrators and victims, and reassures them that good guys win in the end and evil is always punished, encouraging punitive acts against stereotyped villains in real life (or preemptive strikes against perceived threats in a "War on Terror"). Yet some television shows have tragic edges revealing complex sources or consequences — through flawed heroes, culpable victims, and sympathetic antagonists offering various perspectives on good and evil vengeance. These may produce a deeper cathartic effect, evoking insights about violent tendencies in spectators themselves, instead of simplistic scripts for aggressive acts.

Television's accessibility within the home may make its melodramatic thrill ride even more addictive and hypnotic than cinema's.[15] Television involves a secular ritual of watching certain shows each week, while offering the illusory power of tuning into a channel with millions of other viewers. (Shows like *Millennium* also have fan websites for discussion groups.) Spectators not only learn violent scripts and stereotyped characters from television.[16] They become cultivated as a "mass" audience, with their lives ordered through periodic television shows and their highs found in the mesmerism of boob-tube fantasies and celebrity identifications. Of course, television viewers are not always uncritical believers or voyeuristic addicts of media violence. But the "protective frame" of screen fiction (Apter) may lead some viewers to "associate [TV] violence with pleasure" (Briller 40).[17] This can also be explained through the fetishistic mechanism of disavowal that Christian Metz theorized in film: the reassurance that the screen image is not real evokes a greater belief in its fiction, through submission to its painful pleasure as *jouissance* (69–78).[18]

Television, like cinema, inherits the melodramatic structures of nineteenth-century theatre,[19] along with its proscenium and choral frames.[20] These structures define the limits of the spectator's view, yet lure the imagination to participate in diegetic violence or abject horror offscreen.[21] The cinematic frame involves acoustic, not just visual edges of audience participation,[22] which can be traced back to the ancient Greek chorus, as well as the orchestras of nineteenth-century stage melo(dy)dramas and subsequent silent films. Television inherits from film an emphasis on realistic spectacles, which peaked in theatre a century ago, during the early development of film conventions: from moving, panoramic landscapes in stage melodrama to camera pans onscreen, and from naturalistic reproductions of interior rooms in exact detail to movie close-ups of objects as well as faces.[23] Television (or film on video) also involves the viewer, more like theatre, in the live editing of the show, through the power of the "remote control."[24] Television not only reflects, but inverts the naturalism of theatre's box set,[25] creating a "home theatre" where the domestic auditorium becomes the enclosed room in the gaze of the screen,[26] especially through Nielsen surveys, as programmers and advertisers try to perceive and manipulate mass audience desires through the fourth wall.

In their visual, acoustic, and dramatic edges, certain works of television demonstrate the medium's potential for a tragic (or tragicomic) catharsis of violent desires and fears in the mass audience, despite the prevalence of melodrama. *Millennium*, with its 66 episodes in 3 years, offered an extensive melodramatic appeal, and yet occasional tragic challenges to the projection of demonic scapegoats and angelic martyrs, in the apocalypse now of Christian millennialism in the late 1990s. A look at specific episodes demonstrates *Millennium*'s melodramatic thrills and mimetic dangers for the mass audience, but also certain moments of tragic clarity. These may produce emotional contagion through mirror neurons and binary operators in television viewers' brains — or purify limbic-prefrontal filters, encouraging a full Theory of Mind, with compassionate insights about the other's perspectives — depending on how the violence is presented and perceived.[27]

The Millennial, Oedipal Curse

In the fall of 1996, in the midst of public and Congressional pressure on all the networks to institute a ratings system for television violence, allowing parents to control access through the "V-chip," Twentieth-Century–Fox offered a new series with the TV-14 rating (not recommended for children under age 14).[28] *Millennium*, a show about serial killing and apocalyptic visions, became arguably the most violent show ever aired on American television.[29] It was developed by Chris Carter, who had created the highly successful *X-Files* series for Fox. For three seasons *Millennium* took the 10 P.M. Friday night slot that *X-Files* had popularized. But instead of using platonic best friends (FBI agents Dana Scully and Fox Mulder of *X-Files*) *Millennium* focused on a single, hard-boiled, ex–FBI agent Frank Black[30] (Lance Henriksen) and his immediate family: his wife Catherine (Megan Gallagher), and seven year-old daughter Jordan (Brittany Tiplady).[31]

X-Files investigated paranormal phenomena, often involving alien visitors as demonic or angelic figures, through the conflicting views of Scully, the scientist, and Mulder, the believer. Instead, *Millennium* showed the scientific and psychic hunt for violent criminals, involving a supernatural, millennial countdown, through the tortured psyche of one conflicted character, alienated from his family life by his visionary "gift." *X-Files* combined the "buddy film" and detective genres with various extraterrestrial encounters. *Millennium* explored the Oedipal crisis within individuals and families, in relation to apocalyptic fantasies, stimulated by television news stories of criminal cults and serial killers. While both shows involved tabloid fears about evil forces and dangers "out there,"[32] Carter's second series touched a deeper paranoia about the potential violence within each family and viewer around the television set at home.[33] Yet the series also served as a premonition to the apocalyptic logic of international terrorism and its consequent paranoia, experienced by television watchers even more profoundly in the first year of the new millennium, soon after *Millennium* went off the air. God never appears in the series, but a divine battle is indicated, with good and evil forces moving through humans toward a final, millennial climax. This evokes uncertainty as to identities and outcomes, yet a greater significance to mortals suffering in a cosmic theatre. Without God's definite spectatorship, as in medieval drama, but with signs of angels and devils in human groups and actions, plus a scripted fate for the world, various aspects of cinematic deities (defined in the last chapter) enter *Millennium*'s home theatre space: supernatural judgment, compassion, communication, warfare, mischief, and bestiality.

The violent scenes in *Millennium* appear not only through the hero's job as a "consultant" working with the mysterious Millennium Group and local police on homicide evidence, but also through his uncanny ability to glimpse the violence of the crime through the killer's point of view. Experiencing Frank's technique as a profiler, the television viewer gets visual and acoustic fragments of the demonic killer's perspective, seeing and hearing the edges of hallucinatory passions in a perverse or schizoid mind (as with the distorted scenery and sounds of expressionist theatre and film). Because of his familiarity with serial killers and his apparent psychic powers, Frank receives sudden flashbacks as if he were within the killer's mind during the crime, when he nears the evidence and the violent remains.[34] This technique intensifies the experience of violence for the television audience, although the gory scenes are shown only briefly. In fact, each image in the violent montage lasts a fraction of a second, punctuated by flashes of light and the victim's screams or other, uncanny sounds.[35] This offers the effect of being both within the killer's mind and in the victim's horrifying experience — catching the television viewer in the lures of ritual power and mortal vulnerability, while following Frank's heroic desire to solve the crime's mystery.

Usually, the violence appears first through the killer's activities and hallucinations in the episode's teaser (before the series name and initial commercials) or at other points early in the hour. Then the audience is shown some of the same, extremely violent shots again, through Frank's point of view (POV). This not only puts the voyeuristic television viewer into the killer's perverse or psychotic world, but also into Frank's schizoid position of being caught between the legal purpose of stopping further violence and his visionary empathy with it. As he tries to use his "gift" to protect his family and community against further killings, he draws the violence into himself and toward his loved ones. The violence of *Millennium* poses a similar, tragic dilemma for television spectators, as they empathize with Frank and to some degree with the psychopathic killers he pursues — while trying to make sense of such evil in a human or supernatural way.

Frank Black explains his profiling talent in the pilot episode: "I see what the killer sees.... I put myself in his head. I become the thing we fear the most. I become capability. I become the horror. What we know we can become only in our heart of darkness. It's my gift. It's my curse." Is television a visionary gift and millennial curse for the postmodern, schizoid audience — with a show like *Millennium* putting viewers inside a warped, criminal mind? What if the television viewer is not a normal neurotic being entertained, but a borderline psychotic or pervert, being lured to the edge of real violence?[36]

In an episode that aired at the beginning of *Millennium*'s second season (1997–98), Frank kills a man who had stalked his family.[37] That man sent Frank a set of Polaroid pictures of his family to show their vulnerability (as seen in the series pilot). He orchestrated the violence of a serial killer whom Frank worked to catch (in "Paper Dove," the final show of the first season). Then the "Polaroid Man" kidnapped Catherine, Frank's wife, manipulating him into a position of violent rage. Frank kills the kidnapper as a rigged Polaroid flashes and Catherine watches, horrified at her husband's violence.[38] Although Frank heroically saves her, this incident leads to their marital separation. Frank saves his wife, yet loses his family.

In another challenge to gendered, melodramatic conventions, Catherine contradicts the role of the rescued damsel, expressing not gratitude and bonding with the hero, but sadness and separation from him. "I feel like you lost something ... sacrificed to save Jordan and me."[39] (The kidnapper had tricked Catherine into fearing that he had also captured her daughter, Jordan.) Increasingly in subsequent episodes, Frank continues to make that painful sacrifice, fighting with his wife about his involvement in the Millennium Group and losing contact with his daughter, as he tries to save others and perhaps the entire apocalyptic world — by incorporating "the thing we fear the most" and using his visionary gift/curse to enforce the law. But even the law, which Frank reaffirms by catching killers, becomes increasingly ambiguous in relation to the Millennium Group and the sacrifices it demands.

Lacanian theory points to a fundamental sacrifice of symbiotic joy (*jouissance*) in all neurotics.[40] The particular way that one's natural being is sacrificed, in the loss of primal pleasure with the mother's body as the infant enters the paternal (Symbolic) order, sets up personal symptoms throughout life. Freud called such Oedipal sacrifice "castration," the father's prevention of the child from further intercourse with its mother and the law's restraint of primal, polymorphous perversity.[41] The symbolic sacrifice of *jouissance* succeeds, in most cases, in saving the subject from perverse anxiety or psychotic abjection, giving the subject a place in the social and linguistic order. But this sacrificial salvation has tragic side-effects that repeat later in life, through the further sacrificial demands of one's particular symptoms.

The child's brain is structured by particular, dramatic experiences of parental figures as godlike forces, through its nurturing, mimetic interactions with the Mother (or primary caretaker) shaping limbic and right-hemisphere networks. The inner theatre of the child's developing brain is also pruned through a sacrificial redefinition of such Real emotions and Imaginary pleasures by the Symbolic orders of the Father (or the working mother's social connections) via language and laws, especially in the left hemisphere. A culture's supernatural figures thus reflect the particular, primal experiences of mimetic bliss, and its symbolic sacrifice, repeated symptomatically, through evolving connections between brain theatres — and their cosmic projections. In caves, on stages, or on screens, the drama of devils and angels, or other animal and heavenly spirits interacting with human heroes, may reinforce melodramatic conventions of primal nostalgia and symptomatic sacrifice. Or they might provoke changes in neural networks by displaying complex, tragically flawed, role models and groups.

The ambiguity of patriarchal law, with its sacrifice of domestic bliss for the sake of a more independent ego, is shown through the melodramatic dangers of the *Millennium* series plot and its hero's tragic double-bind. Frank saves his wife from the kidnapper's torture and gives meaning to his violent visions by working within the patriarchal order of the Millennium Group. Yet, through this supernatural work, he repeatedly sacrifices the possibilities of a happy family in their "yellow house" — which viewers see in the initial episodes of the series and in the title sequence for each show. During the first season of *Millennium*, tension grows between Frank and his wife over his involvement with "the Group." In the second season, after he kills the kidnapper, Frank and Catherine separate, but they continue to meet and sometimes collaborate on cases. (She is a child psychologist and social worker.) Frank's separation from his wife and daughter intensifies his inner struggle with his gift and his external conflicts with the Group and its members. This also leads to a Christmas episode ("Midnight of the Century") with flashbacks of Frank's mother, who died when he was a child, revealing his continued anger at his father. That episode (explored further below) suggests particular Oedipal sources for Frank's psychic sympathy with demonic killers and the dead throughout the series, while joining and then distrusting the flawed, patriarchal Group.

At the end of the second season, Catherine and Jordan are reunited with

Frank as he decides to leave the Group. They try to escape an apocalyptic plague by moving together to a rundown cabin in the woods, which Frank inherited from his recently deceased father ("The Time is Now"). However, Catherine has already caught the plague virus. Discovering blood on her pillow at night, she leaves Frank and Jordan, walking into the woods to die alone. Once again, the tragic (anti)hero, Frank Black, tries to rescue his family *jouissance* by sacrificing a part of himself, this time his Group identity, yet he becomes further separated from his wife as a result.

Frank's separation from family joys intensifies in the show's third season, not only because his wife is dead, but because Jordan goes to live with her mother's parents. Jordan continues to set an empty plate at the dinner table for her dead mother, while Catherine's father blames Frank for not finding her real killer ("Innocents"). This parental blame touches a specific point of guilt in Frank. As a Millennium Group member he had been vaccinated against the plague (in "The Fourth Horseman"), but then got a syringe with only enough vaccine to save one family member.[42] Catherine helped him to choose Jordan and then realized she herself was dying. Like Sophocles's Oedipus, Frank is both innocent and guilty. He did not kill Catherine but he failed to save his child's mother from a death related (like the suicide of Oedipus' mother) to his gifted detective work. When the third season begins, Frank has partially recovered from this trauma by blaming the Millennium Group for Catherine's death ("Innocents"). He is shown meeting with his therapist and viewing his earlier, more tortured character on videotape. Yet he continues to be haunted by his traumas and ghosts, like Oedipus walking toward Colonus, knowing he is also responsible.[43]

In this way the *Millennium* series, for all its melodramatic violence and apocalyptic fantasy, achieves a tragic dimension that is rare for television drama.[44] The show refracts its violence back upon the tragic antihero, to expose "the thing we fear the most," the capability and horror of violence within us as well. The violence of killers and catastrophes "out there," glimpsed at the edges of the television screen and in the quick cuts of Frank's visions, returns to expose the *hamartia*, the tragic flaw or error in judgment, in the show's hero and the spectator's gaze. When *Millennium* aired, there was a potential catharsis of fear and desire, through viewers' identifications with Frank Black and his family.[45] As spectators gazed (or glanced) at their television sets,[46] tying their inner theatres to it, through personal and social, angelic and demonic, blissful and sacrificial ideals, *Millennium*'s tragic violence evoked a greater awareness of fundamental(ist) fantasies and ego flaws for many in the mass audience. Viewers' symptomatic associations interacted anonymously with the show onscreen, producing subtle, yet exponential effects rippling through millions of homes and minds.

In the overall plot of the series, Frank's visionary gift becomes his tragic flaw — his Aristotelian *hamartia* and Lacanian *sinthome* (an incurable structural Symptom as the linking of Symbolic, Imaginary, and Real) — as well as his returning fate of split subjectivity. Like Oedipus, who saved Thebes by solving the riddle of the Sphinx and found a new identity as its king, initially escaping his horrible oracle, Frank Black bears a superior detective power that saves others and yet tempts him toward the hubris of higher status within the Millennium Group. (Like a mystical sect of Christian fundamentalists, the Group appears able to interpret ancient texts, predicting earthquakes and other apocalyptic catastrophes.) But Frank's gift also becomes a blinding vision, like the knowledge that the blind Tiresias gives to Oedipus, who subsequently blinds himself after realizing his full identity and seeing his mother in his wife's suicide.

As with Oedipus in the backstory of Sophocles's trilogy, Frank tries to outrun his fate, at the start of the series, by finding sanctuary in another city, which is actually a return home.[47] Yet, both heroes cannot avoid the tragic *hamartia* of their brain's inner theatre and its neural rutting along symptomatic pathways. Through repeated hubris concerning his oracles, Oedipus draws a horrifying erotic drive into his marriage and family life, resulting in the death of his wife (and mother), even as he saves Thebes from a plague. Likewise, Frank brings the death drive of serial killing and apocalyptic plague to his family, through his hubris of saving himself and the world with the Group's wisdom.[48]

The Oedipus story may not be a universal family history for all cultures. But its paradigm appears in the collective fantasies of various Euro-American cultures, from Sophocles's *Oedipus* in ancient Greece to Shakespeare's *Hamlet* in Renaissance England (as Freud and Lacan pointed out), to similar works of theatre, film, and television today. Especially in times of patriarchal crisis, Oedipal dramas reveal how rebellion against alpha-male authorities, with a potential return to Mother Nature, or another form of nostalgic bliss, might operate as a tragic lure. They also manifest the perennial temptation of family revenge: to reverse one's separation from the mother by the "Father," who enforces the incest taboo and thus represents the order of society, beyond domestic origins and pleasures.[49] Such godlike, parental figures, involving angelic and demonic temptations, continue to recur in various performance media, even as cultures evolve by transforming them. But they also show the possibility for change in alpha-male role models and rule-governed classes, inherited from the past.

Millennium's apocalyptic plot represents specific crises of patriarchy in postmodern American culture prior to the year 2000, but also after it. Some *Millennium* fans may have believed, like Frank's colleague (and eventual enemy) Peter Watts, in a coming religious apocalypse that would make sense of the random evil of serial killing, despite God's apparent absence.[50] Other viewers probably identified more with Frank's wife and daughter, as they struggled to rebuild the family structure without a father figure in the home. Others focused more on Frank as he faced a patriarchal crisis, not just in his nuclear family, but also in law enforcement, divine providence, and the Group's mystical order. Through these and other perspectives, *Millennium* drew its viewers into a common Oedipal plot connected to the home scene, to current crime news, to the millennial future, and to returning desires and fears from the past.

Like Sophocles's *Oedipus, the King* (as Aristotle valued it), the overall plot of the *Millennium* series increasingly twists around the hero's desire to escape his "curse," peripatetically showing the full recognition of his fate and his family disaster eventually reaching him. The show evoked across its several seasons (as we all approached the year 2000 together) a mass audience's sympathy and fear for the hero's violent family catastrophe — and the world's apocalyptic ruin — leading to a potential catharsis "of such emotions" beyond the television screen.[51] Some episodes were indirectly tied to the overarching plot of Frank's family and the Group. Others gave unrelated comic relief. But each still referred to the show's Oedipal, apocalyptic plot through the title images after the opening teaser, including the family's yellow house and the Group's ouroboros symbol.[52]

The ancient symbol of the ouroboros (a snake swallowing its tail) as logo of the Millennium Group and series appeared not only in the show's title images, but also on Frank's computer screen, during many episodes in the first two seasons, whenever he received a message from the Group with details of a crime. The ouroboros on both the television and computer screen creates a link between history, fictional characters, and the present audience. More specifically, it symbolizes: (1) the cultural heritage of oracles that the Group's members use to interpret our collective, apocalyptic fate, while consulting on violent crimes; (2) the cyclical return of Frank's fatal flaw as his family curse, repeating from the angelic visions and terrors of his mother, when Frank was a child, to his daughter's inheritance of her own psychic gift; and (3) the return of symbiotic identity, with the Group and Frank's maternal ghost threatening to overwhelm his individual ego. The *Millennium* ouroboros also signifies the ritual return of violent visions that the television set gives to its home theatre audience, in this series and many others, relating in some way to the Real *chora* within spectators and to the actual violence of serial killers in society today.[53]

Along with the ouroboros symbol, Frank's computer screen showed him, and thus the mass audience on their television screens, the number of days remaining until the year 2000—a real point in the future that we shared with the show's fiction. Experts told us that computers around the world turning to that "00" date would produce a technological disaster.[54] This millennial point on Frank's screen also showed when the series, if not the world, would presumably end. (The series was indeed cancelled at the end of the 1999 season.) But the "Y2K" date reflected not just our varying degrees of belief in or fear of a coming apocalypse. It revealed the fate of death within each of us and the future point of death that we approached individually as we watched the show together each week. *Millennium* ritually enticed its viewers with scenes of vicarious violence "out there," involving melodramatic forces and figures of good versus evil. Yet the series also expressed a Real death drive, and various reactions to that horrifying gift of tragic knowledge, within the mass audience itself.

Catharsis as Potential "Cure"

Sigmund Freud defined a connection between theatrical and psychoanalytic catharsis, not only by his use of the same term (in his initial work with Josef Breuer),[55] but also in an early essay on theatre.[56] There he reinterprets the Aristotelian spectator's pity and fear—and the catharsis of such emotions—as the pleasurable "release" and "recognition" of unconscious material: "in the manner seen in psychoanalytic treatment, when the derivatives of the repressed ideas and emotions come to consciousness as a result of a lessening of resistance" (*SE* 7: 463–64). Freud refers to the "sacrificial rites (goat and scapegoat)" through which ancient Greek theatre emerged, as being related to the modern spectator's sympathetic identification with the suffering of psychopathic characters onstage (460). The sacrificial hero rebels against "the divine order which decreed the suffering"—or fights against society, other characters, and his own "soul," i.e., against unconscious impulses (461–62). The spectator's usual "resistance," repressing such desires and fears, becomes "distracted" by the vicarious experience of the hero's violent struggle and mental anguish onstage (463–64). The spectator is "transport[ed] ... into the same illness" so that his or her unconscious anguish, paralleling the character's, can reach consciousness (464).[57]

This Freudian model of theatrical catharsis, originating in the ancient scapegoat's revolt against divine order and developing through the modern hero's struggle with society and his own soul, shows a continuing sense of sacrifice onstage and in the audience. The hero's revolt against ritual sacrifice is itself another sacrifice, for the sake of the spectator. Both the character and spectator suffer, however, from the primal sacrifice of Oedipal repression (if they are neurotics), against which their unconscious desires also rebel.[58] Thus, the sacrificial theatres within various minds — of spectator, character, and actor — meet at the stage edge. In that dramatic, transferential meeting, the tragic display of the flawed sacrifice of Oedipal repression may become cathartic for some performers and spectators, "in the manner seen in psychoanalytic treatment" (as Freud put it), and perhaps approach an individual "sacrifice of the sacrifice" (in Lacan's sense of a cure). Whether this happens with a particular performance of *Oedipus* onstage, *Hamlet* (Freud's example), another play, a film, or *Millennium* on television, depends on the way each spectator finds the Other mirrored in the show.[59] For perverse or psychotic spectators, and perhaps some neurotics, the cathartic formula of homeopathic, fictional violence[60] may backfire — and return in the Real — leading to the phenomenon of copycat violence.[61]

Most television viewers experience only small doses of fantasy violence and paranoia. Through mirror neurons and emotional contagion, this may trigger a temporary transport into the psychopathic character's illness, but not a mimetic acting out. How do such doses accumulate in the psyche, especially given the weekly ritual of sacrificial violence for fans of certain television shows (or videogames)?[62] If millions of *Millennium* fans identify with the Oedipal hero Frank Black, seeing what the killer sees and becoming "the thing" we fear the most, will that gradually result in a healthy catharsis of sacrificial violence or bring the screen demons to life?

Lacanian theory describes three goals for psychoanalytic treatment: (1) crossing a fundamental fantasy that conceals one's particular lack of being, (2) identifying with one's structural Symptom as incurable while taking responsibility for it, and (3) experiencing subjective destitution in the non-existence of the Other.[63] More often than not, television offers illusions to screen the death drive and lacking being of its audience. Occasionally, however, the television dream may trouble some spectators so profoundly that it becomes an oracle — collective and particular, apocalyptic and tragic — like the Lacanian analyst.[64] It may provoke some viewers beyond the easy illusions of television violence, to cross fundamental fantasies involving personal, symptomatic identifications. Violence onscreen (or indicated offscreen), in the right context, could inspire subjective destitution — a glimpse of one's own death drive,[65] as with *Oedipus* or *Hamlet* onstage. Given the millions of minds that watch a show like *Millennium*, through conscious and unconscious associations, the effects of television violence and its sacrificial illusions become difficult to analyze. Certain possibilities for tragic catharsis may be found in the plot details and performance moments of specific *Millennium* episodes. But the danger of television violence, as a homeopathic drug, as cathartic medicine and poison, can be seen, too, with the inspiration that its melodrama might give to some of *Millennium*'s psychotic and perverse, as well as neurotic fans "out there."[66]

Despite its apocalyptic framework of angelic and demonic visions, the series at some points encourages cathartic insights, in all three senses of the Lacanian cure. Frank experiences, as might the sympathetic and fearful viewer: (1) a purgation of desire's detours in the crossing of a fundamental fantasy, like that of saving the world through the Group, (2) a purification of drive *jouissance* in the identification with, rather than avoidance of one's primal Symptom, as with Frank's "gift," and (3) a clarification of the Other's lack through subjective destitution, as in Frank's loss of his wife through their separation and her death.[67] Frank also views God as lacking when his daughter almost dies in a third season episode ("Borrowed Time"). Thus, the sympathetic *chora* of some in the mass audience might achieve a new cathartic awareness, from limbic to prefrontal networks. This may occur not only through Frank's psychotic traits of hallucinatory visions and violent paranoia, but also through similar traits in female characters (his friend Lara Means, his daughter Jordan, and his mother Linda), plus the psychotic or perverse traits of various demonic killers and of the Group itself.[68] Neurological research shows that damage to the prefrontal cortex can cause sociopathic traits, such as those presented by *Millennium*'s human devils.[69] But perhaps that area of the brain, as a moral filter in most television viewers, can be improved through identification with Frank and sympathy for, as well as fear of the tragic sociopaths that he connects with.

Violent rites in many cultures teach spectators to submit and conform. Yet they may also evoke a questioning of personal, familial, and social sacrifices — as when the fictional scapegoat, onstage or onscreen, is not just objectified as evil, but reflects the spectator's inner theatre of cruelty and the danger of communal vengeance. Artaud's awareness of a risk in stage violence points to the mimetic danger in screen violence as well. If the violence against the demonic scapegoat is not shown in a way that clarifies tragic suffering, then television may simply mirror the illusory, yet seductive powers of the heroic ego: its victorious, apparently moral *Gestalt* in saving or avenging the victim and defeating the evil one. The violence of ego rivalry at the Imaginary level will be encouraged and focused against villainous stereotypes — instead of revealing the Real fragmentation and lack of being behind the mirror, as the potential for evil in the good hero and the "uselessness" of reciprocal vengeance (Artaud, *Theater* 82). Yet, a more Brechtian catharsis, clarifying the Symbolic dimension in ego sacrifice, whether in the home scene or in other social mirrors, creates another homeopathic possibility: counterbalancing the Artaudian risk of abject, narcissistic regression[70] into primal, mimetic violence.[71]

Bruce Fink describes how "something changes" in the Lacanian analysand during treatment, "something takes place at the border of the symbolic and the real" (*Lacanian* 71). Fink describes this as a change in self-understanding beyond Imaginary (mimetic) meaning, "a process which goes beyond the automatic functioning of the symbolic order and involves an incursion of the symbolic into the real: the signifier brings forth something new in the real or drains off more of the real into the symbolic." The violent signifiers of an Artaudian theatre of cruelty, onstage or onscreen, might bring forth something new in the Real of the spectator. But Brecht's theory of an "alienation effect" might work in the other direction at this Lacanian edge, changing the moral filter between the brain's prefrontal superego and limbic id. Brecht argued, in his own terms, for theatrical techniques revealing the Imaginary lure of mimesis in the hero's ideal ego and sacrificial destiny, thus changing the automatic functioning of the current Symbolic order.[72] The A-effect might bring more of the Real into the Symbolic — with a greater cathartic awareness of Real dangers in Imaginary violence, as spectators change the sacrifices in their homes and society.

Brecht attacked Aristotle's theory of tragedy as valuing a catharsis that encouraged audience conformity, rather than social change. But his critique is even more applicable to melodrama, onstage and onscreen. There is a sympathetic and fearful justification for violence in melodramatic plots: the hero or another victim suffers at the villain's hands and the hero takes violent revenge. This hides the way that repeated, reciprocal sacrifices are of "human contriving" and therefore changeable.[73] Melodrama may show vengeance as fated by supernatural forces — as with *Millennium*'s serial killers acting demonically in an apocalyptic framework. Thus, the mimetic mechanism of melodrama focuses the spectator's sympathy and fear toward a simple, Imaginary catharsis: identifying the rival egos of hero and villain with communal beliefs in good triumphing over evil. This draws the spectator into ritual submission and vicarious complicity at the stage or screen edge, watching not only the striptease of the victim's masochistic display,[74] but also the righteous sadism of the vengeful hero, giving the villain his just deserts. Some spectators might identify with the villain, too, in his initial violence and eventual punishment, sadistically and masochistically. However, tragic plot twists and recognition scenes can shift the viewer's perspective toward compassion for flawed heroes and victimized villains, through horror at their violence and its repetition compulsions.

During many episodes of *Millennium*, as with other television melodramas, its fans may become mimetic star worshippers, or little Frank Blacks, ritually returning to the screen each week to see another strip tease of violence. Perhaps some spectators not only join in this communal trance, which Brecht abhorred in theatre, but also feel inspired, directly or indirectly, to imitate the hero by acting against villainous types in ordinary life. Brecht's response to these mimetic dangers of communal submission and revenge (even more apparent in the Nazism of his time) was to disrupt spectators' empathy through various theatrical techniques. Likewise, certain episodes of *Millennium* present, despite the show's melodramatic framework of millennial apocalypse, a Brechtian tragicomic potential for revealing the mimetic sacrifices of commercial television. Television lures the viewer toward violence onscreen for the sake of catching eyeballs and selling products. This may produce not only desensitization and aggression, but also a sacrificial fatalism and the demonizing of apparent villains (especially those who look like terrorists). Yet, the Brechtian twists and Artaudian cruelties of specific *Millennium* episodes exemplify television's possible cathartic insights: clarifying the edges of the Symbolic and Real in the viewer's brain, in the home theatres of television, and in the social metaphysics evolving through them.

Home-Made Tragedies

According to chapter 14 of Aristotle's *Poetics*, catharsis becomes most effective when the play focuses on a family disaster, "when murder or the like is done or mediated by brother on brother, by son on father, by mother on son, or by son on mother" (1468).[75] *Millennium* often projected the evil of family and criminal violence toward an apocalyptic fate for the entire world, caused by satanic forces and by a mysterious, patriarchal group's ruthless war to interpret and control the future. During its second and third seasons, the show became more melodramatic, with the Group sacrificing individuals, ostensibly for the greater good, and Frank blaming it for the death of his wife and others.[76]

Yet, in some *Millennium* episodes, Frank's visionary gift speaks directly to the home and family scene, intensifying the potential Oedipal catharsis. The show's tele-visionary gift to the audience (through Frank's gift) of sudden flashes of the serial killer's violence becomes insightful, not just fanciful and entertaining, when it hits home. Some viewers might even realize an uncanny parallel between their ritual fascination with the show's violence each week and the return of serial killers to the illusion of power and the ecstasy of violence with each new victim. Typically, the real-life serial killer was a severely abused child who became a late bed-wetter and turned to arson or animal dissection for sexual gratification.[77] Such a victim of abuse may become a perpetrator himself, fighting the terror within and reversing the trauma by finding other human bodies to control, torture, or take souvenirs from — as the pervert develops certain sadistic and fetishistic rites, or the psychotic finds significance in an alternate, paranoid universe and hears voices commanding him to kill. The sacrifice thus repeats, as victim becomes villain, from one generation to the next.[78]

While not offering any depictions of perverse or psychotic killers as childhood victims, *Millennium* does strike home by another route. Its expressionistic flashes of serial killers' visions and actions, through Frank's televisual, profiling gift, reflect various perverse and psychotic traits of serial television. These include star and consumer-product fetishism; remote-control sadism and masochism (even before "reality shows" like *Survivor* and *Fear Factor*); eyeball-catching exhibitionism vis-à-vis channel-changing voyeurism; and millennial, terrorist, or local news paranoia.

The "Pilot" of the series begins, for example, with a strip-tease dancer evoking the perverse and psychotic visions of a serial killer, which the television audience sees. But various spectators, male and female, might participate in different ways, while watching through the glass of the television set and then, with the killer, through the glass of the stripper's cell. During the opening minutes of this episode's "teaser," the killer and television audience

A stripper named Calamity (April Telek) dances with blood running down her body and a lake of fire behind her in the killer's psychotic vision — seen also by Frank Black ("Pilot" of *Millennium*).

see blood dripping down a brown wall behind the stripper — after she says, "Tell me what you want," and he responds with a phrase from W. B. Yeats: "I want to see you dance on the blood-dimmed tide."[79]

Thus, the series began by teasing its initial audience, not only with the stripper's body, but also with the killer's demonic gaze, involving apocalyptic fear and fetishistic desire. "You like to watch my body," the stripper says to him, and thus to the television viewer, who is likewise lured by the fetishized flesh of many other erotic bodies in mass-media situations. After the killer replies (again quoting Yeats), "The ceremony of innocence is drowned," a line of blood runs down the dancer's forehead and nose. (She continues her erotic dance, oblivious to his violent visions.) The killer persists with his apocalyptic mantra: "You'll have your part in the lake which burns with fire and brimstone." Although the dancer keeps her bra on for the commercial television audience, the scene's death-drive eroticism intensifies as flames erupt behind her gyrating, blood-striped torso and face. "I know you like it," the stripper says as fire rises behind her bloody face and breasts. "Tell me what you want," she whispers, letting slip the show's desire to please its mass audience, yet also its potential to expose spectators' unconscious, sacrificial desires.

Male, heterosexual television viewers might be more likely to identify with the killer's gaze in this scene and be teased into an awareness of their unconscious fantasies and desires, as akin to his demonic visions. But female viewers might also find themselves represented subjectively, not just as objects of a violent hallucination. Although several erotic dancers gyrate in a collective performance space before their customers' various windows (reflecting the multiple television windows of the mass audience), the future victim is shown making a phone call to her babysitter. She is not only an erotic object; she is also a mother. This suggests that she performs the subsequent "private" dance for an individual customer (who turns out to be the serial killer) because she needs the extra $200 as a mother. Her overtime labor leads to the killer's violent illusions and subsequent actions upon her body — shown later in the episode through Frank's psychic flashbacks.

After the opening teaser, the *Millennium* title sequence, and the commercial break, Frank Black is introduced, moving with his wife and daughter into their new "yellow house" in Seattle. Female as well as male viewers might identify with this family, in their American dream of home ownership in a new place while escaping the past: Frank's horrifying work as an FBI agent tracking violent criminals for ten years in Washington, DC. Catherine expresses her joy at their new home, and Frank tells a neighbor that they have returned to Seattle to "put down roots." But then he picks up his local newspaper and sees a front-page story about the murdered mother of a five year-old girl, showing the face of the stripper.[80]

Frank's escape with his family and return to their home roots is immediately tainted by this news of local violence (reminiscent of the "Green River serial killer" who terrorized the Seattle area in the 1980s and was not convicted until 2003).[81] Many television viewers, male and female, might sympathize at this early plot twist through their own fears of violence in the local news, as they watch *Millennium* in their home theatres. Although he has left the FBI, Frank is lured into using his supernatural "gift" once again to help the local police. As Frank nears the current victim's remains, but before the black body bag is opened, he receives psychic profiling flashbacks (shown to the television viewer) of the victim through the killer's POV.[82] First, her arms wave to fight off the attacker while she screams. Next, Frank's visionary flashes show the victim's bloody face on the floor with a crushed plastic cup near it, the killer's hands crossing hers over her chest, and repeated glimpses of her body on the floor. Without seeing the victim in the body bag, Frank tells his police colleagues that the killer cut off her head and fingers (but that is not shown). The coroner replies: "The man with the X-ray eyes."

Frank then goes to the strip club where the victim had worked and questions another stripper in the same private dance booth where the killer had been. Frank glimpses through the killer's POV, and the television audience again sees, the blood and fire on the dancing murder victim, as shown in the teaser. Later in the episode, Frank interprets these violent flashbacks through the killer's quotes from Yeats, Nostradamus, and Revelations, which were captured on a security camera videotape of him speaking as he watched the erotic dancer.

Through these oracles Frank reveals that the killer has found his violent, psychotic place in the Symbolic order of millennial apocalypse. His motivations have baffled local police, because he not only killed the stripper but also several gay men. But Frank explains that the killer's quote from the sixteenth-century apocalyptic poet Nostradamus — "The great plague in the maritime city will not stop until Death is avenged by the blood of a just man" — means that he is trying to stop the plague of AIDS in Seattle through a bloody vengeance. Frank interprets another Nostradamus quote by the killer — "The Great Lady is outraged by the pretense" — as revealing his family motivations and psychic structure. Frank tells the skeptical police: "The killer is confused by his sexuality. He feels guilt, quite possibly from his mother. So he goes to peep shows to try to feel something toward women, but all he feels is anger. Anger that fuels a psychosis that distorts and twists his view of reality."[83] With this analysis Frank helps the local police to find another of the killer's victims, buried alive in a coffin with his eyes, lips, and fingers sewn shut. The top of the coffin lid is inscribed with "La Grande Dame" ("the Great Lady" from the Nostradamus quote). Inside the coffin another victim's decapitated head is found in a plastic bag.

Frank realizes that the psychotic killer is "passing judgment ... carrying out his death sentences on the afflicted," after testing their blood for AIDS. Frank explains to his friend, Detective Bletcher, how the Millennium Group has helped him to understand his horrifying psychic "gift" in relation to present criminals and the future apocalypse. After Bletcher leaves him at the police station, Frank discovers that the killer works for the police in "pathology"[84]— and the heroic detective goes downstairs to the forensic lab alone. The serial killer attacks him, confirming Frank's diagnosis about maternal guilt and performing the Wrath of God.[85] Frank uses a dead body as a shield while the killer stabs at him; then Bletcher returns to save Frank's life, shooting the psychotic and ending his mission of divine judgment. Yet the killer's final words offer a tragic prophecy about Frank's millennial gift: "You can see it, just like I do. You know the end is coming. The thousand years is over. You think you're the one to stop it.... You can't stop it."

Catherine told him something similar, earlier in the same episode, about his desire to protect his family and home from the evil and violence of his work: "The real world starts to seep in. You can't stop it." It is precisely this intersection of the neurotic's fear of bloody violence seeping into the home scene and the psychotic's paranoid compulsion to perform a bloody, apocalyptic revenge that might shock the television viewer into cathartic identification with both Frank and the serial killer. As *Millennium*'s violence seeps into what Frank calls the "make believe" of his yellow house, of his safe home and happy family,[86] the show's visions may also seep into the television spectator's home, producing sympathetic terror—for better or worse, cathartically or mimetically.

The rest of this chapter delineates the potential for tragic catharsis or melodramatic mimesis in watching *Millennium*, with its troubled hero, his family, his allies, and his enemies. The latter include various types of human and metaphysical villains: perverse and psychotic parents, violent children, angels of death, family ghosts, and the destructive edges of television and Internet screens. Each of these monstrosities represents a specific social terror and potential catalyst of mass-audience sympathies. Such demonic figures and spaces may have been used as melodramatic evils to catch a mass audience. At times, however, they demonstrated complex, tragic desires, regarding erotic and deadly drives that are not just "out there," but in the spectator as well—in the brain's theatre and the home scene.

Maternal/TV Control

Out of the 66 episodes in *Millennium*'s three seasons, only one showed the mother of a serial killer. Yet through this episode and others akin to it, the series evoked complex sympathies in its audience, for both killers and victims onscreen. It produced a potential cathartic awareness that serial killing rituals begin in the family pruning of a child's brain — in homes not completely unlike the ones around the television sets tuned to this series.

In "Paper Dove," the final show of the first season, a killer named Henry Dion takes the corpse of each female victim on a camping trip, so that he can talk with a woman who does not talk back — before leaving the body under plastic and hidden by fallen leaves. Trained as a nurse, Henry removes the larynx (or crushes the hyoid bone or severs the trachea) of his victims after death, but takes gentle, loving care of their faces and bodies while spending time with each one in the woods. (The details of Henry's violent signature are discussed by Frank and FBI agents; they are not shown onscreen, except for a mark on the victim's throat. But Henry's tenderness with his victims' bodies is shown.) Henry still lives in his mother's home, where she cooks for him and fusses over him, talking at him endlessly.

In the episode's teaser, Henry is dressed only in his undershorts while driving a van. He follows a woman home from the grocery store parking lot and murders her in her kitchen.[87] *Millennium* does not show the murder, just the devilish Henry, who then wears rubber fishing pants, surprising the woman in the kitchen and taking a Polaroid picture before killing her. Later, he is shown driving to a campsite with her dead body, talking to her in his van and in the forest after he has covered most of her with leaves. "It's so good to have someone who listens," he says as he lies with the corpse in the leaves at night. He poses for a photo with her, holding her face close to his, smiling, and the cut in her neck becomes visible. After a scene of Frank and his wife, talking and kissing in bed, Henry is again shown lying in the leaves with his dead companion, telling her about his mother: "When we came down from Quebec, I worked hard to lose my accent.... I wanted to be a DJ. She wouldn't let me. No. I had to be what she wanted. But you know all about that. I reckon you heard my whole life story." Henry's symptomatic anger at his mother, for controlling his life, is revealed just under the surface of his pleasant companionship with the victim's corpse.

In a subsequent meeting with FBI agents, Frank Black profiles this unknown serial killer: "He's silencing someone he knows. His wife, his mother. Once they're silenced, they become a captive audience. They'll listen. Be sympathetic. That's what he wants. That's why he never disfigures their faces. He easily enters middle class homes and confronts them. He's comfortable there. I think that's where he lives." Frank realizes that the killer is silencing his victims and transforming each into a captive, sympathetic audience, controlling them because he cannot control his mother's voice. He reverses and avenges his mother's control of him through a surrogate fetish of her demonic maternal body.

But this episode also reflects the ordinary fetishism of television viewing. Overwhelmed by parental desires and voices, from childhood to adulthood, the television viewer can turn to the voices and scenes onscreen as an escape from the Oedipal drama at home. Using a "remote control" device, the spectator makes the voices and visions of television obey. And yet, the television screen — with its many speakers, actors, and channels — becomes the Other whose desires captivate the spectator for many hours each day or week. Like Henry Dion with his victim's corpse in the woods, the television viewer controls the set: making it a good listener by "muting" it, and making it, as dead machine, into the fiction of a live, interactive companion. But television viewers, like Henry, may become subject to and captured by the ritual sacrifice, through its illusion of interactivity and transcendent power.

The "Paper Dove" episode puts the spectator into the killer's position, not just as a devilish villain, but also as a victim. It does this through the teaser's POV shots of Henry sighting his victim in the grocery store parking lot, through Frank's violent glimpses at the crime scene, and by showing Henry's mother in action. When he comes home from his pastoral bliss with a silent female corpse at the campsite, his mother assaults him with quick, affectionate chatter (in a French Canadian accent), squeezing his cheeks with her rubber gloves in the kitchen. "Oh, little nice cabbage, where have you been? Oh, sometime a mother knows, heh? Yes, my big brute. I know where you go. With those girls, I bet, heh? ... You think you become a registered nurse without me teach you? No.... Now you have license to be nurse but don't work? ... Here are ten little chores for my big hairy boy...." Henry then tells her he went "to the trail." This brief statement incites her *jouissance* even more: "I teach for you that love for nature that God give us so free. Your father, he didn't love that. He didn't want to go hiking, never. Yeah, God took him early to his grave for that...." Henry's mother defines his father's flawed relations to God and Mother Nature, demonstrating how the killer's perversities are not just devilish, but symptomatic of his suffering, his drive to recall the Father's Law, and his desperate desire to find a better mother — in nature and his victims.

Indeed, each of the threads of the mother's rapid speech in the kitchen scene connects with the violent signature of Henry's forest altar and serial killing Symptom. His perverse relation to "those girls" who are his victims shows his response to the mother's overwhelming love and controlling envy, to the pain that her *jouissance* causes in him. His work as a nurse with each victim, his surgical care and affectionate bedside manner with the corpse, responds also to her "teaching." And the ritual "chores" that the serial killer completes on the Appalachian Trail relate directly to his mother's "love for nature" and his father's absence. Henry kills his female victims and leaves them

6. Millennium *in the Home Theatre*   249

on the trail not only to revenge and reverse his mother's vocal dominance, but also to become her "God"—the Other who took Henry's father "to his grave" for not loving nature and never hiking there. Years ago, the paternal function failed to separate the little boy Henry ("little nice cabbage") from his mother's desire and vocal *jouissance*. So now the adult Henry reinvokes the Oedipal taboo through its violation, making love to his silenced mother (and to Mother Nature) with each victim he kills—demanding the appearance of the law to separate him from her.

The law's appearance for Henry, at the conclusion of this *Millennium* episode, occurs just after he has achieved a climax to his serial killing rituals, returning the Oedipal sacrifice to his mother's body. After his many rehearsals with surrogate victims, Henry is finally able to kill—and thus make love directly to—his mother.[88] He also brings the Law of the Father upon him, as a judgmental "God." Frank and the police find Henry in the kitchen, sitting on the floor calmly waiting for them, with his mother's dead body and a bloody bolt cutter nearby. There is blood on Henry's face and the garbage disposal is running. Apparently, Henry cut out his mother's voice box, like he did with his rehearsal victims, and ground it in her kitchen drain, consummating the union of her vocal *jouissance* and his serial killing rite.[89] "She's dead," he says to Frank, who represents the Father whose Law of separation has reached him at last. "The important thing is we finally came to understand each other. Isn't that right, Frank?"[90] Through Frank's appearance as a superego, Henry realizes his goal of Symbolic understanding and achieves a cathartic climax but only after killing many surrogate mothers and finally his own, at the edge between the Imaginary and the Real. He reaches an understanding with his verbose, smothering mother, as well as with the Law and his mother's "God"—by invoking the Father's Oedipal intervention.

Some television spectators might experience complex cathartic insights by identifying with Frank, his wife (who is kidnapped at the end of this episode), and even with the devilish Henry, who victimizes others, yet is also a victim of his own perverse Symptom and his mother's *jouissance*. Television viewers might see snapshots of their own unconscious desires, and of the Other's *jouissance*, in the violence of this episode and its tragic characters (including the mother). Others might experience just a partial catharsis that evokes mimicry, in some ways, of the aggression onscreen, through melodramatic fear and vindictive vengeance—especially through *Millennium*'s cosmic schema of good battling evil. With or without that supernatural framework, the weekly rite of serial television, giving viewers a temporary illusion of transcending their insignificance and mortality, becomes akin to the serial killer's addictive power trip over each victim, demanding more and more sacrifices to reach that high again. But there is also the possibility, with

Millennium's tragic twists, of a greater awareness of television's remote control — and of the Other's violence within oneself — in the theatres of many homes, which may involve perverse fathers, as well as smothering mothers.

Perverse Patriarchal Devils

In *Millennium*'s third season (1998–99), the Group, including Frank's friend, Peter Watts (Terry O'Quinn), becomes a perverse fatherhood, orchestrating the killing of a family of blonde female clairvoyants and of anyone else who threatens the Group's control of the future.[91] Frank blames the Group for the plague outbreak that killed Catherine — although his own life was saved by a vaccine that came from the Group and the vaccine may have saved his daughter's life as well. This repositioning of the Millennium Group as melodramatically evil, as the opposite to Frank's patriarchal goodness, reduces the opportunity for the television viewer to experience a tragic catharsis through sympathy with and fear of Frank, as a flawed Oedipal hero realizing that the evil, "the thing we fear the most," is in himself as well. Yet, certain episodes in the first two seasons offer such tragic insights by dramatizing a perverse father within Frank, as well as "out there," and the evil potential of the Father's Law.

According to Lacanian analyst Bruce Fink, "The pervert seems to be cognizant, at some level, of the fact that there is always some *jouissance* related to the enunciation of the moral law. The neurotic would prefer not to see it, since it strikes him or her as indecent, obscene" (*Clinical* 190). I would argue that *Millennium*, at its best, not only entices its audience with paranoid fears and violent desires, somewhat resolved at the end of each episode. It also shows something more obscene and horrifying: a perverse father's *jouissance* within the normal law of Oedipal sacrifice — like a devilish ecstasy in a judgmental God. This evokes a potential catharsis, a sacrifice of the patriarchal sacrifice (Symbolic castration),[92] in the watching television audience, clarifying the brain's moral filter between limbic and prefrontal networks, and changing how spectators' repeat their symptomatic sacrificial demands.[93]

Millennium, as a television series about serial killers and other apocalyptic devils, maintains the ordinary, ritual sacrifice of the spectator's submission to the screen. Even if God does not appear in the series, *Millennium* offers the illusion of the Law's consistency and the Other's moral existence through Frank's capturing of the serial killer at the end of each episode and apocalyptic hints about sacrificial violence. This alleviates the spectator's fear of random, incomprehensible evil in real life, producing a simple catharsis, as purgative release. But a more complex catharsis might be produced by certain episodes. Through Frank's identification with the Symptom of the serial killer, he functions like a Lacanian analyst, focusing the spectator's transferential identifications toward a sacrifice of the normative Oedipal sacrifice. This might move some in the television audience beyond personal fantasies of Oedipal consistency — toward a realization of the Other's lack, morality's inherent obscenity, and perverse parents in the Law and the home.

In a first-season episode entitled "Sacrament," Frank's sister-in-law, Helen, is kidnapped outside a church after the baptism of her child. Helen and Tom (Frank's brother) had brought their newborn child to Seattle, so that Frank and Catherine could become its godparents.[94] Frank is thus doubly to blame in his brother's eyes — for bringing Helen into danger and for his inability to do his job in finding her — as Tom waits for the lacking Law to act, not knowing how long his wife might live or how she is being tortured. This episode expresses, more than most crime dramas, the tragic suffering of a victim's family members as the consequence of violence.[95]

Frank finds a likely perpetrator, a recently released sex-offender, named Richard Green, who now lives in his parents' home. After studying photos of the suspect, Frank remembers seeing the back of Green's head at the baptism, yet decides not to tell Tom. The police have already put Green under surveillance, but they cannot arrest him or search the house without probable cause. Then Tom Black discovers information about Green on Frank's computer. He takes Frank's gun and confronts Green, threatening to kill him if he won't release his wife. But Green, pulling the gun closer to his own chest, tells Tom to go ahead and shoot. Tom then realizes (with the help of the police who arrive on the scene) that his own murderous rage might mean the death of his wife as well, since Green may be the only person who knows where she is, if she is still alive.

Both Black brothers, Tom and Frank, as Oedipal detectives, realize the tragic errors in their ways. (The "black" in their names also suggests a tragic dimension, contradicting the conventional color coding of heroes in melodramatic movies, especially classic Westerns.) Tom learns to grieve the loss of his wife without violating the Father's Law. Frank fights with his former police colleagues, because they will not let him help on the case as a relative of the victim. But he eventually finds a remedy to the Law's impotence,[96] through a sense of family significance in Green's satanic tattoo. When the police search Green's parents' house, a decayed body is found buried in the family's back yard — with the same tattoo that Richard Green wears on his arm burned into the roof of the victim's mouth. Frank realizes that Green may have hidden Helen in his parents' home. Then Frank and the police (along with Peter Watts) find Helen, barely alive, entombed in a newly constructed wall of the Green family's basement.

While rescuing his sister-in-law, Frank looks up the basement stairs and sees Richard Green's father. Frank's visionary gift reveals (through flashing images) that the evil madness he had glimpsed in Green's mind earlier in the episode — the killer in agony crawling out of a lake of fire and the mark of the devil tattooed on his arm — had its source in the human father, for whom the son procured his victims.[97] But as Detective Bletcher tells Frank at the end of the episode, the law cannot touch the evil father, because forensic evidence is lacking.[98] Thus, *Millennium* shows a Lacanian insight to its television viewers: the primal, obscene double of the Oedipal father cannot be eliminated. The Real devil of the death drive (in the human warping of natural instricts) is repressed by, yet returns through the flawed sacrifices demanded by the Symbolic law — in the violence of certain families and their offspring.

Another episode, "Weeds," focuses on patriarchal guilt like "Sacrament" (both were scripted by Frank Spotnitz), yet this time in an ordinary community of upper middle class families. Someone is kidnapping high school boys, not for perverse pleasure, but through psychotic identification with Christ's sacrifice — and a sense of Oedipal fate. The psychotic kidnapper uses a cattle prod to stun and capture the teenagers, selecting each one because of his father's particular sin. He then forces each victim to drink his (the kidnapper's) blood. This ritual cannibalism recreates in Real terms the Symbolic and Imaginary practice of drinking Christ's blood in many Christian churches. But as Frank explains, this psychotic savior offers his blood to "purify" each kidnapped son from a patriarchal legacy, which he reveals through specific clues left for the boy's father and for Frank, as lead detective in the case. Confetti made from dollar bills is left in one father's mailbox, and then his son's dead body is found, because the father refused to confess a sin involving money. Another father and his wife find the number "331" painted in blood on their kidnapped son's bed, corresponding to the hotel room where the father had an adulterous affair. When he confesses this to his wife, their son is returned alive.

Regarding a third victim, Frank receives a paint swatch that matches the enamel from a minivan involved in an earlier hit-and-run death of another teenage boy. Frank realizes that the kidnapper is telling him that the abducted boy's father is guilty of that earlier killing, a hidden guilt in the community — like that of Sophocles's Oedipus and the other two fathers here — which must be exposed and punished in order to prevent further suffering in the family and society. According to Frank, the kidnapper "believes that the sins of the father are visited on the son." When caught, the psychotic judge-*cum*-savior explains his Oedipal logic to Frank: "You've seen what they'll do to their children. They'll make them just like they are — sick. And these boys. Am I the only one? I saved him [the last victim]. I made his father pay the price for what he did." The serial kidnapper turns out to be a local resident who had organized a community meeting about the missing boys earlier in the episode. After his arrest, the father that he mentioned is found dead, having hung himself at home.

Even regular *Millennium* fans might be shocked by such an episode, as it brings the psychotic killer's logic and identity so close to home. Most episodes involve criminal psychopaths or satanic monsters. But the communal Christ-figure of "Weeds" is a local suburban resident whose psychotic break is triggered by his sense of hidden crimes in the community — greed, adultery, and hit-and-run manslaughter — as tragic flaws that will cause a legacy of suffering, from Oedipal fathers to sons. He creates an absurd imitation of Christ's sacrificial blood offering and a vengeful demand for further suffering and death as repayment for sins. But he also exposes the cruel logic of unconscious signifiers and family symptoms, inherited from one generation to the next, as original sins and Oedipal structures of the psyche. The *Millennium* viewer — identifying with the Imaginary egos onscreen and perceiving the Symbolic dimension of the fathers' hypocrisies through the kidnapper's clues — might experience a catharsis of the Real within the home theatre around the television screen. The viewer might see a related fate or similar repetition compulsions in his or her own family. And yet, a more complete vision of "the thing we fear the most," within our psyches and families, may be evoked by other episodes that show guilty mothers as well as fathers, and even the potential evil of children, in their sacrificial compulsions between heaven and hell.

Mother's Little Angels

Angels have been very popular in American mass culture in recent decades. Yet *Millennium* does not present its angels as biblical warriors (as in *Dogma*), nor as compassionate helpers in the spiritual journeys of humans (like in *City of Angels* or the CBS series, *Touched by an Angel*). *Millennium*'s angels are terrifying messengers — more like the character of Death in the medieval morality play, *Everyman*. Even when shown as traditional cherubs associated with the cuteness of children, these angels are then unmasked to reveal a terrible potential for violence in the primal foundations of family and society. Such unmasking of a violent *chora* in mothers and children, as well as fathers, pushes the usual television melodrama of purely evil villains toward a tragic complexity, thus meriting a detailed look at certain episodes.

"Covenant" begins with the apparent murder of three children and their mother by the father/husband. He has confessed to the brutal crime. His fingerprints have been found on the murder weapon, one of his woodworking tools from the garage. In fact, Sheriff William Garry has already been found guilty for the four murders by a local Utah jury. Frank Black has been called in simply to offer a psychic profile at the sentencing hearing, which the prosecutor hopes will result in the death penalty. The evidence is explained to Frank while photographs of the happy, smiling children are shown, along with the mother's dead face and the four bodies wrapped in bloody sheets. As the prosecutor, Calvin Smith, tells Frank: "The last thing those children saw before they saw the face of God was their father's face, the face of a murderer. I want William Garry to pay for that." But Frank discovers that the case is not so clear-cut. The godlike, yet monstrous face of a murderous father (in the prosecutor's imagination) is challenged when Frank finds a wooden angel that Garry carved for his wife's birthday, left in the garage with other woodworking tools. Etched on the bottom is the message: "love Bill."[99] Frank then sees a woodcut sign in Garry's garage: "If a man fails at home, he fails in life."

Frank plays a tape of Garry's confession while inspecting the family home. He finds other carved angels on a kitchen shelf. Then he walks downstairs while he hears Garry describe, in a very calm, emotionless voice: the stabbing of his oldest son, his wife, his daughter, and then his youngest son. "I'm grateful now he didn't wake up to see me looking down at him. He was only five years old. Once you start something like this you somehow have to finish."

Garry wants to be executed. According to his "religious beliefs," his attorney says, he must die for his sins; "his blood must also be shed" in order to be "forgiven by God." Garry's desire to be sacrificed, after his apparent sacrifice of his wife and children, puzzles Frank. The prosecutor asks Frank if he believes in the death penalty and then introduces him to Mrs. Garry's parents. They ask him, if he is "a good man," to make sure that William Garry "pays the price he ought to pay."

Frank interviews Garry himself. The television viewer hears Garry's voice describe the aftermath of the killing and sees him, with blood on his face, carrying the dead body of his youngest son (covered in a sheet). "There was no pain. No sorrow. Just a kind of dullness after the anger. And the need to finish it." Frank asks why he did it and Garry tells him that he "fantasized about it for a long time." He got to the point where he hated his wife and she hated him. "You know what it's like to scream in silence 365 days a year?" This dialogue involves the television viewer in imagining, through Artaudian cruelty, what Garry means — or distancing oneself, with Brechtian estrangement, from his hate, to consider the broader, social issue of the death penalty. Frank then introduces another, Artaudian and Brechtian twist by placing the carved, wooden angel on the table in front of Garry. Why would the father create such a birthday gift and then turn so violent? Is society right in taking melodramatic revenge on such a man, even if he wants the death penalty? Garry responds that the angel "is a lie."

Although the television viewer saw Garry holding his son's corpse in an apparent flashback, Frank questions this detail as he inspects the house again with the assistant coroner, Didi Higgens. Carpet fibers from the stairs were found on the body, so it was not carried but dragged down the stairs. The lie in Garry's confession and in the television flashback,[100] unmasked by the wooden angel and carpet evidence, is further clarified when Frank discusses with Didi the "defensive" wounds on Mrs. Garry's hand. As Frank and Didi later explain to the judge, the cuts on the hand show a pattern of her stabbing herself, instead of being attacked. (Didi demonstrates this to the judge, stabbing herself with a ruler that leaves dark lines on her own hand where the cuts were on Mrs. Garry's.) Frank also discovers that evidence in the Garry home had been tampered with, that Mrs. Garry was pregnant when she died, and that William Garry may have had a prior affair that greatly troubled his wife. He diagnoses Mr. Garry as "suffering from delusions of guilt, most likely a fixed false belief syndrome, brought on by severe depression." Garry passed a lie detector test, saying that he killed his family. But Frank explains that Garry "answered truthfully" in the polygraph; he said what he believed the killer felt — rage, anger, and hatred — because he felt so guilty for the killings himself. The evidence reveals, according to Frank, that the real killer felt "love and compassion," seeing the children not as victims but as "angels" (as shown in the episode's teaser).

Frank tells the judge it was the mother who killed the children and then killed herself in front of her husband. She stabbed herself with the knife that also cut her hand, after killing all three children with her husband's chisel. When Frank confronts Garry with this theory, the convicted man still insists that he is guilty. "I'm the only one who knows what happened that night, who's responsible. My blood must be shed at the moment of my death. To rob me of my salvation would be sentencing my soul to eternal damnation." Frank responds that he cannot let Garry die for something he did not do. Like a Lacanian analyst, the hero of *Millennium* maneuvers Garry and the legal system (and perhaps the television viewer) toward a tragic admission of a Real lack of retribution, toward a crossing of the fundamental, melodramatic fantasy that demands further human sacrifice for the sake of cosmic and social justice.

Garry continues to believe that he is the villain, guilty for the murders, like most in his community believe and many in the television audience may think during the initial parts of this episode. As Frank presents evidence to the contrary, even the judge has difficulty believing that a mother could kill her three children and then stab herself in the heart. Frank explains: "As mad as it sounds, she saw her children as angels and wanted to keep them that way." Frank then narrates (and the television viewer watches) an alternate sequence of events. Garry enters the house from the garage and finds his dead son William at the bottom of the stairs. He discovers his youngest son Gabe also dead and wrapped in a sheet in the hallway. Then he sees his daughter dead in her bedroom. In this revised flashback, showing Frank's account of the killings,[101] Garry calls to his wife as he finds the dead children: "Dolores" (a name that means sorrow). He then hears a noise in the kitchen and finds her, with blood on her face, stabbing herself. She collapses in his arms.

After the sentencing hearing, Deputy Kevin Reilly, a close friend of Sheriff Garry, who had helped him to change the evidence, confesses to Frank more of the crime's details and motives. "She told Bill it was his fault. She said he made her kill the children. And then she stabbed herself one last time.... She said she couldn't bear the thought of living in a world of adulterers. Men like him." Frank responds: "Instead of looking for the truth, the people of this town were looking for whatever would put the pain and the blood behind them." Frank asks Deputy Reilly to tell the truth now, before the judge decrees the death penalty. But Reilly still wants to keep silent so as not to "betray" his friend. The episode leaves the ending open, not showing what decision Reilly makes, nor what the court decides. This leaves a Brechtian opening for the television audience to complete the show by reconsidering the demand for blood sacrifice in many religions and jurisdictions, including their own. Yet this episode also offers an Artaudian cruelty that may be even more challenging: the suffering of three children caught between two violent parents in a broken marriage—and the suffering of those parents themselves.

Garry tries to save himself with a suicidal lie—to find his "salvation," even if that means tricking the law, as a former lawman, and hiding his wife's crime. With the wooden angel and other evidence, Frank unmasks Garry's melodramatic guilt: his "fixed false belief" in himself as the murderous villain, due to his adultery, as his wife's dying words encouraged him to think. Frank also reveals the community's violent vengeance in making Garry pay for the murders. Crossing the abyss of the mother's abjection, Frank exposes a horrifying tragedy in the Garry family and in the surrounding town, beyond melodramatic villainy and revenge. The television audience is asked to consider not only the judge's question: "Why would a mother kill her children?" But also, how could she kill them with "love and compassion"?[102]

Like Euripides's Medea, Mrs. Garry killed her children out of vengeance against her adulterous husband. Unlike Medea, she does not have a voice in this television drama, except as recollected through men (Reilly's account of what Garry said he heard her say). But Frank reveals a tragic *chora* of Medea-like abjection in the "love and compassion" of the killings, in Mrs.

Garry's love for angels, and in her hidden pregnancy. Dolores Garry was a perverse mother who loved her children to death, making them into objects of her desire as "angels." She kept them with her in death, away from their father's adulterous world, which they were already growing toward as they separated from her body and her loving care. She drew them back into her religious fantasy and physical womb, as her angels and babies, by stopping their progress away from her. She also, by killing herself, kept the new angel in her womb from leaving it.

Frank's diagnosis of her as the true murderer, killing with maternal love, opens a tragic mystery in this drama. Through this Lacanian breach,[103] William Garry, his community, and the television audience might reach a cathartic clarification of the sacrifices already made and of the demand for a further death sentence. It is much easier to blame the father in this case as an inhuman (Caliban-like) devil and to justify a vengeful, legal triumph through the bloodshed of a firing squad. But a more tragic catharsis is offered through Frank's discovery of evil within the good, of the mother's love and compassion as possessive and sacrificial.

It is very difficulty for parents to let go of their children, especially for the mother, who carries each child inside her body for nine months, who then gives birth in excruciating agony, yet finds a bundle of joy in this product of her body. The view of children as angelic, and the depiction of angels as cherubic, masks an abject passion in maternal love — the potential for violent control with children as objects of the (m)Other's desire. As Lacan puts it: "The mother's role is her desire.... The mother is a big crocodile, and you find yourself in her mouth" (qtd. in Fink, *Lacanian* 56, from Lacan's *Seminar XVII*). As Fink points out, this monstrous role of the mother's desire comes from both directions: "both the child's desire for the mother and the mother's desire per se" (57). In the Garry household, with both parents having strict views of marital fidelity, the father was flawed — at home and at life — as the sign said in his garage. The Name of the Father, as substitute for the mother's desire, no longer protected the children "from a potentially dangerous dyadic situation." Crocodile jaws closed upon them, through the mother's desire for her children as angels.

Witch-Hunt in the Garden

Another episode, shown early in the second season, focused like "Covenant" on child abuse, involved angels, and explored the temptation of violent revenge — but with different tragic and melodramatic twists. In "Monster," Frank himself is accused of violence toward his daughter, Jordan. He is shown early in the show losing his temper with Jordan in a department store. Catherine takes Jordan to a dentist after she spits blood from her mouth while brushing her teeth. The dentist suggests the little girl may have been severely disciplined. Catherine, who is now separated from Frank after seeing his brutal violence in saving her from a kidnapper, begins to suspect the father — especially when Jordan admits that he got angry with her in the store. Frank learns that he is being investigated for a possible assault on his daughter, while he investigates an alleged abuse of children at a daycare center in Arkansas.

Prior to that, Frank meets a new Millennium Group colleague, Lara Means (Kristen Cloke), when they realize that they were each sent to Arkansas separately by the Group — and that they both have psychic visions, Lara of angels and Frank of devils. Rumors have circulated throughout the small Arkansas town about children being abused by Miss Penny Plot, even though she has run the daycare center for over 30 years. Public fears intensify when a bite mark is discovered on the back of a boy at the center. He is the son of a policeman, who finds it difficult to believe Miss Penny is guilty, since she also took care of him as a child. Then another child, Jason, dies at the daycare. He suddenly stops breathing while Frank and Lara are visiting — after Frank sees the Gehenna devil (which also appears in other episodes) and Lara sees a luminous angel. The coroner defines the cause of death as an acute asthma attack. But when the children from the center are interviewed at the police station by Frank and Lara, one of them, Danielle Barbakow, says she saw Miss Penny get angry at Jason for wetting his pants. After Miss Penny took him into the bathroom, Danielle heard "a smack," she says, with Jason crying and wheezing, and then heard him scream: "No, don't touch me there!" Danielle also says she saw Miss Penny, a few days later, talk to Jason on the playground. "She had a doll and I heard her say: 'If you ever tell....' And then she twisted the arm and broke the head [of the doll] and threw it into the bushes."

Miss Penny is arrested and appears agitated in the interrogation room. She admits that she spanks the children "when they're very bad," but she has done that for 36 years, she says. Then she jumps up and yells at the cop, the father of the boy she is accused of biting, who is videotaping the interview: "I wiped your ass when you were five and your brother's. How dare you?" Even as Miss Penny claims she is "a good person," the show encourages the audience to feel she may be a monstrous, maternal figure like Mrs. Garry. At a town meeting, angry parents demand the closure of Miss Penny's daycare center, especially since she has been released on bail. But the cop rises and tells his community that although his own son has "a mark" on him (and will not say who did it), he cannot believe Miss Penny is responsible. "In my heart I know she's a good person." Yet the rest of the community starts chanting: "We believe the children" (though only one child had been shown talking about Miss Penny as violent). A mob of angry parents, still chanting that slogan, paints it in red letters on the white picket fence around Miss Penny's daycare center — and sets fire to the fence.

But Frank and Lara feel that Danielle is responsible for the preschool violence. Lara, a psychologist, suggests the idea to Frank, but still finds it difficult to believe herself. "How can anyone know about, let alone commit murder, at five years old?" She also asks broader questions about social metaphysics. "Has our culture bred this possibility? Is it violence on television? Is she Damien? Is this girl an evolutionary mutation? And are there more of these kids, these people coming?" This episode draws on audience fears of child monsters, inspired by news accounts of real violence by young people in schools and homes, and by fictional devils and alien creatures onscreen — like those seen in other *Millennium* episodes as well as in movies. Frank historicizes this fear in describing how he experiences evil. "It feels like a force, like gravity, like the wind. It has blown across Cambodia, been a cyclone in Nazi Germany. It gusts throughout Los Angeles." He calls the little girl, Danielle, "a pre-storm breeze of an approaching hurricane."

At this point, the series drew a melodramatic picture of evil in the twentieth century, as a supernatural force that would hit the world with even greater villainy in the year 2000. Yet it also showed the tragic delusions of a small town mob, seeing evil in a woman they had trusted for 36 years. That force of mob violence was also the wind of evil that Frank described in Cambodia, Nazi Germany, and Los Angeles. *Millennium* showed, to those willing to see the tragedy behind the melodramatic vision, how evil might arise through the fear of evil and the scapegoating of others as devilish.

American psychoanalyst Joseph H. Smith has theorized the source of human evil in the child's earliest fears of inner danger. "The cry is initially a crying against inner danger and only later a call for the mother" (35). Smith finds a primal violence in the infant, prior to Lacan's mirror stage: "the subject is always already a subject of aggression, already divided and dispersed, and in that sense alienated from itself" (37).[104] Smith notes that guilty feelings can be a defense against inner danger: "being guilty, whether directly suffered or denied, is more tolerable than danger and loss." (This would help to explain William Garry's clinging to his guilty fantasy.) Smith then describes the social "bias and prejudice in favor of one's own kind ... [as] replay[ing] the function of the mirror image," through the collective masking of lack and focusing of primal aggression toward other groups or persons (39). In cases of "radical evil" (depicted in many *Millennium* episodes), primitive guilt and inner danger become projected on others. Thus, the primal fear of evil and aggression can turn into a radical evil, in violent acts by certain humans against others, especially through "crowd psychology" (44).

Millennium touches on the primal fear of evil in its mass audience. But episodes like "Monster" and "Covenant" also reveal the tragedy of mob and family violence. This may help some spectators to "own" the original evil (or fear of evil) in themselves, beyond primitive guilt, clarifying their personal potential for violence, rather than justifying personal or group vengeance as a defense mechanism, as a preemptive strike or retaliation against the other's evil. According to Smith: "since no one ever completely and forever overcomes primitive guilt, everyone, at times, invokes such defenses; in dire circumstances, anyone might be capable of evil ..." (44).

When Frank and Lara visit Danielle's home, she is staring and smiling at the television set, as it shows a monstrous spider and offers melodramatically stirring music, suggesting that its images of evil may be absorbed by Danielle (like Lara wondered). Frank then talks with Danielle in her room, while Lara talks with the mother, who says the girl has "always been an old soul." When Frank asks Danielle if she liked Jason (the boy who died), she turns to face him and Frank again sees flashes of the Gehenna devil. Danielle screams: "No, no. Get away from me. Don't touch me. Get out." Frank and Lara leave the house. Then the mother hears a thud, goes upstairs, and finds Danielle fallen to the floor, her face cut and bruised. A pewter angel is shown on a low shelf in the girl's bedroom before and after her mother finds her on the floor. Here the gray, metal angel seems to reflect the evil in the child, or the force Frank described as a fierce wind and saw within the child as a devil.

In the next scene, Lara tells Frank about her occasional, frightening visions of an angel as "portent of evil." She tells him that in the Old Testament "angels are everywhere" and that today "more people believe in and see angels than in any other time in history." Lara's terrifying sensation of an angel as messenger of evil and destruction parallels Frank's sense of evil as a gust of wind or hurricane. It also relates to the biblical angels and devils of death and separation, especially those who force Adam and Eve to leave the Garden of Eden — separating them (like the Oedipal superego) from mythic, immortal bliss with the (m)Other in paradise — as shown in the medieval *Adam* and Wakefield Cycle. Partial, erogenous pleasure replaces the lost symbiosis of Paradise and personal guilt masks the primal danger of original evil, with the Serpent as satanic tempter.[105]

Frank and Lara are likewise separated, just when they start to commune, as Group members, through their shared psychic knowledge. Agents of the Law break into the hotel room and arrest Frank for hitting Danielle and breaking her jaw. Frank is also informed that he is under investigation in Seattle for assaulting his daughter. Eventually, however, Lara becomes like a Lacanian analyst for the community (and Danielle's parents), encouraging a cathartic clarification of the witch-hunt. She compares an infrared photo of Danielle's severely bruised cheek with a pewter angel, which she finds in the girl's bedroom. The mark of the bruise exactly matches the angel's metal wing. With this evidence Lara convinces Danielle's parents to speak to Frank, because the little girl might have injured herself with the angel, to pin guilt on an innocent man.

At the police station, Danielle's father calls Frank a "monster." Frank responds: "You're seeing what you want to see, what you can see…. It's like when you look in a mirror." Frank appeals to the girl's father to see the monster (or Lacanian Real) not only behind his own face in the mirror, but also in the scapegoating of Miss Penny and himself. Mr. Barbakow scoffs at this; so Frank turns to Mrs. Barbakow. He tells her about the joy he felt at his own daughter's birth. "I realized I'd forgotten that I was born, thought I'd manufactured myself. And the gift she gave me was from that day on I could look at every man and see the child in them." He then appeals to her for the truth, "because there's a witch-hunt in progress over here and they're not going to understand." Her husband continues to scoff, but Mrs. Barbakow suddenly offers a tragic statement about the mystery of original evil within a child's angelic joy and apparent innocence. She says she always believed she would see the "gift" Frank described in her child's eyes, but she had lied to herself. "I only see something that I don't understand."

Then Mrs. Barbakow tells the authorities that she knows Frank did not hurt her daughter, because she "heard her do it to herself." Frank is reunited with his own daughter at his release, because Catherine flew to Arkansas with Jordan to help Frank. But the episode ends with a more tragic reminder, after this melodramatic triumph of Frank Black against the witch-hunt and wicked girl. The cop and his family (including the boy with the bite mark) are shown leaving the town where he grew up, a "For Sale" sign on their home — as the voice of Miss Penny is heard reading the conclusion to the story of the three little pigs and the big, bad wolf. Danielle then appears at a dinner table in a new family, her hands folded to say "Grace." The Group ouroboros appears on a computer screen behind her, signifying its mystical role in handling her evil.

This episode allows the television viewer to see a villainous devil in the violent, lying, malicious child (like the possessed Regan in *The Exorcist* or Caliban in Prospero's view). She not only injured other preschoolers and herself, sadistically and masochistically, but also ignited a witch-hunt in her hometown. Yet the devil in Frank's psychic view of Danielle, the "monster" whom her father saw in Frank, the "something I don't understand" that her mother saw in her, and the metal angel Danielle used to break her own jaw, along with the town's vehement turn against Miss Penny, all point to a more tragic sense of evil shared by each of them, and by the television audience as well. The extreme vulnerability that each person experiences as a child, and the necessary alienation from the (m)Other, may turn from an inner danger and fear of evil into evil in action. This is exemplified, too, in the sacrificial rage of the mob attacking Miss Penny's daycare, showing the hypocrisy of melodramatic vengeance. But the episode offers the audience a more tragic awareness: the owning of potential evil in oneself, so as not to project it and enact it on others. For the tremendous success of our species in evolving bigger brains, through language, art, and technology, has a tragic flaw. The experience of primal aggression in our lack of natural being, through the prematurity of human birth, the vulnerability of infants, and the pruning of neural networks by particular family situations, leads to a fear of evil in others, which may produce a radical evil performed by individuals or groups, as if possessed by ancestral devils.

Angels as Masks of the Death Drive

Good and evil ghosts haunt various *Millennium* episodes. For example, in "The Curse of Frank Black" (a Morgan and Wong Halloween episode in the second season), the show's hero meets the ghost of a man he remembers from his childhood, who tells him to leave the Millennium Group because the devil is going to win anyway and will then reward him. In the third season, after Catherine's death, she also visits Frank as a ghost. However, the ghost story that reveals the most about Frank's childhood and his current family tragedy, regarding angels as well as devils, aired in the second season at Christmas time.

As "Midnight of the Century" begins, a flashback shows the five-year-old Frank Black drawing an angel. Then, in the present day, the adult Frank receives a Christmas card with an angel on front. Inside it states: "It's the Midnight of the Century." The card's envelope arrived in the mail with a typed address and no return, stamped by the Seattle Post Office with the date: Dec. 24, 1946. As the viewer learns later in the episode, this is the day Frank's mother died, after helping him to draw the angel. For him, every Christmas since then was about birth and death, about the nativity and yet also his mother's death on a Christmas Eve midway through the century. The millennial year 2000 that the hero and his audience approached together, during the run of the series, was shown in many other episodes to have grand, apocalyptic meanings. But here it becomes very personal — with no serial killer except the death drive itself.

When Frank goes shopping for a Christmas gift for his daughter, he flashes back to a scene (in black and white) of himself as a small boy, toy shopping with his mother, as her voice quotes *Macbeth*: "tomorrow and tomorrow and tomorrow." Frank also glimpses a strange young man, reflected in the present-day shop window, who whispers the same line, then disappears. This stranger turns up again outside the small theatre of a church hall, where Frank is attending his daughter's Christmas play. Frank sees the shadow of a figure behind a window in the back of the auditorium and again hears the line from *Macbeth*. So he leaves the play and goes outside into the church's graveyard. There he meets the stranger, who tells Frank about "fetches." These are ghosts, or rather "the souls of those destined to die in the coming year," who walk the churchyard on Christmas Eve, "in search of those who are soon to be their companions." The stranger also asks Frank: "Why put off till tomorrow what should be done today? It is, after all, the midnight of the century." He then disappears from the churchyard. Was this ghostly messenger threatening Frank, when he also said: "There's no telling whose face you might see [as a fetch]"? Or simply giving him a common-sense warning about not procrastinating?

After the Christmas play, Catherine shows Frank the drawing of an angel that Jordan said she made with the help of her grandmother, his deceased mother, which is identical to the one he drew with her as a boy. Frank flashes back (in black and white) to his father a half century before, pointing at the viewer and saying: "You're just a kid. You didn't see nothin.' It's all in your head." Frank reminds Catherine that their daughter has a psychic "gift" like his. "You can't suppress it," he says. But she responds: "Your gift gave you a nervous breakdown. Your gift makes you see horrible images. It's turned you away from your family, from your daughter." She also tells him that his "gift" might be lying to him.

Frank continues his hunt at the toy store, after buying two gifts that he then discovered his six-year-old daughter already had. "I have less than 12 hours," he tells the store clerks. They send him "upstairs," because he still insists on a "traditional" gift. There he finds an angel doll on an upper shelf, saying to himself: "She's gotta like that." It looks new as he brings it down, but when he turns it over to see its wings and then looks at its face again, it appears old and grotesque. A close-up of the doll's face shows its eyelids as blue and its skin as darker. Frank drops it on the floor. Next, he is at home looking through memorabilia in a box from the closet. He finds his mother's obituary and the angel drawing he made with her on her last day in this world. A flashback memory shows her hand over his, drawing the outline of the angel. At the bottom of the drawing, which Frank also holds in the present, is written: "Frank drew this on December 24, 1946."

Various personal, apocalyptic twists are given so far in this episode: the Christmas card mailed in 1946, the stranger's warning about fetches and procrastination, Jordan's drawing of an angel with her dead grandmother, Catherine's concern about the "gift" that father and daughter share, and Frank's own hallucination with the angel doll (like his psychic profiling flashes in other episodes), plus his flashback memories. All of these draw Frank, and the sympathetic television viewer, toward facing the death drive in the family and in the communal experience of Christmas joy — especially in the gift buying pressures shown in this episode and rehearsed in the commercials interrupting it.[106] Through Jordan's inheritance of her father's psychic gift,[107] and through the gift he found in the upper floor of the toy store, Frank starts to unmask a tragic conflict within his own family and history. He begins to excavate a painful sense of the "angel" as master signifier, masking an abject *chora*: not only the loss of his mother when he was five, but also her psychic gift and death-drive *jouissance*, with his connection to that visionary abjection being negated by his father. As Frank struggles to remember what his father said he did not see, the television viewer follows him in crossing a fundamental fantasy that estranged him from his father for half a century.

Lara visits Frank at his request and looks at both Jordan's angel drawing and his own, drawn with his mother on a Christmas Eve 51 years ago, the same day she died. Frank thinks that his daughter's gift may be more like Lara's than his own. Lara tells Frank about her first experience of an angel, as an uncanny "presence," when her father brought a business associate home and she had the premonition, as a little girl, that the man would die soon. She saw "an intense, beautiful light" behind the man; then told her mother that he was going to die. "He had a coronary in his hotel room that afternoon." Lara also tells Frank that the angel's presence "gets more intense as I get older." She then describes his appearance as an overwhelming, painful pleasure — in Lacanian terms, a full-body, death-drive, Other *jouissance* (a feminine, mystical alternative to partial, phallic *jouissance*).[108] "When he appears all my senses are heightened. Everything is clear and uncluttered in my mind and my heart. I feel like nothing can stand in my way. But I would have done anything, anything not to have this in my life." Lara then reminds Frank that biblical angels are messengers and that there is "only one place" for him to discern the meaning of his angel drawing. Another black and white flashback is shown, this time with the father in another chair, as little Frank sits on his mother's lap drawing with her. Then, in the present, Frank's Christmas tree falls over, with a tinkling bell sound. Lara remarks, parodying the idealistic view of angels in the classic film, *It's a Wonderful Life*: "Oh, great. Another angel just got its wings."

Later, Frank goes to his father's house and hallucinates a lamp lighter at the streetlight outside, before seeing that it actually has an electric bulb. When he climbs the steps to his father's front door, there is a black and white flashback of his mother's bedroom door, inside the same house, with the drawing of an angel on it, closing after he climbs the steps to it, shutting him out as a little boy. Then the father's voice (Darren McGavin) is heard, saying to a barking dog: "Quiet. You'll raise the dead."[109]

When the father, Henry, lets him in, Frank has another flashback of following his mother upstairs as she leaves the living room, but she stops half way up the stairs and touches his head. He sits on the steps. Thus, Frank's flashback memories appear to be contradictory: he both sits on the steps as she goes up to her room and he follows her to the door, which she then shuts in his face. But this ambiguity shows a more honest way of rendering memory than is usually done in film and television flashbacks. Rather than a complete and consistent view of what really happened, Frank's black and white flashbacks, like his psychic profiling flashes in color, show fragments of what might have been — bits of memory reconceived each time they are envisioned.[110]

Henry, who apparently lives alone, shows that he has been estranged from his son and granddaughter for a long time. He asks if Jordan received his photo and if she knew him. He then asks Frank to sit down: "We got a lot of years to talk over." Instead, the Oedipal detective demands "the key," telling his father that he needs "to see her room." Henry barely masks his disappointment, saying he thought Frank came to visit him at Christmas and show him pictures of his granddaughter. Frank heads up the stairs and Henry calls after him: "You know you're not supposed to go up there. You're not allowed up there." But Frank finds the key on the doorframe ledge, unlocks the door, and enters the room alone.

He sees angel drawings covering every wall. As he touches a drawing on the wall, Frank has another flashback of his mother. This time she opens some wrapping paper and finds a small statue of an angel inside, as the young Frank looks on. She says to his father: "They're not coming back, Henry. They're gone forever." His father takes the angel from her and places it on a glass stand. His mother then rises from her chair to go upstairs. In the present, Henry talks to Frank about the angel drawings when he comes downstairs to the living room. "What you just saw pushed us away from each other, even though nothing made us sadder." Frank's response reveals a fundamental fantasy about his parents: "Sad enough to let her die alone?"

Frank's memory of his mother leaving him on the steps or behind the door of her room, to die in there "alone," after drawing the angel with him — and of his father telling him that he did not "see" like her and that he was "not allowed" in her room — form a fantasy that helped Frank make partial sense of his traumatic loss for 51 years, but at the price of estrangement from his father. Now he is haunted by his psychic gift like his mother, which he sees even more in Jordan. Yet his gift is a connection with her behind the door, against his father's Oedipal "No."[111]

The angel, as an image of his mother's desire and of her Other *jouissance*, is also a barrier between Frank and his father. Frank must traverse his fantasy and split this signifier of the Other's lack[112] to reach his father once again before it is too late, before his father joins the fetches at midnight on Christmas Eve. Frank is haunted by two ghosts now: his dead mother and his living father. Yet Henry holds a "key" that can help Frank open another door beyond the angel drawings and death room. The little statue of the angel, to which Henry also clings as a master signifier and point of fantasy contact with his wife, will inspire a catharsis in Frank of his psychic gift and painful memories — perhaps unlocking similar signifiers and Christmas ghosts for sympathetic viewers.

First, Henry points to an old flag in his window that has a star on it. He explains that it hangs there in memory of Frank's Uncle Joe, his mother's brother, who died in the Allied invasion of Normandy. Henry describes how Frank's mother, Linda, knew on June 5, 1944, that D-Day had begun, even though it was still a "big secret," as yet unannounced. "She burst into tears, huge weeping tears," and ran upstairs to the room, knowing also that her brother "died in the first wave, on the beachhead at Normandy." After that night, the father says, "She took to that room. She moved out of our bedroom, out of my bed. And she started drawing those little angels." Linda also told Henry about her "vision" of Joe's impending death, which "scared me to death," the father says, because it seemed like "a ticket to the nut house." He also saw a similar "ability" in little Frank. "It was taking my family away."

According to Henry, Linda told him that she had "seen" she was about to die. "Well, that was the last straw," Henry says to Frank. He repeats his words of the past: "You don't see nothin', Linda" — as Frank's earlier flashback of his father, at a younger age, pointing at the television viewer is shown again. Henry then tells Frank: "I said that to you, too." This clarifies Frank's memories of his father's negation of the psychic gift, in himself and his mother. Frank can now see beyond the fantasy that his father simply prevented him from joining her in the bedroom and that he made her "die alone." It was her own desire to be alone in the room with the angel drawings, knowing that she would die that night. She drew the angel with little Frank, before leaving, as her only way to express this. Her Other *jouissance* of angels and psychic visions gave her a feminine, mystical joy and pain, apart from the father's bed. But the cause of her solitude and suffering was not the father; it was the "thing we fear the most," the death-drive visions in Frank as well, which he struggled to subjectify in his own way.

Henry explains the angel statue that is still on the glass stand (as in Frank's memory). It was a gift he gave to Linda on her last Christmas Eve, as a way of reconnecting with her. "She said that she would move the figurine

6. Millennium *in the Home Theatre*

to show that she was waiting for me on the other side," where they would someday be together and both see real angels. Henry describes (somewhat differently from Frank's earlier flashbacks) how Linda then kissed him goodbye, hugged and kissed her two boys, and then went upstairs. "With a most serene smile on her face, she walked up to that room alone, as she wanted." He tells his son: "It was the only way it could have been, Frank." But Linda's promise to act as a ghost was not fulfilled. "And that figurine didn't move an inch in 51 years."

Frank then shows his father the angel drawing made by Jordan. Henry suddenly takes the angel figurine off the glass stand, in a shockingly un-ritual gesture (or Brechtian gest), and gives it to his son. He tells Frank that he got the "sign" after all — but not in Jordan's drawing, as Frank thinks. Instead, Henry sees Linda's presence in his son's visit after so many years away: "She brought you to me." Frank gives his father a photo of Jordan, hugs him, and asks: "Why did it take so long?" As Frank leaves, Henry puts the photo of his granddaughter against the glass stand where the angel had stood for half a century as a promised fantasy, master signifier, and dead end.

The episode ends with Frank giving the angel figurine to Jordan on the steps of the church at night — the special Christmas gift, which he had searched for and failed to find in the toy store. Without her father explaining, she says, "Grandma wants me to have it." Then she and Frank both see fetches walking toward the church. Among the group of ghosts-to-be, Frank sees his father. Thus, the episode offers its viewers a sentimental conclusion to a lovely Christmas story of family reconciliation, confirming the hope of life after death with loved ones. Yet it also reveals the other side of that hope: the death drive in the characters and the spectator, expressed through criminal violence in other episodes, but here as the violence of a symbol. The angel was a "sign" not only of impending death, but also of the Other's *jouissance*, of a painful, ecstatic joy beyond the patriarchal order (in Lara and Linda). It was a signifier blocking Frank from his father for half a century — until he faced it, split it, and crossed its fantasy in his own way.

Millennium offers various possible viewer identifications here: with the women or the men, with the mother or granddaughter, with the son or the fathers (and grandfather). Whichever path the television spectator takes during this episode, through his or her personal memories and transferential visions, a tragic loss may be addressed, beyond the angel signifier and mother's abjection. For the ghosts in the audience are also fetches,[113] visiting the television shrine at the midnight of the century — remembering lost loved ones and approaching death in their own ways, on Christmas Eve.

Profiling the Sacrificial Audience

Millennium's Christmas episode in its second season revealed the family heritage in Frank's "gift" of profiling serial killers. But this supernatural gift in the series plot is also a tragic flaw: a painful, repetitive, Other *jouissance*. It is Frank's psychotic trait, connecting him with his mother's ghost, yet also turning him "away from his family," as his wife tells him, like it had pushed his mother and father apart a half century before. While his parents may find each other again in the afterlife, and Frank may have reconciled with his father just in time, the tragedy of his failed marriage continues—until he is reunited with Catherine at the end of the second season, only to lose her again as she dies of the plague. Although Frank crossed a fundamental fantasy about his father forcing his mother to die alone, he then shifts his Oedipal anger and paranoia toward the patriarchal cult of the Millennium Group in the show's third season, after Catherine dies alone. And yet, through the shifting melodrama of the series plot—with Frank's battle against various serial killers and demons, against his father, and against the Group itself as a collective villain—*Millennium* shows both Artaudian and Brechtian possibilities for tragic (or tragicomic) catharsis. This also reflects the consolidation of good and evil stereotypes through the mass media, and their possible further evolution in the neural networks of the audience.

While the Group becomes evil and villainous in Frank's view, his former friend Peter Watts maintains his loyalty to the Group even when it kills off various rivals. A spectator's sympathy with Watts as an ally to Frank, who has helped him to solve cases and save his family members in past episodes, is thus complicated in a tragic direction (even as the third season turns more melodramatic): showing the danger of a good person becoming evil, by collaborating with cult violence through the collective, apocalyptic fear of evil. This Brechtian twist in the series plot distances *Millennium* fans from Watts and evokes a more critical view of Frank Black as he turns against his friend. But it also parallels the tragic awareness that might be aroused in viewers of earlier episodes, with the Artaudian cruelty that Frank and other characters discover in themselves and their families—a cathartic clarification of the potential for violence, even in good people who fight against evil.

Not all *Millennium* episodes followed the series formula of Frank Black solving a serial killer crime or supernatural mystery each week. Some featured other characters with cosmic ties. In "Anamnesis," for example, Lara and Catherine work together (without Frank) to investigate five teen-age girls experiencing religious visions of the Virgin Mary or of Mary Magdalene. They are unable to stop an assassination attempt on the group's leader, though someone else is killed instead—with Lara explaining that there had to be a "sacrifice" to set certain events in motion. In another episode, "Somehow, Satan Got Behind Me," four old men talking in a donut shop transform into grotesque devils, discussing their projects to tempt human souls. They also recall, with surprise, how Frank Black seemed to see their devilish "essence" inside the old man disguises.

But the most insightful *Millennium* shows, in a Brechtian sense, were those that followed the usual formula of Frank's melodramatic battle against various serial killers, yet reflected certain aspects of the television viewer's sacrificial submission to the screen and idolization of its good and evil characters. In "Loin Like a Hunting Flame," Frank tracks a serial killer who drugs his victims and leaves their dead bodies in ideal fantasy scenes as lovers (including a scene that mimics the Garden of Eden). Frank realizes that the killer is trying to capture his victims in a perfect moment of lovemaking, on video while they are still alive and then again in a death-scene fantasy. This episode of *Millennium* might evoke a cruel Artaudian intimacy, and then Brechtian distancing of the spectator, provoking personal sympathies and yet critical thoughts about screen fantasies and society's use of ideal bodies as commercial lures.

Another episode, early in *Millennium*'s first season, "522666," began by showing a quote from Jean-Paul Sartre onscreen: "I am responsible for everything ... except my very responsibility...."[114] This episode shows a domestic terrorist, a serial bomber, who masturbates near the bombsite as he waits for the explosion. When Frank follows the clues and contacts him by phone, the killer calls himself a "star" and says he will make the detective a star, too. But Frank knows that the bomber's thrills and fame must increase with each violent event. Frank himself is almost killed in another bomb blast, but he is saved by a worker in the building. That savior becomes famous on television, although he is actually the serial bomber. Later, he tells Frank that he wants everyone to know his name. He controls the final scene with the fake threat of an apparent bomb detonation device, so that the police will shoot him unnecessarily and make him an even bigger star after his death. This episode ironically reflects the media's melodramatic fetishizing of villains and heroes, showing the bomber not only as a vulgar onanist, but also as Frank's savior and as a martyr to the mass audience. This irony may increase the tragic awareness (and Brechtian critique) for television spectators who share the killer's ordinary trauma of anonymity, the alienation of being no one who counts in the postmodern world, unless one is known, by name or image, onscreen.

A Halloween episode in *Millennium*'s final season presented further ironic gests of ideal, erotic death scenes, with the serial killer, victim, and detective as stars, in a Brechtian, tragicomic critique of the show itself. "... 13 Years

Later" begins with a beautiful actress and her director in the midst of sexual play, squirting stage blood on each other. While she takes a shower in the hotel room, the director is killed by a mysterious intruder. The actress is then stabbed in the shower, *Psycho*-like, her real blood mixing with the stage blood she was washing off. Frank is also shown early in this episode lecturing on a similar crime at the FBI Academy. When he visits the current crime scene, his psychic profiling vision reveals faces with star-shaped makeup like the performers in the rock band, "Kiss." He then visits the movie set where the star actress and director had been working before they were murdered. A new director is continuing the film, making "magic" about a famous Frank Black case from 1985. The character of Frank in the scene being filmed boasts that he invented profiling. When Frank tells the director this is not true, the filmmaker replies: "It *is* true, just improved ... [for] all those people you fly over on the way to New York. They want to be entertained." Frank watches the moviemakers replicate a murder scene from his past casework. It is "improved," however, by putting a nude woman in a swimming pool for the killing. Then a finger from one of the real victims is found in the new director's sandwich and Frank shuts down the movie set, realizing that the killer may be someone involved in making the film.

Frank and his FBI colleague, Emma Hollis, watch slasher movies on videotape in his hotel room, because Frank thinks the killer is recreating film murders "to drive me insane for the third time in my life."[115] They hear a scream on the roof of the hotel and find a female victim hanging upside-down with Kiss-like star makeup on her face. (The famous rock band also plays in the show's finale.) Later, Frank discovers that movies listed in a current *TV Guide* match the M.O. of the real killings. He mentions that the serial killer Ted Bundy described murder as being "like watching himself in a movie." When another killing occurs, writing on the wall of the movie set is found, graffiti written in blood from the new victim: "MOVIES KILL." When Frank at last catches the killer, he turns out to be the actor who was playing Frank in the movie. He says, "I wanted to be like you, Frank. See what the killer sees ... [but I] found one important thing about killing ... I liked it." In the end this character (Bianco) lectures to fellow inmates in prison, like Frank Black earlier in the show with students at the FBI Academy.

This episode is not only entertaining, with reflective complications added to the typical *Millennium* plot; it also provokes the audience toward a Brechtian awareness of a social problem. Melodramatic violence sells well in film and television, as an easy thrill and superficial catharsis. But it can set precedents for real killers to follow or exceed — as they model themselves after the star killers and detectives seen onscreen. Another episode, from the show's second season, presented a more direct, Artaudian sense of the mass media's cruel influences, bringing the audience inside a violent, psychotic mind. "19:19" begins with a man living in a trailer, listening to different news broadcasts on many radios and television sets simultaneously. He makes furious notes on his linoleum floor with a felt tip marker, adding further details to a complex picture of interconnecting words in black and red lines, which focus on a central empty spot, where he writes: "ME." He then steps outside the trailer, holding his Bible, and sees a nuclear explosion.

After this opening teaser, Frank Black and Peter Watts investigate a missing school bus in Broken Bow, Oklahoma. The bus is found and pulled out of a river, but the children and driver are still missing. Frank touches a child's backpack and receives a vision just like the nuclear apocalypse in the teaser, sharing a glimpse (with the television audience) of the psychotic man's hallucination. Meanwhile, that man and his accomplice, the kidnappers of the school children, discover that one child who usually takes the bus is missing. The psychotic kidnapper says to the captured bus driver, who will not reveal the name of the missing child, "Worthy is the Lamb who is slain," and asks if he is "worthy"—threatening him with a large wire cutter. Then the bus driver is heard screaming offscreen.

Through various clues, Frank catches this kidnapper trying to take the missing child from her home. He realizes that Matthew Prine sees himself as being on a mission in the apocalyptic war between good and evil—like Frank, Watts, and the Group. But Prine sees his mission through his own visions of Revelation, items in the news media, and the complex diagram that he drew, which Frank also discovers. As he looks at the diagram on the floor of Prine's trailer home, with numerous television sets and radios chattering on different channels, Frank asks Watts for the help of Lara Means in this case: "We can use her friends in high places. We're sitting in someone's brain."

Frank and Lara reproduce the diagram from the trailer home on the floor of Prine's jail cell. Prine traces the word "ME" with his fingers and then points to various other terms—"king assassinated," "Rwanda," "Libya," "Syria"—and says: "This is wrong." He corrects the diagram, which Frank and Lara had intentionally copied wrongly to get him to speak. "Gog is not China; it's the Russians. So that makes Magog America."[116] Frank thinks that Prine has kidnapped the children to prevent World War III, but the kidnapper replies that the war has already started. Yet he confirms Frank's guess that one of the children is a "peacemaker."

When Frank, Lara, Watts, and the police take Prine to find the place where he and his accomplice hid the children and bus driver, a tornado hits. This is the "sign" that Prine was seeking through his biblical interpretations. He leaves the police car and walks toward the funnel cloud, his handcuffed wrists raised over his head, disappearing into it. Frank and his friends are at a loss now as to how to find the children — since Prine's accomplice is dead, killed in a gunfight with the police. But miraculously, the children emerge from behind a hill, from an underground bomb shelter unearthed by the tornado, where they had been put by Prine and his accomplice. The tornado destroyed the school where the children would have been. They were saved through Prine's apocalyptic vision that one of them would save the world. Yet this episode also reveals that the (melo)drama of the daily news media can exacerbate the paranoia of certain viewers and listeners toward violent visions and psychotic actions — for good or evil, depending on one's perspective. As Frank says of Matthew Prine, when he begins to understand his tele-visions: "To him it's a fight between good and evil, and we're the ones with the horns and pitchforks."

One more episode will be considered in this chapter, because it not only involves demonic threats, with the Artaudian cruelty of psychotic violence and the Brechtian revelation of mass media alienation, but also points to the rites of sacrifice around a new millennial screen. As "The Mikado" begins, three teenage boys are surfing the Internet for sex sites. They find the live image of a woman tied to a chair (apparently drugged), wearing only a bra and panties. A number is painted on the wall behind her, "37122." As the number of "hits" on the site, also shown onscreen, nears that number, a man in a black hood and robe appears behind the woman and slits her throat with a large knife. Blood gushes from it and the shocked teenagers print the image from their computer.

Frank Black and Peter Watts discuss the unique problems of this postmodern murder case: the website has disappeared and no authority can claim jurisdiction because the place where the crime occurred is unknown. Watt says: "The law just hasn't kept up with the technology." He also suggests that it could have been staged as some "sick performance art." Frank believes the Net killing is real, "though the elaborate nature of the murder shows that he's performing for us." Frank asks the help of a computer hacker, Brian Roedecker, who "deconstructs" the woman's face from the printout to create a "digital abstraction" that he compares on his computer to faces in the national database of missing persons. He finds six matches, but only one was a heavy Web user, online for hundreds of hours each month. Roedecker calls her "my kind of woman, a real Net chick." Frank is certain she is the victim: "The Web has become his [the killer's] world.... He stalked her, then made contact in cyberspace before abducting her in real life." Frank also feels that the killer is "trolling for his next victim."

With the help of Watts, they enter the Millennium Group's room of high-powered computers, which Roedecker calls "a hacker's wet dream." He hacks into the victim's e-mail account. The masochistic titles of messages seem unusual for a librarian at a music conservatory in Sheboygan, Wisconsin. But Roedecker explains: "In the anonymity of cyberspace people are free to experiment." He says that he has often changed his name, "appearance, sexual orientation, even gender," and he sticks out his tongue impishly, like a devilish trickster on the side of the law.

The team sets up video-cameras and sound communications in three cities where the victim's most frequent online friends live, so that simultaneous interviews can be done without any of them tipping off the others. Frank holds the controls in Seattle, watching three screens, while Watts and his colleagues in those other cities investigate. But the screen medium frustrates Frank's profiling gift. So he tries to voice the killer's thoughts to Roedecker, who is still with him in the computer room. The shocked face of the hacker, who had been so enthusiastic about the high-powered screens and mutable Internet identities, then reflects the Artaudian effects of this show and the series, as Frank speaks to him. "You slaughtered that young girl.... You hate yourself, but you're God at the same time. But maybe the bitch had it coming. I watched the life drain from her body. And now she's mine forever. I showed her to the world. And now I have her in a safe place where I can see her anytime I want."

Frank first puts Roedecker in the role of the Net killer, then shifts into that position himself, suturing the television viewer into a *chora* of violence at the edges of the computer and television screens. The multiple subjectivities of the Net and various video channels reflect one another in a hall of mirrors — between *Millennium*'s screen and others in the home theatres of the mass audience. The rites of sacrifice on both types of screens come together, just before the commercial break, when police on the San Jose video break into a cemetery shed and find two severed heads in plastic bags. One is from the Web-death victim and the other from the suspect they had planned to interview on video. They are real, abject, body parts, from a killer and his victims, beyond the control of tele-video cops and the lacking computer law.

Inside the shed the detectives also find a series of numbers painted on the ceiling. These numbers refer to another of the killer's homepages, but this time the chair is empty. Roedecker tries to use his high-powered computer to trace the geographic location of the website, but its origin is very Derridean, always already displaced as soon as he comes near it in the mapping of phone lines on his screen. There is so much "code" in the Net, the hacker says, that "no one knows what all's in there." (This also seems like a premonition about toxic assets in the recent international banking crisis.) The killer in the Web who cannot be traced presents a postmodern model of primal violence in the social unconscious, as a *chora* of virtual creation and Real destruction.

Frank finds a clue, however, in the new number on the wall of the "Web death room." It matches the case file number from a serial killer known as Avatar, whom Frank tracked and nearly caught 12 years ago. Watts is unsure, because the current crimes are so different, as "murder on display." But Frank says it is "a new medium to vent the same old rage." And he adds: "From the slip-stream of electrons, a world as real as life or death can disappear in the blink of an eye. The devil has a new playground."

Another clue is found when minute differences between two visually identical ciphers posted on the website produce a sound file, part of a song from Avatar's favorite operetta, "The Mikado," with the lyrics: "Defer, defer, to the Lord High Executioner." Avatar also puts a new victim in the chair, first showing a blank screen, "a curtain, just like on a stage" (as Frank says), then close-ups of her body parts: fingers, breasts, and knee. But this woman's face is covered to prevent identification. Frank sends his own message to the killer via the Internet (through a news group he may be monitoring), a quote from Henry James that Avatar had used a dozen years ago: "Our doubt is our passion, our passion is our task, the rest is the madness of art"—with the last word changed to "pain." The killer responds by showing his new website victim with the word "pain" cut into her forehead. Thus, *Millennium* represents its own precarious position in a not-so-new, rage-venting medium and devil's playground, between art and entertainment, madness and ritual, violence and law, fiction and pain, as the show itself cuts into the lives and minds of the mass audience.

The television viewer is shown close-ups (with a heart-beat sound) of Avatar zipping up his black robe, of his eye through the hole in his hood, of his hands binding a victim's arms and breasts to the chair with rope, of his knife at her throat, and of her eye. Then her scream is heard. These images and sounds turn out to be just a dream that Frank has while asleep. But after he wakes, he and his colleagues create a similar theatre of cruelty, to try (like Artaud) to avoid the real act of violence. They reproduce the victim in the chair, with an actress wearing a similar short slip, with ropes binding her and a cloth covering her face (below the "PAIN" cut, painted on her forehead), along with an identical wall and number behind her. Roedecker then switches the image on the website to their restaging of it, so that the number of hits by voyeuristic/sadistic computer viewers will not increase to the kill number.

As they create this substitute theatre of cruelty to gain time and find the killer, Roedecker shows his own desire for the actress in the chair in the room with him, offering to give her a massage during the one-minute interval between updates, to get her "circulation" back. But she gives a voice to the unknown female victim, while distinguishing herself from that character in a gestic feminist twist,[117] distancing both Roedecker and the voyeuristic television viewer. "Is this some kind of a sick come-on? I'm not enjoying myself.

I don't get tied up in real life." Another distinction is shown then, too, as the actual victim in Avatar's chair (being monitored by the team on another computer screen) has her throat cut by the Lord High Executioner in the black robe.

Although Frank eventually flies to San Francisco and finds Avatar's "Mystery Room" on the stage of a closed theatre, he must still "defer" to the serial killer. In fact, he almost shoots the last victim, who was not actually killed as shown onscreen. First, Frank finds the empty chair onstage and sits in it himself, appearing on the website that Watts and Roedecker are still monitoring. Then, in a hallway backstage, someone shoots at Frank and he chases the shooter, returning to the stage where he finds a figure in Avatar's black hood and cloak pointing a gun at him. But Frank's finger pauses before firing at it, as the hooded figure freezes also. Frank walks toward the person with the gun pointed at him, unmasks the figure, and discovers it is the female victim, still alive, shuddering with fear.

This suspenseful twist and melodramatic happy ending, with Frank saving the last victim, offers a simple emotional release for sympathetic television viewers. And yet, it also offers a complex, tragicomic clarification of perverse desires. The law fails to triumph over Avatar, in the *chora* of the Web. But he is not the only villain here. As Watts said earlier about Internet users who visited the snuff site, raising its hit number toward the kill score: "Their actions now practically make them accessories to the crime."

Like its serial killer, with his Web audience, this episode of *Millennium* arouses the voyeurism and sadomasochism of the television audience, displaying erotic, violent fantasies on various stages, videos, and computer screens. Yet it gives only a partial conclusion, with the hero saving one victim, but not others, and failing once more to catch the devilish villain. "When he finds that next thing," Frank says, "that next weakness, that's when he'll be there again." In this way *Millennium* reveals the tragic edges of the Real, within and around cinema, television, and computer screens, perhaps provoking some spectators to cross fundamental fantasies within themselves and in the new media of a postmodern society — rather than being "accessories to the crime." Thus, the devil's playground of mass media screens may evolve in therapeutic and artistic, as well as moneymaking, social networking directions.

Various *Millennium* episodes ironically suggest a parallel between the angelic position of the serial viewer (or Web surfer) and a devilish serial killer. The television spectator's habitual channel-surfing, then intense focus on certain shows, gives a temporary sense of transcendent immortality and significance, through screen identifications and commercial buy-products (as offscreen objects to possess). Likewise, the serial killer's trolling for victims — to capture, torture, and kill — reverses his childhood trauma with addictive rites of power over someone else's life, yet repeats the objectifying of a victim.[118] At some level, we share with serial killers a primal trauma, in the human lack of natural being, with remnant instincts of fear and aggression transformed by particular experiences and family models, which prune our brains and shape our sacrificial symptoms.[119] Like the serial killer with his ritual victims, television and the Internet are ways to temporarily fill our lack of being through the illusion of controlling other bodies. We briefly overcome our primal alienation and mortal vulnerability with the power of virtual violence and its sublimated *jouissance*. We identify with and against the monstrous idols onscreen, especially through the predictable rite of melodrama, in order to experience the terror of evil "out there" and enjoy the hero's ultimate victory. But we also become victims of the television habit, returning each week for another hit of the same show, sacrificing our bodies in the serial rite, while learning aggressive scripts, related to real life (even more today with competitive "reality show" identifications).

Most shows in the *Millennium* series fit this habit-forming pattern, reflecting the fears in the news media about perverse and psychotic killers as melodramatic villains, while extending their evil into cosmic, apocalyptic plots. The series was prophetic, in some ways, about the 9/11 tragedy involving Islamic terrorists and the subsequent melodrama of a war against such militants, plus their Afghan, Iraqi, and Pakistani relations. They have become the apocalyptic villains of the new millennium, believing (according to Western news reports) that they attain transcendent sainthood, in a Jihad heaven of virgins, by destroying thousands of innocent lives. But Americans, too, are tempted to destroy others, as enemies or collateral damage, through revenge and fears of vulnerability, which the news media perpetually restage.

As mass-audience angels, we enjoy the immortal voyeurism of watching, with judgment and compassion, the mortal lives onscreen. Yet, we also become accessories in television's fetishizing of "Most Wanted" criminals and international terrorists as our current, evolving devils. Their violence is rewarded with media fame and their mass-killing martyrdom with stardom. Each of us contributes to some degree — like Avatar's fans raising his website hits toward the kill number — if we vote with our remotes for stations and networks that script the news, or further fictions, with melodramatic polarities of good, innocent, and evil objects onscreen. In the network of our brain theatres, as in the devil's playground of the Internet, simplistic plots and objectified characters of screen melodramas encourage real-life stereotypes of vengeful violence in the wars on terror and crime, instead of a more tragic catharsis with compassion for victims on both sides and with terror at our own acts as seen through the other's eyes.

6. Millennium *in the Home Theatre*

Millennium presented its angels, devils, heroes, serial killers, and victims within the melodramatic framework of a detective series and apocalyptic countdown — focusing audience anxiety on particular phobic objects.[120] Yet, the tragic complications of certain shows, beyond demonic fantasies, exemplified television's potential cathartic effects upon the mass audience, as art and therapy, even in the devil's playground. This challenged the commercial priorities of network television, contributing perhaps to *Millennium*'s cancellation after three seasons.[121] For example, there was a Red Lobster Restaurant commercial with people enjoying a beach scene, right after the psychotic kidnapper, Richard Green, crawled out of a lake of fire, shown through Frank's supernatural profiling "gift." This Brechtian juxtaposition of water scenes might not appear as a gift, or a good investment, to Red Lobster executives. Yet *Millennium* was an artistic gift to receptive spectators in the late 1990s — broadcasting its tragic insights in an era when PBS television and the NEA faced cuts in funding, with pressures toward self-censorship, for being too subversive.

Children may need the V-chip and family viewing times, plus melodrama's masks, to be sheltered from too much traumatic violence on the home theatre screen or too much Real fear of a mean world "out there." Yet mature spectators who want to be challenged by the artful complexity of tragic or tragicomic insights not only benefit from a series like *Millennium*, but also contribute to the social metaphysics evolving in postmodern culture. As the episodes analyzed in this chapter illustrate, there is a homeopathic potential in television violence to change the angels and devils in the audience, as a network of brain theatres. While reflecting the devil's playground of various screen media today, *Millennium*'s edges of Artaudian cruelty and Brechtian alienation alter the usual melodramatic formula of television as commercial entertainment, questioning the moral triumph of good over evil stereotypes. This creates new possibilities for angels and devils as extensions (and repeated symptoms) of the human brain, through further neural connections between Symbolic, Imaginary, and Real areas, personally and collectively. With God's uncertain role in the apocalyptic events onscreen, the mass audience of *Millennium* was implicated as its godlike Other, watching the characters and yet moving beyond them, into the future. Depending on how we watch such a show, with its oracles of the communal unconscious, we may change the serial sacrifices of television and other screen rites to be less addictive and destructive. This is all the more vital as we become interactive accomplices, not just spectators or potential victims, in the high-tech desires, social networking, and global terrors of our new millennium.

Conclusion: More Morality with Humans as Gods?

> *God is dead.... And we have killed him.... Who will wipe this blood off of us? ... Must we not ourselves become gods simply to seem worthy of it?*
> — Friedrich Nietzsche, *The Gay Science*, 1882

From its ancient and medieval to early modern, modern, and postmodern eras, Western culture altered its collective sense of a cosmic theatre. Instead of multiple gods or one God watching from above and sometimes acting in the human realm, they, as well as angels, devils, and other supernatural creatures, became less significant as ruling, warring, or devious spirits.[1] Humanist ideals developed as Guiding Identity and Purpose Principles (GIPPs), sometimes wearing divine masks, but often with more ordinary characterizations, onstage or onscreen. Yet they, along with the Romantic ideal of a sublime power in nature, operated as transcendent functions, extending the inherited structures of the human brain. Reason, rights, and the sublime gave life a greater meaning, beyond natural survival and reproduction, as certain societies grew larger and more powerful technologically, spreading their religious and secular ideologies across the globe. Eventually, the watching gods were supplanted not just by human rulers, but also by the "mass audience" as the mysterious Other that orders each brain, transforming natural values into cultural fashions.

In the West and elsewhere, humankind has become the metamind Spectator of itself, rewriting the scripts of its nature, through various cultural drafts and performance experiments, while striving for the ethical wisdom to handle its Pandora's Box of contending neural networks. Democratic, consumerist, mass-media technologies have positioned the human audience as a godlike Other, with shifting angelic and demonic values. The mass audience acts like a god: giving oracles in opinion polls, demanding sacrifices for the "immortality" of fame, and altering moral rules with evolving drives, desires, and fantasies. Likewise, First World luxuries cause Third World sufferings — as if selfish, whimsical gods were taking advantage of weaker mortals, or making them dependent on their benevolence.[2]

Yet, as *Millennium* predicted, the twenty-first century resurgence of religious fundamentalism involves terrorist acts and counter-terrorist measures performed not only for a global audience, but also for a certain transcendent Being.[3] In daily life, many people believe that God is watching, or that angels and devils might influence their performances, as in medieval drama and *Dr. Faustus*. This is reflected by the various God, angel, and devil movies of the last several decades. Such believers may find a greater purpose to life, with freedom from existential despair and loneliness, plus the thrill of participating in a supernatural struggle for souls, through the continued sense of a cosmic theatre.[4] On the other hand, atheistic scientists, philosophers, and pundits (such as Richard Dawkins, Daniel Dennett, Sam Harris, and Christopher Hitchens) argue that empirical discoveries should free us from religious superstition and political demonizing — through a shared sense of human evolution, beyond any nation "under God." And yet, how moral is human nature with its godlike power of higher-order consciousness?[5] Do we need more morality through the cultural evolution of our brains' inner theatres, involving divine forces watching and interacting with us — or new ideals of the mass audience? What is the moral role of today's stage and screen media in this light?

Good and Evil Spirits in Human Nature

Nature has its own rules, which sometimes seem immoral to us.[6] Species compete and cooperate for survival within a given environment. Sexual creatures compete within a species (and in social group hierarchies) for males to be selected by females for reproductive opportunities, with males fighting to control females for that reason. Infanticide is committed by various mammals, especially when a new male takes control of a female with infants from another male. Among our primate relatives, baboons and gorillas frequently perform such killings (as considered in chapters 3 and 5 here). Male chimpanzees and orangutans regularly beat up or "rape" females that refuse to mate (Wrangham and Peterson 132–46). During raids across the borders of their territory, chimpanzee "gangs" may attack, severely wound or kill, and even drink the blood of an enemy chimp, although he used to be a member of their group, perhaps a sibling (13–18). Male chimps, in the wild and in captivity, have been known to tear out the testicles of a rival (de Waal, *Our* 44–48). Females in certain rodent and larger mammal species will kill weaker offspring or abandon an entire litter when resources are low, or with the possibility of male infanticide, so as not to waste their nurturing investments (Hrdy). But humans have gone much farther with rape, murder, mutilation, genocide, slavery, and other cruelties, especially in times of war — and with infanticidal acts by mothers, as shown in the *Millennium* show and the real-life tragedies of Susan Smith and Andrea Yates (Cheryl Meyer).[7]

At other times, animals seem to exhibit generous, ethical behaviors through "kin selection" (for genetic relatives) or "reciprocal altruism" (when returns can be expected) — as considered in chapter 2. Yet humans transform such drives, carrying self-sacrifice for kin, adopted family, religion, or country far beyond the genetic logic of animal altruism. Rhesus macaques demonstrate self-sacrifice for a kinship group, starving themselves to avoid harm to others (chapter 3 here). But in humans this instinct is radically extended, beyond kinship generosity, as with the destructive pride of martyrs, which may involve a righteous "holy war" against others perceived as evil. Capuchin monkeys reward each other for cooperation in getting food, and yet insist on fairness in food distribution to the point of wild tantrums, throwing away food when others get more or better rewards (de Waal, *Our* 214–20). Yet humans show much greater generosity, envy, and rivalry, for good or evil, in various cultural traditions. Chimpanzees may share food because they anticipate resentment in others and want to avoid vengeance (219–20).[8] But humans go much farther in anticipating others' views and reactions, especially through theatre. They perform characters and plots far beyond their own lives, experiencing the thrill of vicarious conflicts — with melodramatic (good versus evil) aggression, or tragic consequences of vengeance, or comic relief and renewal, or satirical challenges to the status quo.[9]

Both physically and behaviorally, humans exhibit "neoteny" in relation to other apes (de Waal, *Our* 240–41). Like young apes, adult humans have round heads and flat faces, while often showing a playful inventiveness and curiosity.[10] Indeed, we may need our many forms of play, from childhood to adulthood, in sports, games, theatre, movies, television, and today's interactive media, because of our enlarged primate brains. Human dreaming is inherited from our mammalian ancestors as a playful workspace of memory sorting, reproductive desires, and threat rehearsals while asleep. But we extend such biological functions, magnified by our brain's tremendous capacity and plasticity, into fictional performances while awake, on various stages and screens of shared cultural fantasies. The exponential powers of our brain's inner theatre, with an individual's dreams and meta-representations connected to others' through language and technology, create vast transformations of our behavioral identities and physical environments.[11] We continue to need not only dreams and play, well into adulthood, but also the super-parental guidance of gods, or godlike ideals, through various theatrical media, due to the heaven and hell that we make on earth — even as science frees us from past superstitions and reveals new orders of emergent complexity.

Our highly imaginative brains are built on the practical foundation of millions of years of trial and error learning, which honed the structures we inherit. Brainstem and limbic circuits store traumatic and desirable experiences, rehearsing them in dreams with painful or pleasurable emotions and coded images, beneath the level of waking consciousness, yet influencing it and sometimes appearing on its stage — through the brain's inner theatre and outer performances. Such repetitions of pain and pleasure circuitry cause more "drama" in life, as we create iconic dangers, egos, and superegos, from the family influences of childhood and later networks in stressful cultural environments. Humans stage a social metaphysics of kin compassion and alpha competition through the Nurturing Parent or Strict Father metaphors (Lakoff and Johnson 318–27, 559), plus others we sometimes embody, such as the brash warrior or iconoclastic trickster. In the last several centuries of Western culture, older ideas of God have been replaced, for many people, by Enlightenment, Romantic, and modern ideals — involving Universal Moral Reason (319), the beautiful and grotesque sublime, and the libidinous id. Current theatrical media continue recombining prior supernal principals (or GIPPs) to express and contain our collective neural networks. Today's re-engineering of fictional supernatural species often appears as a playful postmodern pastiche. But there are real world consequences. Stage and screen media may confirm the certainty of good and evil types, with entertaining threats and heroic victories, exercising primal emotions, from virtual scripts to real tendencies.[12] Or they may create a more complex awareness of moral ambiguity within rational, uncanny, and libidinous brains — changing future perceptions and actions.

Emotional Contagion, Mirror Neurons, and Psychic Structures

Various types of human evil gain prominence through the news media today, often in a melodramatic context: terrorism, serial killing, mass murder, infanticide, kidnapping, vigilantism, and mob violence. Yet, as shown by *Millennium*, television drama may expose the tragic complexities of such evils, involving spectators' emotions in cathartic reflections about family and society. Films may likewise engage the mass audience in tragic twists that illuminate the angels and devils within human nature, even if they appear as supernatural figures onscreen. There is a vast potential for these mass media to evoke collective forces of good and evil, or individual acts of generosity and violence, from moral compassion[13] to neurotic, perverse, and psychotic aspects of sacrifice, especially through two features of human nature: emotional contagion and mirror neurons.

Emotional contagion is shown by many animals in flocks, herds, and packs, when individuals react together by picking up subtle cues from one another.[14] But it gains a new networking force with human brains and their mass-media technologies.[15] The theatrics of good and evil may produce contagious emotions of righteous idealism, fear mongering, and collective rage against scapegoats, magnified by mass communication (as in Nazi Germany and Rwanda).[16] Yet, compassion may also circulate in the social metaphysics of neural networks, with charity extending worldwide through news media coverage of humanitarian crises.[17] Such threatening, nurturing, protective, controlling, judgmental, and compassionate systems, from nature to the enlarged human brain and its expanding communities, were staged in the past as evil and good forces of the cosmic realm.[18] Now, such super-natural systems appear through names, images, and contagious emotions in our screen media — with emergent desires and fantasies reshaping our inherited survival, kinship, territory, and status drives.

The mysterious Real of emotional contagion between humans, focused by Imaginary and Symbolic, good and evil identifications, involves a mirror neuron system in each brain, with unconscious effects rippling through adjacent bodies and current screen media. Through mirror-neuron mimesis, human brains, like those of other primates, simulate doing what they perceive others doing, as a way of intuiting others' emotions and goals (Gallese and Goldman; Iacoboni 67). Yet, unlike in monkey brains, human mirror neurons fire at a pantomimed gesture, without an object being present (Iacoboni 26).[19] This shows a key difference in our evolution — toward unconscious and conscious theatricality.[20] Canonical neurons have also been discovered, which fire at the sight of an object that might be handled in a typical way. These and other mirror neurons may fire in a "logically related" chain (14, 24–26, 77). Such communication includes facial, as well as gestural muscles, triggered by mirror neurons, automatically mimicking the other's expression to interpret emotions, which are shared or repressed (111).[21] Mirror neurons, even for simple grasping actions, may fire differently depending on a scene's meaningful context (6–7). There are also neurons in the cingulate cortex that mirror the other's pain, just with an abstract signal (122–25). Thus, theatrical gestures, props, facial expressions, social scenes, and stylized signifiers communicate across many watching bodies, often at an unconscious level, with simulated muscle actions, desires, and emotions. This involves both sight and sound, affecting the Symbolic dimension of word meaning, through echo-mirror neurons (in Broca's area) relating heard speech to the auditor's own inner speech, including tongue and lip movements at a sub-threshold level, with mimicry circuits active as if, but not to the degree of acting likewise (102–05; Rizzolatti and Sinigaglia 168–69).

Such neurological discoveries confirm the Lacanian (phenomenological and psychoanalytic) notion that one's desire is the desire of the Other.[22] The mirror stage continues throughout one's life, through mirror neurons and emotional contagion. Yet, initial mirror-stage experiences are crucial in structuring particular neural networks that internalize facial expressions, gestures, and sounds to repeat or repress the Other's desires, while also distinguishing one's own as ostensibly independent. If there is not enough "alienation" of being, usually given by the Father's Law and by the (m)Other's desires moving beyond the child as her object of *jouissance*, the infant's brain may develop a psychotic structure (Fink, *Clinical* 86–89, 179).[23] Emotional contagion may overwhelm the child and, later in life, turn into hallucinatory voices and visions (82–86). The psychotic might seem possessed by demonic forces of emotional contagion from elsewhere — as shown in the violent missions of various *Millennium* characters and in devil-possession films.[24] Mirror neuron networks may also respond perversely if there is not enough "separation" from the (m)Other's desires, through the Father's naming of lost objects, which normally gives meaning to such sacrifices of *jouissance* and significance to the subject in the social order (175–76, 191–93).[25] Such a brain structure might mirror social norms perversely, repeating voyeuristic, exhibitionistic, sadomasochistic, or fetishistic rites that objectify the subject and others, thus invoking the Law (180–92) — as with *Millennium*'s villains, recent film devils, and prior stage characters. Even with the "normal" passage of most brains from alienation to separation to neurotic symptoms of obsession and hysteria, the traits of psychosis and perversion may emerge, reflecting foundational networks. These uncanny effects are magnified today as mass audience brains become possessed by mimetic imagery and emotions, perpetuating unconscious patterns — or raising them to cathartic awareness — through stage and screen media.

Media Mind Control

Theatre, with its various screen offspring, offers special spaces for observing fictional others, while imagining new worlds in the mise-en-scène and its diegesis. Our interaction with live, filmed, or virtual characters, vicariously through mirror neurons and motor circuitry (more so in environmental theatre and videogames) may immerse us in the emotional contagion of melodramatic thrills. Or it may create reflective spaces of tragicomic awareness, through Artaudian cruelty or Brechtian distancing. In many brains, this might alter the ventromedial filters[26] between limbic passions and prefrontal (right/left) intuitions and controls,[27] toward a more complex moral compassion.[28] Despite the usefulness of the left cortex's binary operator in our ancestors' evolution, distinguishing the good and evil, nurturing and threatening forces of nature, today's emotionally contagious, mirroring, and symbolic brains produce even greater dangers through such identifications — making man a wolf to man.[29] We need our stage and screen media not just as escapist entertainment, but also as artful therapy, while we generate new angels and devils in the real world.

A recent study (Hasson, Nir, et al.) involving fMRI scans of five subjects watching parts of the Clint Eastwood movie, *The Good, the Bad, and the Ugly* (1966, Sergio Leone), showed synchronized activity in the primary visual and auditory cortices, secondary somatosensory areas, multimodal regions of the frontal gyrus, and limbic areas in the cingulate gyrus (as if by emotional contagion). The movie created a collective "control" of face, object, and spatial recognition areas of spectators' brains and in their eye movements (Hasson, Landesman, et al. 5). In contrast, a video clip of an unstructured real-life event (an outdoor concert in New York), with a single camera and no editing, produced less correlative, synchronized activity, beyond the basic visual and auditory brain areas (7–8). Eye movement was more varied also (13). Viewers were then shown four different versions of the Charlie Chaplin films, *The Adventurer* (1917) and *City Lights* (1931): original sequences and three other versions with 36-second, 12-second, and 4-second parts "scrambled" in these silent films (10). Sensory brain areas showed greater "inter–subject correlation" (ISC) across all versions; but cognitive areas varied more because they "accumulate information across shots and process the movie as a whole" (12). Comparisons were also made as subjects watched an episode of Alfred Hitchcock's 1961 television show ("Bang! You're Dead"), the Eastwood movie mentioned above, an episode of Larry David's 2000 cable television show, *Curb Your Enthusiasm*, and the real-life video of New York (14). The correlative "control" of audience brain areas was 65 percent, 45 percent, 18 percent, and 5 percent, respectively.[30]

Entertainers often try to engage audiences through a collective focus on key details of a thriller, western, or comedy (the examples in the brain-scan study). Yet, artists have also developed cathartic approaches to expose the dangers of collective mind control. Such a catharsis would not just stimulate and purge emotions, creating synchronous brain activity by confirming good and evil stereotypes.[31] It could involve a tragic Artaudian or tragicomic Brechtian challenge to spectators, developing new and different views, through emotional contagion, mirror neurons, and more complex meanings. It might change limbic and prefrontal (PFC) networks through a clarification of cruel passions or by an alienating awareness of the social metaphysics that persist within and between our brains — affecting various brain areas that are synchronic or not, while watching the stage or screen.[32]

More in this vein, Hasson and his colleagues found that correlative (ISC) activity of subjects' brains in the dorsolateral prefrontal cortex, involving "higher cognitive functions," was relatively low, but then increased during a two-minute period, two-thirds of the way through the Hitchcock television episode (20). Along with the high correlation of limbic areas throughout the melodramatic thriller, this evidence may indicate certain tragicomic twists reaching a collective, cognitive catharsis, clarifying emotions at that point — and perhaps later, in retrospect, after the show ended. Such cognitive, cathartic experiences are all the more crucial today, given the prevalence of violence on many interactive screens and evidence from both neurology and sociology about its potential effects.

Mirror neurons fire in humans (but not in monkeys) with video as well as live performances, although less actively (Iacoboni 161). Research has found not only motor actions being imitated through observation, but also "various forms of complex mimicry" changing test scores when subjects are asked to think about certain types of people (200–01). This involves mirror neurons that fire more when the subject acts than when the subject observes, but are still influenced by images and ideas of the other (202–03). The placebo or pricey taste effect, via a doctor's or advertiser's influence, shows the power of the Other's authority to shape perception and action, through the PFC (Lehrer 146–48; Lipton 137–41). This also relates to media studies showing that children get more aggressive after watching violent movies, with long-term effects lasting for decades (Iacoboni 205–07).[33] Media violence triggers arousal systems in the brain, which may reduce the way that "super mirror neurons" inhibit the imitation of observed behaviors (211). I would add that such aggressive imitation, plus an increased appetite for media violence due to repeated arousal (in the brain's adrenaline and pleasure systems), is more likely to occur when the violence onscreen or onstage does not produce a higher-cognitive, PFC catharsis, through tragic or tragicomic plot twists with reflective, ironic insights.[34]

Researchers have found that mirror neurons fire more with video clips of social relationship actions, involving the spectator's identification with others, creating a feeling of intimacy and group belonging (Iacoboni 228, 257). This stimulates the brain's "default state network," with regions that are usually active only when a person is at rest, doing nothing or daydreaming (252–53, 257). A spectator sitting passively, watching social relations onstage or onscreen, is doing nothing, and yet also actively daydreaming with others in the audience and with those who have created the performance. The spectator's mirror neurons fire with intimate identifications, emotional and cognitive affiliations, and a sense of communal belonging with or against the heroes, villains, angels, and devils in the theatrical fantasy. Depending on the focus of attention (and sense of self through super mirror neurons), the spectator's brain simulates being in the body observed. Thus, the power of audience sympathy with fictional characters, through emotional contagion and mirror-neuron simulation, creates the possibility of extended neural networks, perhaps even more so with interactive videogames and websites.[35] These networks may evoke demonic passions that produce catastrophic acts, individually and collectively, unless fundamental fantasies, with godlike egos and superegos, are challenged and revised through a better cathartic awareness.[36]

Unlike other primates with mirror neurons, humans create theatrical artworks that represent dramatic conflicts in a "what if" situation. Depending on the script, performance, and audience, this may improve our higher-order consciousness of primal passions, encouraging better ways to deal with them in life. It may also reveal a continuing bio-cultural evolution in the human species: a further cathartic distancing (or alienation of being) from the animal's immersion in nature.[37] And yet, the "moral" ordering of limbic emotions and subcortical instincts by each brain's prefrontal lobes, learned through a particular culture's pedagogical arts, can twist these animal drives into monstrous forms — by focusing pity, fear, and revenge into good and evil projections, through the brain's binary operator. The primal drives of survival, reproduction, and social status, which fuel the moral ideals of compassion and autonomy, as well as the ethics of awe and cosmic meaning (Haidt, *Happiness* 188–206; McGeer), create extraordinary examples of human altruism and self-sacrifice. But they also produce greedy, competitive, insecure, and destructive egos, along with superego evils, through the righteous demonizing of others as perverse mutations.

Divine Traces

This book has considered the evolution of gods — made in the image of humans and animals — through brain structures that we share with our ancestors.[38] It has traced the judgmental, compassionate, warrior, rebel, trickster, and bestial aspects of gods, angels, and devils from ancient to postmodern stages and screens, reflecting left, right, and limbic brain areas. Whether or not such divinities (or a monotheistic God) exist, they represent super-natural symptoms of the human brain and specific cultural networks — even today as spiritual realms become "virtual realities."[39] Thus, aspects of the cosmic theatres of the past still structure our social metaphysics and our brain's inner performance spaces.[40]

The prehistoric art of Europeans tens of thousands of years ago shows a possible ritual theatre of animal and human forms interacting through the cave-wall membrane. Ancient Egyptian and Greek dramas reveal animal-human or immortal human deities as competing influences within mortal brains and cultures. A desert devil (Set) or the vampire-like Furies or rebellious Prometheus or trickster Dionysus compete against alpha sky gods and earthly rulers for territory, moral identity, symbolic objects, and sacrificial offerings. In medieval European theatre, Lucifer or Satan (as a snake in the garden or lord of the underworld) battles with Almighty God and His Son for the supernatural territory of human souls, using various devils and angels (or the figure of Death) as warriors, tempters, and messengers — redefining primal passions and transcendent aspirations. Renaissance theatre further transforms the animal nature and higher cognitive morality of humans toward the godlike potential of a rationally ordered society, despite lingering devils, angels, ancient deities, and nature spirits. Enlightenment sentimental comedy also idealizes the perfectibility of human nature, with the "man of sentiment" as a noble, rational, and moral force, emerging as distinct from his baser brothers. But Romantic drama turns again to nature's sublime forces, beyond civilization and within the human animal, as beautiful yet grotesque, through social outcasts, dreams, and conflicts of passion and honor. In popular melodrama, evil and good take obvious forms through human villains and heroes, their victims or love objects, and comic sidekicks, battling toward a happy ending of poetic, providential justice. This provides moral certitude with binary opposites, at the loss of tragicomic complexity.

With modern, "slice of life" naturalism and psychological realism, theatre attempts to be more scientific in exploring social problems, like a naturalist studying the human species in its urban environment, or a psychoanalyst discerning the monsters in the repressed unconscious. Yet this also involves an optimistic faith in the godlike power of human spectators as world makers — to experience, study, interpret, and change society through art. This faith continues with modern theatre's various anti-realist styles, including Artaudian cruelty and Brechtian distancing. Such ideological beliefs in a transcendent theatre experience and the audience's godlike power continue to influence the visionary worlds of postmodern plays, as do many aspects of prior periods, with ethnic gods returning and with diverse political figures fighting against the white, patriarchal, capitalist Other, as principles of good and evil. The popular melodramas of film and television involve, at times, more explicit angels and devils, with God as a distant spectator, occasionally interacting with humans (giving targeted help or challenges), and as a mystery, an absent presence and providence, even at the world's end.

While humans may be in charge of the world more than ever before, according to current beliefs in science and technology, we still need godlike principles and devilish alternatives — as cultural extensions, restraints, and mutations of our inherited brain structures. We also need supernatural figures for such neural networks, onstage and onscreen.[41] We need them not just as moral ideals and their binary opposites, with melodramatic good triumphing over evil, but as ethical tragicomic complexities, with renewed challenges to transform the gods in our prefrontal lobe's cross-filtering of compassion and reason.

The ultimate Metamind may exist beyond us, as a beginning and end to life, as prime mover and omega point. It may function not only as a distant Author/Spectator, but also as a more intimate Director — giving meaning to our mortality and a framework for morality. Or it may exist not as a power over us, but within us, changing and suffering through our animal to human evolution as earthly Actor in all of life.[42] The staging of self-awareness by nature has led (by chance or design) to the tragicomic success of our species, with its forces of good and evil. Even if we are alone as Metamind mortals, without gods in other realms, we create many angelic and demonic performances in this world, shaped to some degree by what we see, hear, and question onstage or onscreen. As we acquire more godlike powers, through language, art, mass media, and other technologies, we also need a neuro-cosmic wisdom, through the further evolution of selfish and compassionate networks, with complex identifications in our brain theatres and social metaphysics.

Notes

Introduction

1. My approach is somewhat akin to Michel Foucault's critique of "biopolitics," as further developed by Giorgio Agamben: "the growing inclusion of man's natural life in the mechanisms and calculations of power" (119).

2. Cf. MacIntyre on the Aristotelian and Thomistic philosophy of humans as rational, dependent animals.

3. See Kagan 302, on how children older than 6 begin to form moral ideals by noting characteristics that are praised by their community and inferring more perfect forms of those (and their opposites) as difficult to achieve.

4. See d'Aquili, Laughlin, McManus, et al. 171, on the human need for myth and ritual due to the "adaptive advance" of abstract, causal problem-solving, in the brain's evolution of higher cortical functions, and yet the corresponding awareness of mortality in an "unpredictable world." They call this existential anxiety the "curse of cognition." Cf. social psychologist Leary on the "curse," as well as adaptive success, of human self-awareness.

5. "Environmental theatre" is a term coined by Richard Schechner to denote an alternative tradition to the Western convention of proscenium frame and stage, using the given environment instead. See his *Environmental Theater* and *Performance Theory*. Here I am using the phrase in an even broader sense: the entire environment in which humans become aware of their mortality and then play (or work) at transcending it.

6. Democracy and terrorism are not mutually exclusive. Consider, for example, the ancient Athenians who voted in their assembly on the fate of the people on the island of Melos, who refused to pay tribute and join the Delian League. In 416 BCE, they were starved into submission through a winter siege. Then the entire male population was executed and the Melian women and children were sold as slaves. Aristophanes makes a brief reference to this in his comedy, *The Birds* (281, 386).

7. Cf. Hobson on REM (rapid eye movement) dreams, showing "the impressive level of consciousness ... achieved by the brain with no help from the external world" (391). REM dreams simulate problem-solving activities through creative associations and non–REM dreams refine recent learned behaviors through long-term memories.

8. I will not explore the racial issues of the "black Athena" debate about whether Greco-Roman culture actually began in North Africa (Bernal). Yet, I am including ancient Egypt as part of the circum-Mediterranean world that developed into Western culture. See also Price 16–17, on the origins of Greek gods and myths in India and the Near East. Cf. Zhang for research on European readings of fairy tales, as more objective, individualistic, analytic, and horizontal in contrast to the Asian approach as contextual, communal, holistic, and vertical (cosmic).

9. See Jablonka and Lamb on the four dimensions of evolution, from animal to human: genetic, epigenetic, behavioral (social learning), and cultural (symbolic communication emerging only with humans). See also Lipton, who extends epigenetics, via quantum physics, to theorize how the mind pervades all the body's cells. Cf. Durham, who gives an anthropological view of biological and cultural "coevolution."

10. See also Hayden 127–29 on the geographical advantages of Europe during the Upper Paleolithic period, tens of thousands of years earlier.

11. See also McConachie, "Falsifiable." He

attacks Lacanian theories as non-empirical, while misrepresenting key terms, such as the gaze, Imaginary, and Symbolic. My approach is more inclusive and, I hope, more accurate.

12. Cf. Barash and Barash 2–3. They use evolutionary psychology to find "[u]niversal human nature" in various literary works, but they often reduce "socially constructed" complexities to simple biological functions. See Hogan 197–213, for a critique of such reductionism when evolutionary psychology is applied to art and culture. See also Gibbs 251–59, on the embodiment of emotions, involving cognitive universals and cultural variations.

13. Such a multidisciplinary view of the human brain and its diverse performances was suggested decades ago by anthropologist Victor Turner, toward the end of his life, and by his colleague, theatre practitioner Richard Schechner — although without reference to Lacanian theory (*Future* 230, 239–40, 25 0–51). Cf. Walter Freeman 109, for a neuroscientist's brief reference to Lacanian theory.

14. See Cheney and Seyfarth, plus chapter 5 here. Charles Darwin suggested in one of his notebooks that human metaphysics could be understood by studying our primate relatives, such as baboons.

15. Cf. Holland. See also Sacks, "Sigmund," on the Freudian foundation of current neurology, especially regarding regression and drives. Cf. Doidge 233–34, 379–80. He cites studies in 2001, 2004, and 2006, involving interpersonal psychotherapy, cognitive behavioral therapy, and psychoanalytic treatment — all of which show a physical "normalizing" of brain pathways. He also quotes Nobel Prize winning neurologist Eric Kandel: "'There is no longer any doubt ... that psychotherapy can result in detectable changes in the brain.'" See Kandel, *In Search* 385–88, on neurological evidence for the Freudian unconscious. See also Kandel, *Psychiatry*; Corrigall and Wilkinson; Green; Kaplan-Solms and Solms; Siegel 334–35; and Vaughan — plus Panksepp, "Emotions," and the commentaries that follow that article.

16. See Durham 170–71, 188–89, on the use of "culturgen," "symbol," or "meme" as the gene-like unit of cultural replication and transmission. For a further development of meme theory, see Blackmore. See also Aunger who distinguishes his "neuromemetics" and "electric meme" theory from Dawkins's biological view (331). Cf. Fuster on the "cognit" as a neural network representing an item of knowledge.

17. See Reid for a new theory of biological evolution that stresses emergent structural orders, not just random mutations and external selection pressures. This would also relate, in my view, to Lacan's structural orders regarding the human brain and its various emergent cultures (with each being a Lacanian Symptom, a particular way of tying together the other three orders, in individuals or collectively). See also Margulis who argues that "symbiogenisis," as cooperation between species, from bacteria onward, explains "punctuated equilibrium"— the sudden emergence of new life forms (8–9).

18. Cf. Tomasello, *Origins*, on "shared intentionality" as the evolutionary distinction between great ape and human communication. See also Dehaene 146, on the "cultural evolution" of mathematics, as a "mental Darwinism" building upon the earlier "phylogenetic evolution" of space, time, and number representations in the human (and monkey) brain.

19. A half century ago, Lacan began theorizing the "mirror stage" of children, at the age of 6–18 months, when they can be observed sitting up and taking notice of their whole form, as an image reflected in a mirror. Lacan argued that the physical "prematurity" of humans at birth, in comparison with other primates, means that the infant experiences its body as "fragmented" (*Écrits* 75–81). Yet, the child gradually finds an illusory wholeness through the mother's interactions with it, reflecting and shaping its image of self through various physical and emotional senses. Like the visual self in the mirror, which differs from the fragmented body, the Imaginary self of the child is alienating. Eventually, this fragile illusion of a whole ego (or imago) evokes paranoia about loss of face and aggressive competition with others. Cf. Schore 28–36, on recent research that shows a crucial mirroring between mother and child, starting in the second month of life, with "affect synchrony," "attuned interaction," and "emotional communication." Cf. also Nadel 57, on the "imitative" body language of preverbal children, mirroring each other via "meta-cognitive" (and theatrical) capacities that they start to possess at 18 months, such as attributing intentions to the other, taking turns, and switching roles. Likewise, Asendorpf states that "mirror self-recognition and synchronic imitation develop in close synchrony during the second year of life..." (67). Yet infants as young as 3–5 months start to discriminate between videos of their own face and that of a peer (70), perhaps showing an early form of the mirror stage. See also Hurley and Chater 22–24; Iacoboni 128, 133–35, 140–41, 162; and Rochat 87.

20. See Feinberg on the unity of the self in the brain as "a nested hierarchy of meaning and purpose" (149). Cf. Hinde 81–82, 92, for a cognitive view of religious figures of transcendence forming from parental paradigms. Freud also suggests this in *The Future of an Illusion*.

21. I thank David Bashor, a neuroscience researcher, for his advice on these parallels.

22. See Winkielman et al. on research about general affects and specific emotions in conscious and unconscious brain systems, from limbic to

other areas, including the emotions triggered by subliminal images.

23. See Cozolino, *Neuroscience of Human* 24–25, on the triune model of the human brain, with reptilian brainstem, paleo-mammalian limbic system, and neo-mammalian cortex, theorized by Paul MacLean a half century ago. "The model of the triune brain serves the valuable function of providing a connective metaphor that encompasses the artifacts of evolution, the contemporary nervous system, and some of the inherent difficulties related to the organization and disorganization of human experience." Yet, this model now appears too simple for the "brain's many complexities," which do not have such clear delineations (25). For a further critique of MacLean's theory, regarding the limbic system, see LeDoux, *Emotional* 92–103 and *Synaptic* 35, 210–12. But see also Panksepp 70–72, 341; he updates MacLean's model and counters LeDoux's critique. Cf. Butler and Hodos 114, 630.

24. There is evidence of hemisphere asymmetry in human infants, even fetuses, but also in various types of animals (including monkeys, birds, and frogs, whose vocalizations are left-brain forms of communication). See Trevarthen and also Ehret 51.

25. Hemisphere asymmetry occurs in the secondary "association" cortices which integrate signals from various modality-specific areas of the primary cortex. See Solms and Turnbull 24–25, 82, 241–72; Kaplan-Solms and Solms 278–79; Newberg, d'Aquili, and Rause 65, 69; Ramachandran and Blakeslee 135–47; Ramachandran, *Emerging* 41; Geschwind and Iacoboni 53, 62; Edwards-Lee and Saul 306, 314; Cozolino, *Neuroscience of Psychotherapy* 110–20; Siegel 197–98; and Goleman 12.

26. A problem with phoneme-recognition due to damage in Wernicke's area is known as Wernicke's aphasia: patients can produce language but not comprehend what they hear (Solms and Turnbull 248). In normal brains, phonemes are also connected to images, creating meaningful words, in the left occipito-temporal area (249). In the left parietal cortex, near Wernicke's area, the syntactical structure of speech is analyzed, including the quasi-spatial grammar of word order (250). Patients with Broca's aphasia, with a lesion in the left premotor cortex, closer to the front of the brain, may analyze or initiate speech with individual words, but cannot string words together in grammatical sentences (252). Cf. Corballis, *Lopsided*, on the left cortex's generative assembly device (GAD).

27. Ninety-nine percent of right-handers have language areas and other "dominant" functions in the left brain. This may be due to the evolution of verbal language from the right-handed gestures in our hominid ancestors, since the left side of the brain controls the right side of the body (and vice-versa). See Geschwind and Iacoboni. See also Wolf 144–54 on the language and visual areas for reading in the left and right hemispheres.

28. See Solms and Turnbull 228–29. See also Seigel 191–92, on gender differences in hemisphere asymmetry and its possible source in "cognitive crowding," plus the evolutionary roles of males and females. Cf. the recent study by Cela-Conde et al., which finds that men use the right parietal lobe more in appreciating the beauty of nature and art, while women use left and right together more. This may be because the right parietal specializes in more precise, "coordinate spatial relations" and the left processes general, "categorical" relations, regarding positions or parts of objects, such as above/below, inside/outside — with males having evolved as hunters, needing such precision, and women as gatherers, using categorical relations more for object contents and locations (3849–50). Also, the right parietal involves global attention and the left local (3851). Brain scans show, too, that men are more left-brain oriented during speech, while women are more bilateral (Panksepp 235; Clements et al.)

29. There may also be an Imaginary dimension in the Symbolic order of the left hemisphere and a Symbolic dimension in the Imaginary order of the right. See Aziz-Zadeh et al. on the congruence between visually presented actions and their literal phrases in the mirror neurons of the left hemisphere's premotor cortex. See Bonda et al. on the involvement of right temporal lobe areas and the limbic system (including the amygdala) in perceiving signs conveyed by emotionally expressive body movements. See Winkelman, *Shamanism* 139, on the "visual-spatial system of symbolic presentation [in the right hemisphere] ... normally inhibited by the dominance of left-hemisphere verbal representational systems," but allowed further expression in dreams and shamanic visions.

30. See also Cozolino, *Neuroscience of Psychotherapy* 116–17, and Bolte Taylor.

31. Right hemisphere damage may also impair "autobiographical memory and self evaluation" (Decety and Sommerville 529). See also Decety and Jackson.

32. Throughout this book, I use the Aristotelian term "catharsis" to mean the clarifying of emotions, especially sympathy and fear, toward a more complex awareness through tragedy or tragic twists in plays of other genres. Elsewhere, I have argued that melodrama provides a simpler form of catharsis, as the purging of such emotions (*Theatres*). See also Belfiore and Bennett 81–91, on six historical interpretations of catharsis, including "clarification," though I do not view this interpretation as just "intellectual" (89–90).

33. Cf. McConachie, *Engaging* 111: "Crying as a means of helping to modulate our physiological thermostats may be the closest that cognitive science can come to the Aristotelian notion of catharsis."

34. See also Panksepp, "Emotions" 31, on the ventromedial PFC and other brain areas related to Freudian therapeutic issues: anterior cingulate areas regarding depression and obsessive-compulsive disorders, or lateral and medial temporal lobes and amygdala zones, plus frontal areas, regarding anxiety and anger.

35. See Nadel 58, who relates her own research on children's preverbal imitative language to Donald's theory of a mimetic stage in human culture. Rizzolatti and Sinigaglia also relate their mirror neuron research to Donald's theory (162). Cf. Blackmore 76, 98. She argues that memetic replication drove hominid tool use and brain expansion 2.5 million years ago, but she distinguishes this from mimesis in Donald's theory.

36. Neural expansion occurred especially in the prefrontal cortex, now occupying 29% of the human brain in comparison with 17% in chimpanzees and 11.5% in macaque monkeys (Murphy and Brown 132).

37. See also Donald, *Origins*.

38. See Geschwind and Iacoboni 52–53 and Edwards-Lee and Saul 306–07, on the right hemisphere's comprehension and output of language prosody, expressing the emotional content of speech. See also Niedenthal et al. Cf. Rizzolatti and Sinigaglia 158 -59, on the double evolution of human communication, with "emotive vocal expression such as exclamations, yells, and so on" (linked to language prosody and music) involving deeper brain areas, as with non-human primate calls, while human gestural and verbal languages involve higher cortical areas.

39. Cf. Dunbar's argument that religion emerged not with the early hominid, *Homo erectus*, but with later *Homo sapiens* ("Social" 179). Dunbar also relates this to his theory of gestural and verbal language evolving (as gossip) from hominid grooming behavior — as a crucial aspect of the "social brain." See Dunbar, *Grooming*.

40. See also Cozolino, *Neuroscience of Psychotherapy* 80.

41. Cf. Robert K. Logan on the evolution of human language (through reciprocal altruism) as an "extended mind." He refers to Donald's stages, but not to Lacan, nor to neuroanatomy.

42. See Kristeva, *Revolution* 25–28, 46–51, for her theory of the *chora* as a maternal, womb-like or ritual space with abject power in the human psyche, art, and culture — producing violent, creative and destructive eruptions in the symbolic, patriarchal order. It is "the place where the subject is both generated and negated" (28).

43. Donald cites the argument of evolutionary biologist Terrence Deacon that, with the expanding brain size of *Homo erectus*, the executive system of the prefrontal cortex "invaded many brain regions that formerly dominated the control of action in primates," creating a "major evolutionary change in hominid physical self-consciousness" (*Mind* 271). See Deacon 343–44.

44. Cf. Greenspan and Shanker 172–73, who find traces of artistic abilities in earlier hominids, as far back as 300,000 BCE, showing the evolution of "affective signaling." Taçon presents similar evidence for this, in relation to religious ritual and burial practices (63–65).

45. Late in his career, Lacan theorized a fourth order called the Symptom (*sinthome*), as the particular interlinking structure of the Symbolic, Imaginary, and Real, in a certain subject's mind or the intersubjectivity of human culture — as explored recently by philosopher Slavoj Žižek.

46. Referring to Donald's earlier work, Mithen says he is following in "Donald's footsteps" while also correcting his errors of oversimplifying the archeological data and underestimating the cognitive ability of living apes, in comparison with early hominids *(Prehistory* 10, 227n4).

47. Cf. Fauconnier and Turner on their linguistic theory of "cognitive blending," especially their critique of Mithen, though they fail to consider his full argument (174). See also Mithen, "Out," especially his theory about supernatural beings as "unnatural" to the human mind's domain-specific foundation, and yet as useful with its evolving fluidity (130). Through their material symbols, supernatural beings become cognitive "anchors," shared between brains, as "an extension of the mind."

48. Cf. Colin Renfrew, who finds evidence for earlier forms of artistic "patterning," but not art per se, in pierced shells as beads, and carefully shaped blocks of red ocher, in southern Africa from 70,000 years ago (77–78).

49. Cf. Goldman, whose cognitive "simulation theory" ties inner mental re-creations to external mimicry, behavioral imitation, pretend play, and "drama" (9). Goldman also relates his theory to mirror neurons (14) and to research on how "internally rehearsed" motor imagery shares "the same neural substrate" as actual movement (11). For that research, see Decety. Goldman also points to research showing that patients with prefrontal lesions "compulsively imitate gestures or even complex actions performed in front of them," because the resonant inner image that we all generate when watching actions, with neural pathways for performing the same actions ourselves, are not inhibited by the prefrontal lobes of those patients, as they are in the rest of us (14). See also Buckner et al. on the "default network" of the human brain, with extra frontal lobe activity during introspection: "self-relevant men-

tal explorations — simulations — that provide a means to anticipate and evaluate upcoming events" (2).

50. See Ryle's critique of the "ghost in the machine"; Dennett, *Consciousness*, on the Cartesian theatre; and also Murphy and Brown 27–29.

51. See also Ramachandran and Blakeslee 152; Gazzaniga, *Mind's Past* 21, 73; and Baars et al.

52. Cf. Minsky, *Emotion* 125–26. He uses Baars's theory to consider "multiple levels of representations and different short-term memory systems to keep track of various kinds of context." See also Minsky, *Society*.

53. Cf. Edelman, *Wider* 57–58, 61, 116, on animals' basic ability to create a scene within the brain as part of their "primary consciousness" — and humans' higher order abilities.

54. See Glenberg on the human skill of suppressing environmental influences upon the brain's conceptualizing, which increases its powers of prediction, memory, and language.

55. Mirror neurons, discovered in monkeys' brains, respond to a specific hand motion in the same way, whether it is performed by the self or the other. Unconscious motor commands signal the miming of such movements. They may be inhibited, yet are "used to interpret what is seen," especially by mirror-stage infants rehearsing self and other movements (Churchland 108–10). See also Rizzolatti and Arbib. Neurologist V. S. Ramachandran argues that this discovery in primates today may explain the "great leap forward" in hominid evolution 40,000 years ago, as a few of our ancestors invented new tools, which were then imitated by others using their mirror neurons, rapidly spreading the cultural transformations (*Brief* 80–81 and "Mirror Neurons").

56. Cf. Solms and Turnbull 150–67, on semantic, procedural, and episodic (autobiographical) types of memory. See also Schacter 66–67, 266, on memories as constructs of fragments and fantasies, not whole experiences.

57. See also Boehm, *Hierarchy*. Cf. Everett 110–12, on the egalitarian aspects of today's Pirahã, Amazonian hunter-gatherers who use spirits, ostracism, and food-sharing rules, rather than chiefs, for social control.

58. Cf. LeDoux, *Emotional* 239–46, for a neurological view of Freudian repression.

59. Neuroaesthetics includes fMRI studies to map spectators' brains while they experience a movie (Hasson et al.; Pessoa) or a dance (Calvo-Merino et al.), plus applications of neurology to visual art, literature, and music (Ramachandran and Hirstein; Zeki, *Inner* and "Neural"; Jourdan; Levitin; and Sacks, *Musicophilia*). See also Levy, *Neuroethics*, and Murphy and Brown. Neuroeconomics also involves moral issues (Zak). Cf. Smail, who uses the term "neurohistory," while critiquing evolutionary psychology, in favor of "deep" cultural history (141–45).

60. Cf. Nettle's evolutionary typology of four dramatic genres: tragedy, heroic tragedy (melodrama), love tragedy, and comedy, regarding negative and positive outcomes of survival and reproductive drives. See also McConachie, *Engaging* 175–77.

Chapter 1

1. Cf. d'Aquilli, Laughlin, McManus et al. 351–54, on "science as a ritualization phenomenon."

2. Cf. Boyer's notion of "cognitive constraints" to religious ideas.

3. Cf. Edelman and Tononi 102–10, on the "primary consciousness" of some animals, in relation to human "higher order" consciousness.

4. Burkert considers the human sacrificial test, through an "ordeal," as akin to divination (163–64).

5. Cf. Wise, who sees the technology of writing as crucial, beyond ritual, to the birth of theatre in ancient Greece.

6. Cf. Ciompi and Panksepp 43. "In sum, affects [in the brain and collectively] appear as elementary forces with Janus-like properties."

7. Evidence for the left brain's executive control over the right, like a war-room "general" denying the contrary views of a "scout," comes from anosognosia patients, who deny their obvious paralysis (on the left side) after a right hemisphere stroke (Ramachandran and Blakeslee 132–37). Ramachandran has done experiments with such patients to prove that the "left hemisphere is a conformist, largely indifferent to discrepancies, whereas the right hemisphere is the opposite: highly sensitive to perturbation" (141). He also cites brain mapping research for this (141–42). With damage to the right parietal lobe, confabulations and denials tend to involve body image (142). With damage to the right ventromedial frontal lobe, "denial is broader, more varied and oddly self-protective."

8. See also the recent argument by Haught that theology should accept Darwinian science, but anticipate an evolution toward God, which gives purpose to suffering and death, while providing the basis for morality.

9. See also Groopman's critique of neurotheology for mixing science and religion. He accepts the possibility that human beings may be intrinsically wired for spirituality, as an evolutionary blueprint, but argues that science cannot measure the soul or photograph God.

10. Cf. Stewart Guthrie on the evolution of religion through the cognitive bias in humans to "animate and anthropomorphize" their environment. See also Shermer's theory of how humans evolved a "belief engine" in the brain.

11. Cf. McNamara, "Religion" 248–49. He argues that religious practices "promote devel-

opment of the frontal lobes," especially the right frontal cortex.

12. See also Saver and Rabin's "limbic marker" hypothesis that numinous religious experiences are "similar to those of ordinary experience, except that they are tagged by the limbic system as of profound importance, as detached, as united into a whole, and/or as joyous" (507).

13. Cf. Brugger 209, on the right cortex making remote, less prototypical associations with diffuse semantic activation, in contrast to the left, and how this relates to creating patterns out of random configurations — with belief in paranormal and magical phenomena. See also Weingarten et al.: "bodily sensations, alimentary sensations, sense of fear, déjà vu experiences and the psychical illusions or hallucinations ... can all be reproduced by electrical stimulation of the depth structures of the temporal lobe" (215).

14. Cf. Horgan 95. See also Persinger, "Experimental" 286.

15. For more details, see Pizzato, "Nietzschean."

16. See Siegel 181–82, on the right brain's "reading of social and emotional cues from other people" and its "external expression of affect" (more on the left side of the face, controlled by the right brain) — through representations of gut reactions and changing bodily states. Though not using the term "mimetic," Seigel calls this the non-verbal "language of the right hemisphere" — processing and expressing "primary emotional reactions ... before the involvement of ... [left-brain,] rational, linear analysis." Seigel also specifies the "positive," social emotions of the left brain, and its "display rules" for emotional expression, versus the "negative," basic emotions of the right, such as "sadness, anger, fear, disgust, surprise, interest/excitement, enjoyment/joy, and shame" (184–85) — again supporting the Apollonian versus Dionysian distinction.

17. See also Dissanayake on the "mutuality" of mother-child attachment and communication, across various human cultures, as an evolved foundation for artistic belonging and expression.

18. Cf. Gough and Shacklett, who develop a scientific argument for the existence of spirits as "metaphysical 'bodies'" linked to the physical world by "interpenetrating patterns" (48).

19. See also Panksepp 235, on the "tendency of females to use both hemispheres in speech while males tend to use only the left side of their brains." Cf. the recent research of Clements et al., confirming this and finding that males are more bilateral and females more right-brain oriented with visuospatial tasks.

20. Joseph also describes how differences in male and female limbic systems relate to sexuality and how "seizure activity within the amygdala/temporal lobe may result in bizarre sexual changes" (145–46).

21. Joseph argues that death is a gradual process, with some cells and tissues "living for hours or even days or weeks before the body decays" (288). He then speculates: even after death, so long as the body (or at least the limbic system) lives, one's sense of a personal soul and identity remains intact. This personal identity is perceived, after the death of the body, as an out-of-body experience. Moreover, this personal identity, the energy field associated with the dying body, may be perceived by others as a ghost, spirit, or departed soul. Presumably, this ethereal after-death existence and sense of personal identity remains tethered to the body (or limbic system) until the body completely dies and decays.

Here, Joseph presents a theory that is similar to the after-death scene in a modern American play, Thornton Wilder's *Our Town*, where the dead wait in their graves while the earth part of the self is burned away and the soul is weaned toward its eternal destiny. "As the body is consumed, perhaps so too is the sense of individuality, freeing the soul, ... thereby becoming One with the Great Spirit and the Gods" (Joseph 288). Cf. Churchland 395–96, who theorizes, as a neurophilosopher, that near-death, out-of-body experiences (NDEs and OBEs) might be produced by the lack of oxygen (anoxia) and endogenous endorphin (opiate) release in brains under stress. See Blackmore 179–82, who makes a similar argument. See also Coveney and Highfield 343, on the use of computers and complexity theory to explore NDE and OBE. Cf. Moody's interviews with those who have had near death experiences and his list of possible explanations. See also Sabom, who analyzes the data on "autoscopic" and "transcendental" NDEs. Both Moody and Sabom are medical doctors, each surveying over 100 patients, but the latter conducts a more scientific study.

22. Joseph also goes much farther than Persinger or Ramachandran in his neurological speculations about brain anatomy in the evolution of human spiritual experience. Persinger points to the common structure of human brains "for thousands of years," as the basis of similar themes in God Experiences across widely diverse cultures and religions (*Neuropsychological* 34). Ramachandran considers the surprising evidence that earlier hominids, Neanderthal and Cro-Magnon, hundreds of thousands of years ago, had average "cranial capacities" larger than those of modern humans, suggesting a "latent potential intelligence" that may have been greater than ours today (Ramachandran and Blakeslee 191). Cf. Mithen, *Prehistory* 11–13, 147–50. However, Joseph uses this evidence to argue that those prior hominids, as primal savages (the short, cannibalistic Neanderthals) and godlike giants (the six-feet-tall, fire-making, technologically superior Cro-Magnons), left traces in the mythic sto-

ries of the ancient Sumerians and in the Book of Genesis (15, 21–41). Cf. Jane Renfrew on the evidence for Neanderthal cannibalism (54), and yet also their care of the sick, burial of the dead, bear rituals, and artistic activities.

According to Joseph, the existence of early hominids with bigger brains than ours, along with other flaws in Darwinian theory, shows that life on earth must have come from elsewhere — with DNA landing here from another planet, either by chance or by alien interference (44–46, 197–205, 254–75). Like Persinger and Ramachandran, Joseph focuses on the limbic system as demonstrating the human evolution of divine apprehension. Unlike his fellow neurologists, however, Joseph sees the limbic system as evidence for a divine presence that has always been latent in the astronomical gift of DNA — calling that part of the brain a "transmitter to God" (74, 187). He argues that LSD produces supernatural limbic visions — like the dreams, trances, and drugs of religious rites — by enabling "the brain to continue processing information that is normally filtered and suppressed," which might include the perception of "gods, demons, or angels" (277–78). He asks: "Why would the limbic system evolve specialized neurons or neural networks that subserve the capacity to dream about, experience, or hallucinate spirits, angels, and the souls of the living and the dear departed, if these entities have no basis in reality?" (279). He considers the possibility that the limbic system evolved its ghost and god experiences as a Darwinian "adaptive illusion" in individuals with a "religious-moral conscience capable of redirecting and controlling the more dangerous limbic impulses"— with these individuals then surviving and breeding more successfully. But Joseph uses the example of reported "after death" experiences by patients who died but were resuscitated —floating above the body and moving through a dark tunnel toward a bright light — to counter this Darwinian logic (280). What, he asks, "is the adaptive significance of these [limbic] neurons firing and creating a hallucination even after death? A capacity such as this would represent a degree of limbic-evolutionary foresight that is almost too incredible to accept without positing some guiding intelligent force behind its design."

23. Ramachandran also refers to, but discounts, Christopher Wills's theory of the "runaway brain": general brain growth producing specific leaps in human talents (Ramachandran and Blakeslee 192). Ramachandran speculates instead that the "esoteric talents" of music, poetry, drawing, or math evolved randomly in some savants through the abnormal expansion of the left (mathematical) or right (artistic) angular gyrus, then became perpetuated by natural selection because they were "sexually attractive ... as an externally visible signature of a giant brain" (196–97). See also Wills 56–57, for his incorrect guess that DNA evidence would prove that Neanderthals were direct ancestors to modern humans.

24. See Ramachandran, *Brief* 96–97, where he lists five characteristics of the common human sense of "self": continuity, unity, embodiment, agency, and awareness. But he says that stimulating the right parietal cortex with an electrode will create an "out-of-the-body experience," temporarily dissolving the embodiment of self. Furthermore, all five aspects can be "selectively affected in brain disease." See also Wegner.

25. It has been argued by Dawkins himself, as well as his critics, that the structural unit of the meme is unknown, unlike the gene's DNA. But I would suggest Lacan's notion of the Symbolic signifier as the meme's unit of replication (genotype), through conscious and unconscious languages, within the Imaginary order of cultural reality as meme product (phenotype), using our brains as Real vehicles. Dawkins coined the term "meme" from the Greek word for imitation, *mimesis*, which Aristotle saw as the source of theatre within human nature. Cf. Blackmore 43, 53–58, 63–66, on the terminology of memes, as replicators with unknown units, involving products and "vehicles."

26. See also Dawkins, *Selfish* and *Extended*. Cf. Microsoft Word inventor Richard Brodie's explanation of the meme as a "virus of the mind" and Aaron Lynch's theory of popular memetic beliefs as "thought contagions." See also Edward O. Wilson 136. For critiques of Dawkin's theory, see Atran 236–62; de Waal, *Good* 14–18; and McGrath, *Dawkin's* 119–38. Cf. Dennett's answer to Atran's critique in *Breaking* 379–84.

27. Cf. biologist Edward Wilson who finds a specific sense of "free will" in the brain's neural complexity — despite the "self" being an illusion, as an "actor" in a "perpetually changing drama," lacking full control, while decisions are actually made by the unconscious brain: "strings dancing the puppet ego" (119–20). See also Satinover 6–7, 129, on the "quantum brain."

28. Blackmore says that morality will be the result of living the truth of no self: "you stop inflicting your own desires on the world around you.... This lack of self-concern means that you (the physical person) are free to notice other people more. Compassion and empathy come naturally" (245–46). On Blackmore's personal experience of enlightenment through Zen Buddhism, see Horgan 106–19. See also Varela et al. for a detailed application of Buddhist teachings to cognitive science and evolution.

29. Cf. Durham 431, 459, for a contrary view that human choice, more than "nature," determines meme selection.

30. Cf. Schroeder 154–55: "Every particle is

an expression of information, of wisdom. The self-awareness we experience is the emergent offspring of that wisdom.... The same elements that build our brains and bodies form stars and galaxies. In that case [if these elements have the trace of consciousness] our minds are but a part of a vast and conscious universe." But see also Margulis 120, on Gaia as "neither vicious nor nurturing in relation to humanity," and 124, about Lovelock eventually giving up his initial notion that Gaia (geophysiology) is teleological. Cf. Davies, especially 259, for a more recent theory of scientific teleology, from astronomy to quantum physics.

31. See also Anthony Freeman's cognitive view of "God-consciousness" as being like the property of liquid realized in water molecules and human consciousness in brains: "a higher-level property still, caused by and realized in the physical-and-mental-totality of human beings" (153).

32. Cf. Atran 182–86, for criticism of this theory from a cognitive view.

33. Cf. Jane Renfrew, who finds that Neanderthals were "capable of symbolic behaviour" as well as burial of their dead (59). But Lewis-Williams argues that Neanderthals, unlike early humans, had only primary consciousness, not higher-order consciousness (in Edelman's sense), and therefore could not remember their dreams, nor conceive of a spirit realm or an after-life ("Of People" 144). So, they left no grave goods at their burial sites. See also Pizzato, *Ghosts* 27–28.

34. Cf. Lakoff and Johnson on the "embodied" binaries across human cultures.

35. Cf. Hogan, who briefly mentions a cognitive "hierarchy" of genres, from melodrama to tragedy, with the former as most emotionally expressive, but the latter as altering emotional orientation with more restraint (171–73).

36. See d'Aquili and Newberg, *Mystical* 53–55, 150–54, on Western binary, causal religions (Judaism, Christianity, and Islam) versus more holistic Eastern religions (Hinduism and Buddhism).

37. Studies have found this binary distinction at the level of specific neural cells: "we know that certain neurons in the left orientation area respond only to objects within arm's reach, while others respond only to objects beyond" (Newberg, d'Aquili, and Rause 28). See also d'Aquili and Newberg, *Mystical* 33–37.

38. Cf. Schore, *Affect* 300, on the two circuits of the limbic system, involving reward functions and autonomic-visceral aspects of emotion, in relation to the sympathetic and parasympathetic systems. Both circuits mature in the infant by 18 months of age (during the Lacanian mirror stage) and respond "to changes in socioaffective information emanating from an emotionally expressive face."

39. Richard Schechner made a similar argument over a decade ago, using the earlier work of d'Aquili, as well as Victor Turner, about ritual performances that involve "the ergotropic [arousal] and trophotropic [quiescent] systems of the brain," possibly leading to "a 'rebound' or 'spillover,' simultaneously exciting both left and right hemispheres" (*Future* 240). See also Schechner, *Performance Theory* 276–77, and d'Aquili, Laughlin, McManus, et al. 174–75.

40. See Newberg, d'Aquili, and Rause 86–87 on the hippocampus, as "diplomat" of the limbic system, regulating neural flow and thus producing the inhibitory effect of deafferentation in the parietal lobes, through hyperquiescence or hyperarousal of the autonomic nervous system, "blurring the edges of the brain's sense of self." The hippocampus, located in the temporal lobe, also links emotions to images, memory, and learning (45). It is crucial for the autobiographical sense of self with its directories of episodic memory storage (Solms and Turnbull 97, 162–63).

41. Cf. Mithen, *Prehistory* 20, 108–11, 140–41. He details the emergence of *Homo erectus* 1.8 million years ago, with language evolving about 250 thousand years ago — if not in the prior *Homo habilis*, with a possible left-brain structure (Broca's area) for language 2 million years ago. Mithen also says that archaic *Homo sapiens* "are likely to be descended from *Homo erectus*" in Asia and Africa, starting about 400,000 years ago (20, 25). They probably evolved into anatomically modern humans, *Homo sapiens sapiens*, in Africa about 100,000 years ago. Cf. McNamara, *Mind* 68–69, and Banyas 99.

42. Cf. Bloom on research with children today showing that the human brain has evolved with a natural bias toward dualism, with the intuitive belief in a soul as well as body.

43. See also Laughlin, McManus, and d'Aquili, on shamanic myths and rituals involving the "theater of mind," neurology, evolution, and phases of consciousness (or altered states of consciousness).

44. Cf. Nelson 46 on the emergence of the human mind as an "exaptation," not adaptation, like birds' feathers. Nelson also describes culture as an exaptation or "spandrel," a by-product in the evolution of the neocortex with intelligence and language as adaptations for survival, producing "complex cognitive potentials that were not confined to the solution of the original problems," but constructed "new cultural environments" that promoted group survival (56).

45. Cf. Teilhard de Chardin's notion of "Christian 'pantheism,'" with God as both Prime Mover and teleological "Omega" point of nature's evolution (171, 240). See also Paul Fiddes's theory of an alienated Christian God, suffering through the free will of all creation (225–26),

and physicist Freeman Dyson's idea of God learning and growing as the universe unfolds (119).

46. Cf. Hayden for the archeological evidence of religion in various prehistoric periods. See also Montelle, *Palaeoperformance*, on the emergence of theatre from "theatricality," as evidenced by prehistoric cave art (2–3).

47. "Ancestors" are meant here in a general sense. As to why cave art has been found mostly in France and Spain, see Mellars, who explores climatic and ecological factors that led to larger and less nomadic human groups there with abundant food (especially reindeer but also a diversity of sources). He argues that population density produced sharply defined territorial groups, promoting ethnic, totemic symbolism, religion and ritual to reduce competitive survival anxieties, and powerful shamanic specialists (223–24). He also theorizes how climactic changes led to the end of the cave art period (225–27).

48. See also Raphael 9, where he finds in Paleolithic cave art "the first idea of catharsis, and the germ of the chorus." He also sees scenes with "the grandeur of Aeschylian tragedies."

49. See also Lewis-Williams, "Constructing"; "'Meaning'"; and *Mind*; plus Lewis-Williams and Pearce, especially 224. On the difficulty of making such ethnographic comparisons, and yet the value of this particular theory, see Curtis 126–27, 140–41, 217–27.

50. See also Lewis-Williams, "Harnessing." Cf. Halifax (especially 11, 54–56, and 82) and Eshleman. For a critique of the theory that Paleolithic cave art resulted from "regular" vision quests, see R. Dale Guthrie 36–40. He argues that the lack of "lampblack" (deposits from burned materials), litter, and trails in the deep caves proves that very few people used them.

51. See Ryan for the use of historical shamanic rites, visions, cosmologies, and artifacts, in various parts of the world, plus Jung's theory of collective unconscious archetypes, to decipher prehistoric cave art.

52. Cf. Schechner, *Performance Theory* 68–69.

53. See also Pfeiffer 227: "the artists had a fine sense of drama and illusion."

54. Cf. Montelle, "Mimicry," for his view of Donald's theory. See also Donald, "Roots," on prehistoric cave art.

55. See also Montelle, *Palaeoperformance* 188–89, where he finds "origin" myths being communicated in the larger cave galleries, and "revelation" myths in the less accessible ones, though without considering gender implications. Cf. Hayden 193–94, on female "goddess" figures created in the Upper Paleolithic period and becoming more prevalent in the subsequent Neolithic, although animal figurines continued to be the most common.

56. According to Clottes, *Cave* 121, this painting is in an area of Lascaux "prone to extremely high rates of carbon dioxide, which may cause hallucinations and acute discomfort." Thus, the imagery may depict trance visions.

57. Cf. Ramachandran and Blakeslee 239, with a modern example.

58. A child's footprints were left here, as in Pech Merle, but in a side cavern not open to the public (Clottes, "Sticking" 198).

59. Cf. Clottes, *Cave* 194, who interprets the bison's position instead as lying on its side, dead with "five arrow-like signs."

60. Paul Bahn speculates that the Bedeilhac cave was "decorated from the entrance inward, but nothing has survived until about 350m in, where you encounter some red dots on the left wall" (121).

61. There is evidence of early human habitation in some of the caves with wall art, including one I visited, Bedeilhac (Clottes, "Sticking" 199).

62. See also Clottes, "Sticking" 208–09, on the prehistoric pieces of bone, shell, or stone stuck into cracks in caves with Paleolithic art, perhaps as a way "to pierce the veil separating the world of the living from that of the spirits."

63. Lewis-Williams finds evidence for this, too, in the rare images of humans within the caves (*Mind* 284). See also Curtis 133. "It is another version of the biblical Fall of Man.... *Homo sapiens* acquired forbidden knowledge and came to believe they were somehow distinct from other animals. The paintings express the guilt, the regret, and the triumph that came with the belief in that separation."

64. Cf. Clottes, *Cave* 23–25, who says that parts of the theories prior to his and Lewis-Williams's "are still valid."

65. The cave paintings were probably more communal activities, with greater planning required and division of labor (with some workers holding torches or lamps, others perhaps making the paint or scaffolding, as well as the actual artists). But the etchings, single-line sketches, and clay moldings on walls might have been made more directly in relation to individual visions, during altered states of consciousness, deeper inside the caves. Cf. Lewis-Williams, "Of People," 146, 150.

66. Cf. Winkelman, "Shamanism" 389, on the activation of both sympathetic and parasympathetic nervous systems during the altered states of consciousness (ASC) in shamanic rites.

67. See Kirby's theory (in the 1970s) of shamanism as the origin of theatre. Such a connection is more prevalent today in non-Western cultures. See Lee 41 on the shamanistic elements in Korean folk theatre and Kister 112 on comic catharsis in Korean shamanic rites. Cf. Shim. See Lemoine 112 on private and public exorcism per-

formances of Hmong shamans. See Dooley on shamanic elements in Japanese noh theatre. But cf. Matthews and Matthews on the Western tradition of shamanism, magic, and mysticism.

68. Cf. Clottes and Lewis-Williams 17 about research in the 1970s on subjects' hallucinations being "projected onto surrounding surfaces. Western subjects liken these projected images to ... 'a motion picture or slide show.'"

69. See Halifax 14, on the Ainu shaman of Hokkaido Japan: "Only the shaman is able to behave as both a god and a human. The shaman then is an interspecies being, as well as a channel for the gods." See also Kirby 9–22, on the magic acts, ventriloquism, torture ordeals, and supernatural masks of the shaman as theatrical devices.

70. See also McNamara's collection of scientific articles about religious and mystical experiences (*Where*).

71. Cf. d'Aquili and Newberg, "Neuropsychology" 246 on the possible neurology of Apollonian and Dionysian aesthetics.

72. Cf. Kubiak and Reynolds, on Machiavellian Intelligence in primates with regard to "transversal" theatricality, Shakespeare's *Hamlet*, and postmodern theory.

73. Cf. Turnbull and Solms 64: "We adults *project* our expectations (the products of our previous experience) onto the world all the time, and in this way we largely *construct* rather than perceive (in any simple sense) the world around us." On the psychoanalytic theory of projection as a defense mechanism, in relation to evolutionary psychology, see Nesse and Lloyd. See also Ramachandran and Blakeslee 60–61, 104–12, 152–55.

Chapter 2

1. See Burkert, *Greek* 200–01, on the opposition of sky and earth gods in Greece and other ancient cultures.

2. For a full translation of the remnant text, see Gaster 377–99. For excerpts, with helpful notes, see Jerome Rothenberg 134–41, 455–56. According to Rothenberg 456, the text of the Ramesseum Papyrus comes from the Twelfth Dynasty (circa 1970 BCE) but its contents may be dated back to the First Dynasty (3300 BCE).

3. See Read 94, for a picture of a "helmet mask" of the jackal-headed god Anubis, the only extant example of such a mask that might have been used in performances. The Ramesseum drama includes priestly embalmers "masked" as monkeys and wolves (Gaster 396).

4. See Kernodle, who calls the text a "production notebook," and connects some of its speeches to the (Old Kingdom) Pyramid Texts of 2625 BCE, and says it "confirms the continued performance of that drama in the Middle Kingdom" (21, 27–29).

5. See Wilkinson 10; Meeks and Favard-Meeks 18.

6. There was also a *proagon*, a procession showing scenes from the play prior to its full performance.

7. See Armstrong 58–61, 187–88, on the Anthesteria festival in honor of Dionysus at the time of new wine in spring, and on the "hint of rebellion" in Dionysian religion. See also Padel, *In* 182–83, and Jameson 57–63.

8. Cf. Burkert, *Greek* 64–65, on gods taking animal form in myth (such as Zeus or Dionysus as a bull), animals associated with them in iconography, and animals sacrificed to them.

9. Cf. Heinrichs 31–43 on the Otherness of Dionysus.

10. See also Hayden 50 on Osiris and other gods "based on shamans or supershamans."

11. In the first century CE, Plutarch of Chaeronea likewise finds Osiris to be "identical with Dionysus," based on parallel rites in their worship and both being "the lord of the nature of moisture" (qtd. in Marvin Meyer 169).

12. Cf. Gibbons 49, 70. See also Segal 23, on the *Bacchae* chorus's "delight in raw-eating" (*omophagia charis*), and 48–49, on Dionysian cult *sparagmos* as incarnating the god's mythic dismemberment by the Titans.

13. Cf. Segal on the doubling paradoxes of Dionysus (27–36), as the "animal sexuality" within Pentheus (32)—a "return of the repressed" in the Freudian sense (21–22).

14. Cf. William Scott 345, on the animal/human nature of Dionysus versus the distinctions made by Pentheus "between human aspirations and the short-range, instinctive goals of the animal."

15. Cf. Kihlstrom on the cognitive science of multiple personalities, in extreme cases and in everyone (without reference to Lacan's theory): "each individual possesses a number of context-specific selves" (463).

16. Cf. Segal 21, on Dionysus in the *Bacchae* as "a god close to the free passage of instincts and the open expression of emotions," becoming for Pentheus, "who has blocked that," the return of the repressed.

17. See Kalke on Pentheus's eventual "transformation into the thyrsus of the god" (413).

18. Cf. Barrett on the "metatheatrical manipulations" of characters within the drama and the "metaphorical sparagmos" of its audience (349, 354).

19. Cf. Cozolino, *Human* 305, on this and other "planes of neural integration" through narratives. Cf. also Jaynes.

20. Cf. Marks, Review, who critiques the way that Wrangham and Peterson present evidence of chimpanzees as "demonic males." See also *What* and *Why* for his broader critique of evolutionary reductionism.

21. "They can mate dozens of times a day; ... they manipulate each other's genitals with hands or mouth; they adopt an impressive variety of copulatory positions; [and] their genitalia, male and female, are proportionately larger than humans'" (Wrangham and Peterson 213).

22. See also de Waal, *Our Inner Ape*.

23. See Panksepp 286, for a comparison of rat and chimpanzee "rough-and-tumble play," with their "ingrained ludic impulses," to human "ritualized dominance sports."

24. See Gallup 123. Gallup points out that "chimpanzees reared in social isolation are incapable of recognizing themselves in mirrors." So the mirror stage relation of mother and infant, in Lacan's terms, may be needed to set up such self-awareness. See also Byrne, *Thinking* 100–18. Cf. Ramachandran, *Brief* 121, who questions whether "the Gallup mirror test [is] really a valid test for awareness of self." But cf. also Rochat 87, who argues that behavioral changes in front of mirrors show "reliable cognitive changes in the perspective of both phylogeny and ontogeny." He points out that human infants do not pass the mirror test of self awareness until the middle of their second year. This corresponds to the end of the Lacanian mirror stage of 6–18 months.

25. Geoffrey Miller theorizes the "Dionysian effect" of the enlarged human brain, evolving as seductive ornamentation, like the peacock's tail—though he does not consider the tragic associations of this Greek god in such a brain's flawed success (Ridley 339).

26. To match other apes, humans should be born after an 18–24 month gestation (Shlain 8; Ridley 340; Cozolino, *Human* 21).

27. See Cozolino, *Psychotherapy* 12, 107, 140, and *Human* 69–72. See also Solms and Turnbull 175–76, 280–82, on the executive system of the prefrontal lobes that enables "free will" by inhibiting the repetition compulsions of primitive brainstem structures and emotional memories. (The prefrontal lobes are the anterior parts of the frontal lobes.) The prefrontal lobes "continue to develop throughout the first two decades of life" and thus are "literally *sculpted* by ... parental (and other authority) figures" (282).

28. See Taylor on the "tending instinct" in humans, a nurturing drive that is often stronger in females due to hormonal differences, especially in times of stress, yet still forms the basis for group bonding and sacrificial altruism in males.

29. Cf. Marvin Meyer 3, on the view of Prodicus of Ceos, in the fifth century BCE, that the Greeks personified the good things of life as gods: bread as Demeter, wine as Dionysus, water as Poseidon, fire as Hephaistos, etc. See also Hecht 8, on the similar idea from Democritus of Abdera that earlier Greeks had invented gods as personalities for what they feared and admired in nature.

30. See Rehm, *Play* 160–61.

31. See Mossman 62, on the "horse-breaking" terms used for Prometheus's bonds, with Zeus reducing him to animal status. Io is described in a similar way, when she appears onstage, as half-human and half-cow (63). See also Burkert, *Greek* 126, on Zeus as rain, storm, and lightning god.

32. Cf. Lakoff and Johnson 560–61, on Strict Father (survival of the fittest) or Nurturing Parent metaphors of evolution.

33. "An animal with primary consciousness sees the room the way a beam of light illuminates it. Only that which is in the beam is explicitly in the remembered present; all else is darkness" (Edelman, *Bright* 122).

34. Cf. Pizzato, *Theatres of Human Sacrifice*.

35. For a detailed answer, see Sober. He argues that altruists interacting together benefit group survival and that human compassion, even beyond kin, is a side effect of mother-child nurturing, although that is also biologically competitive, as are groups.

36. See Boehm, "Egalitarian," 358–59, for a summary of the debate on the evolution of altruism and an argument for the "group effects that support so-called 'altruistic' behaviours such as co-operation, sharing, or patriotic self-sacrifice." See also de Waal, *Good* 12, 24, 135–36, for an explanation of reciprocal altruism, as theorized by Robert Trivers in the 1970s, which de Waal sees (along with kin selection) as an evolutionary step toward the Golden Rule of human morality, as shown today in non-human primates.

37. Cf. Dawkins's argument in Barlow 218–19 that meme theory is a better explanation of "cultural evolution" than kin selection and reciprocal altruism.

38. See also Collins, *The Language of God*.

39. Even in the fifth century BCE, Xenophanes of Colophon suggested this idea, "claiming that if oxen and horses and lions could paint, they would depict the gods in their own image" (Hecht 7). And he pointed out that Ethiopians described the gods as black and Thracians pictured divinities as blue-eyed and red-haired like they were.

40. See Scully and Herington's introduction to the play (17). Cf. White 108.

41. Cf. Schore, "Human" 39–43, on the importance of primal, mother-infant attachment communications, involving the limbic system and right brain, to set up an "internal working model" guiding the child's future social interactions throughout life, especially concerning "empathy" as a "moral emotion"—after the "oedipal" development of repressive "competition" in the left brain's dominance over the right, from age three and a half onward.

42. Cf. Nussbaum 388–91, on alternative translations of Aristotle's *katharsis*, not as "purg-

ing," but as purifying or clarifying emotion, "since the word obviously has a cognitive force" (389). See also Pizzato, *Theatres* 7. Cf. Rifkin on the empathic potential of the human race and global economics, through changes in the brain.

43. See Rehm, *Play* 165–66.

44. Atreus was Orestes's grandfather, who served the bodies and blood of his nephews to their unknowing father, Thyestes, in a vengeful, cannibal meal. Atreus's grandfather, Tantalus, played a similar trick on the gods, serving them his son, Pelops, in a meal, to test them. The gods resurrected Pelops, but Demeter had eaten his shoulder, since she was distracted by her mourning for Persephone, her daughter, who had been captured by Hades, god of the underworld. So Demeter replaced Pelop's consumed shoulder with an ivory one.

45. I have explored the ghostly aspects of *The Oresteia* in a previous work (*Ghosts* 30–41). See also Rehm, *Play* 78, on the significance of the *skene* (scene house) as the home where Agamemnon and then Clytemnestra are killed.

46. See also Burkert, *Greek Religion* 197–98, on the Erinyes as an embodiment of "self-cursing" when a false oath is sworn.

47. In Aeschylus's *Oresteia*, Orestes describes the Furies as "Gorgons ... with their serpents knotted" and with "blood-drops from their eyes" (152). But Pythia, priestess of Apollo's shrine at Delphi, has a vision of them as "wingless" (161). Cf. Euripides, *Orestes* 36, for a reference to the Furies as "three women ... like the night," although they do not appear onstage in that play. (The Furies were also daughters of Night, like the Fates.) In Euripides's *Electra*, the god Castor warns Orestes about the Furies as "hell-hounds," but they do not appear there either (174).

48. Megaera comes from *megairō*, meaning "to begrudge someone of something (because you think it is too great for him)," probably also related to *megas* (great). In the passive tense, *megairō* means "to be envied." Alecto derives from *a* (not) and *lēgō* (to leave off). Tisiphone stems from *tisis* (revenge) and *phonē* (murder). I thank my UNC-Charlotte colleague, classicist Dale Grote, for these insights.

49. Cf. Girard, *Violence*, on "reciprocal violence" and the cathartic "scapegoat." Emotional contagion is a term in cognitive neuroscience, considered in the conclusion of this book.

50. Cf. Victor Turner's use of the anthropological term "social drama" in relation to neuroscience.

51. As the Chorus of Trojan women, Agamemnon's captives, report to Orestes in the second play: "he was mutilated, did you know that? / Cut apart, / disfigured? And she who buried him like that / did it to make / his death impossible for you to bear" (Shapiro and Burian 123).

52. See Padel, *Whom* 80, who calls this "one of the loneliest lines in tragedy," epitomizing the Greek sense of madness as alienation — in seeing too much truth, not just a false delusion (95–96).

53. The Furies persecute those who break nature's laws and would even stop the sun god, Helios, from changing the course of the sun in the sky, according to Heraclitus (Dodds 7–8). On the different moralities of the Furies and Apollo, see Helm 49–42.

54. Tartarus was a place "as far below the earth as heaven above it," where Zeus imprisoned the Titans when he overthrew them (Lefkowitz 19).

55. See Burkert, *Greek* 181, 331–32, on the *daimon* as "a special being [that] watches over each individual ... person," as an interpreter moving "between gods and men," and as good or evil — since some are "filled with greed for blood and sexuality."

56. According to Rehm, the audience then has a double vision, like Pentheus under Dionysus's influence in the *Bacchae*, of two suns in the sky: Medea's chariot and the natural one (*Play* 268).

57. Whether a raised stage separated actors from chorus in the fifth century is debatable. See Rehm, *Greek* 36, and David Wiles 77. But the stone ruins of fourth-century theatres suggest such a division of performance spaces.

58. See also Obbink 83–85, on the ancient theory of Prodicus that Dionysus was a personification of the natural power of wine.

59. Cf. Rehm, *Greek* 105, on details in the play that undermine Apollo's argument, so that it has "less than ... Aeschylean approval."

60. See also Schore, *Affect* 238–39: "In the middle of the second year ... the father first begins to become an emotional object on a par with the mother, thereby shifting the infant from primarily dyadic to triadic object relations. His increasing role ... influences ... the experience-dependent maturation of the dorsolateral system" and its ties to the left hemisphere. Cf. Pizzato, *Ghosts* 196–98, for Lacanian and Kristevan parallels with this neuro-psychoanalytic, object-relations theory.

61. See Cozolino, *Psychotherapy* 192: "The building of the social brain between 18 and 24 months is driven by the attunement between the right hemisphere of the parent and the right hemisphere of the child.... [Thus] the unconscious of the mother is transferred to the unconscious of the child."

62. See Pizzato, *Ghosts* 35–38, for further details on Apollo and the Furies, with the latter as his obscene underside.

63. Zeus swallowed Athena's mother, Metis, who then gave him a headache as she made Athena's armor inside him. Hephaestus split

Zeus's head to release Athena, fully armored at birth — akin perhaps to the left-brain defense mechanisms that Ramachandran describes.

64. See Bacon 54, on the Eumenides as "resident aliens" (*metoikoi*), as outsiders who become insiders, like Orestes. See also Caldwell 153, on the Erinyes as castrating females in relation to the psychoanalytic sense of a phallic mother.

65. See Kay 80–82, 171–72, on Slavoj Žižek's political elaboration of Lacan's theory of the Symptom (*sinthome*): "the manifestation of the subject's enjoyment which he ... should embrace as 'what is in him more than himself'" (Kay 80).

66. See LeDoux, *Synaptic* 322, on the "imperfect set of connections between cognitive and emotional systems in the current stage of evolution of the human brain." See also Deacon 437, who makes a similar point, specifically regarding the human awareness of death.

67. See also Newberg and Waldman 78, for a more precise location of the abstractive operator "at the junction of the superior temporal lobe and the inferior parietal lobe in the left hemisphere."

68. Cf. Detienne, "Forgetting," on the many "couplings" of Apollo and Dionysus, as similar or opposed, in various ancient sources, beyond Nietzsche's paradigm.

69. See Seigel 183, on the "even-keel emotional states of mild interest and calm" in the left hemisphere, whereas the right produces states of "high arousal, ranging from intense joy to rage."

70. In Aristophanes's *The Clouds*, the character of Socrates uses rational arguments in trying to get his neighbor, an old farmer, to question the existence of Zeus, as the force behind rain and lightning. See Hecht 12, who also claims that by the previous century, "educated people commonly held that traditional belief in the Olympian gods had been fully discredited" (10).

71. See Sissa and Detienne 210–14, on various versions of Athena, including as "Mother" (211).

72. See Ramachandran and Blakeslee 179–82, on patients with a "temporal lobe personality," who experience a mystical communion with God during limbic epilepsy seizures — as mentioned in chapter 1 here. Cf. Padel, *Whom* 213–17, who interprets the madness caused by ancient Greek gods as "conflicting divinities" within the self, especially with Orestes's schizophrenic double-bind.

73. Cf. Padel, *Whom* 133–34, on the "revel-dance" of the Furies and Bacchae, in contrast to the ordered measures of other choruses. See also 128–30, on the *ekphron* (out of mind) state caused by the Furies versus the *entheos* (enthused, in-godly) condition of Dionysus's Bacchae — in relation to the Greek view of the mind's mad wandering like the hysterical womb in the woman's body. But she also finds that Dionysus drives Pentheus *eksteson phrenon*, "out" of his mind (193).

74. For a critique of Winkler's argument, see Rehm, *Greek* 26–27, 151 n11.

75. Cf. Newberg and Waldman 207, on the disorienting "randomness" of Pentecostal glossolalia (speaking in tongues), which "disorients the practitioner," while "the music encourages the brain to have an integrated and spatially realistic experience."

Chapter 3

1. Cf. Pagels 39–44, on a shift in the Jewish tradition, in the sixth century BCE, from "the satan," as angelic messenger and servant of god, to a demon representing certain Jews considered immoral. In the second century BCE, Satan became personified, through apocryphal versions of Genesis, as a fallen, rebellious angel and as the cosmic force behind such immorality, even in God's chosen people (48–51), perhaps with other demons assigned by God to different nations to lead them astray (54). The four gospels then presented the non–Christian Jews as demonic to distinguish the story of Jesus from other Jewish rebellions against Roman rule and to connect Christ's earthly suffering with the cosmic battle of God versus Satan (6–12). Early Christian writers under Roman persecution, such as Justin and Origen, also developed this idea, viewing the Roman gods as devils, natural catastrophes as caused by demons, and themselves as warrior-martyrs in a cosmic struggle (116–41). Yet, the Roman writer Celsus, in 180 CE, criticized such a dualist view as blasphemous against the pagan sense of one God or the collective gods (141–43). See also Wink 32–35.

2. See Perry and Schweitzer 5–7, on the difference between and yet lineage from medieval religious anti–Judaism to modern secular anti–Semitism and racial genocide.

3. Some scholars view these villains as also representing Lollard heretics. See Dox 192n2, for a summary. See also Chemers on the "threat of Islam," an evil that can "leak in," with the Jews' curses in this play invoking Mohammed (38).

4. See Kobialka, on the Eucharistic doctrine of transubstantiation in relation to medieval theatre — with God's body and blood "truly contained" in the changing substance of the bread and wine at mass, according to the Fourth Lateran Council of 1215.

5. See also Enders, *Death* 124. Cf. Enders, "Theater," on the Christian woman who sells the Host to the Jew in this play and then is burned at the stake, confessing also to killing her own child, after being raped. This, along with the Host turning into the Christ Child, ties the Jew to another medieval stereotype, that of Jews as baby-killers. Cf. also Enders, "Dramatic," and Perry and Schweitzer 46–56.

6. I am referring to the biblical story of Yah-

weh's command that Abraham sacrifice his son Isaac. But cf. Perry and Schweitzer 47: "no trace of human sacrifices existed in Jewish rituals." Talmudic rules require the blood to be drained from meat before it is kosher, yet medieval Christians accused Jews of blood-thirsty cannibalism, especially with Christian children.

7. See Clark and Sponsler on "racial cross-dressing" as a means of scapegoating in both of these bleeding Host plays. They find that both involve a "complex othering of Jews [that] is indeed anti–Semitic," with a more "grotesque exaggeration" of physical Jewishness in the Croxton play and a "less embodied," more performative otherness in *Sainte Hostie* (79).

8. I thank Dr. Dox for introducing me to the Croxton play in the 1990s, when she was a graduate student, studying under Michal Kobialka at the University of Minnesota, while I was teaching at the University of St. Thomas. My friendship with Prof. Kobialka and his passion for medieval theatre influenced my writing of this chapter as well.

9. Cf. Beckwith, *Signifying* 27, on God as a concept that "underwrites" medieval culture, whose "very impossibility" of representation determines an "imaginary signification" that generates "creative mythologies through which that society looks at itself." See also Beckwith, "Ritual," on the threat that Host plays posed to clerical control.

10. See also John Cox, *The Devil* 40–59, on the related figure of Vice (or vices as personified deadly sins) with a similar anti-social function onstage, from medieval to Renaissance England.

11. This Latin and Greek opening was also used in the Chester and York cycles, performed earlier in the 15th century (Burns 3; *York* 1). For clarity, I will focus on the Wakefield Cycle throughout this chapter, with brief references to other cycles in the notes. However, parts of two plays in that cycle, *The Creation* and *Abraham*, are incomplete in the original manuscript. So I am assuming, with editor Martial Rose (*Wakefield* 54), that the missing parts were similar to corresponding lines in the York and Brome cycles. I will also use the modernized English in Rose's version, rather than the original, Middle English terms and spelling (or the Latin stage directions). Cf. Cawley for the latter.

12. Other cycle plays begin this way as well, with an "archetypal anti-social act, which is also demonic"— as Lucifer and his followers rebel against "the undivided community of heaven" (John Cox, *The Devil* 19). In the Chester Cycle, there is a similar debate among the angels about Lucifer's coup. But God returns, after Lucifer sits on his throne, and expels the conspiring angels from heaven (Burns 5–7). In the York Cycle, Lucifer's vanity and greed for power are briefly expressed—without the angels' debate and then with a more sudden fall (4).

13. Satan probably appeared on the medieval stage dressed in black and with black makeup, as one of his compatriots, the 1st Devil, complained when thrown out of heaven with him: "Now are we made as black as coal / And ugly, tattered as a foal" (63). That devil also described Lucifer's face as "dark that once did shine."

14. Cf. Penchansky, chapter one, on the "insecure" and thus monstrous God of Genesis.

15. See also Armstrong, *History* 130, on Anselm's 12th-century theology, which "reinforced the Western image of a harsh God who could only be satisfied by the hideous death of his own Son, who had been offered up as a kind of human sacrifice."

16. Cf. Armstrong, *History* 148, on the Islamic tradition of "Satanic" verses expunged from the Koran.

17. See also Rizzolatti and Arbib, especially 192, on the imperative structure of canonical neurons and the declarative structure of mirror neurons, as providing the initial grammar and "mimetic capacity" for human language to evolve. See also Corballis, *Lopsided*, on the generative assembly device (GAD) in the left cortex and "special talents for music and art" in the right, as unique to humans (273). He contrasts the "vocabulary-based, generative mode" in the left with the "earlier, holistic mode of representation" in the right (309). And he points to recursion, embedding phrases within phrases, or allowing thought about thought, as a key property of GAD and human self-consciousness (312).

18. Cf. Staub and Hussey 181, on the medieval performance space being "unframed" like current cyberspace, with the computer game user as an "Everyplayer." Their comparison elides the different cosmological and technological frames of the "mystical and sacred space" in each case.

19. This line was cut, along with other choral parts, in the spectacularly bloody movie version of 2007, directed by Tim Burton and starring Johnny Depp. Sondheim's 1979 musical, *Sweeney Todd*, was based on the 1847 melodrama by George Dibdin Pitt, *The String of Pearls*.

20. Cf. Burkert, *Greek* 82, on the *pharmakos* as *katharsion*.

21. See Leonard Katz 1–77, for further details on this position, as presented by Jessica Flack and Frans de Waal, with various responses and critiques by others, plus their counter-argument. Cf. Prinz for a philosopher's challenge to such evidence that "apes have anything like human moral attitudes" (263), while arguing that morality is a biocultural byproduct of human evolution, with "some" moral rules "informed by biological tendencies" (274).

22. For specific details on the neuroanatomy of human morality, see Greene and Haidt. See also Greene, "Secret," for his critique of Kantian

moral theory, with various cognitive experiments showing that "deontologist" principles, though ostensibly rational, are based on evolved emotions. Elaborate "confabulations" are produced when the irrational must be rationalized. Cf. Moll et al., who argue that morality is based on certain mammalian emotions (attachment, aggression, and social rank dominance or self-esteem), plus further cognitive mechanisms in humans (4, 7). In contrast to Greene's "dualist notion" of reason versus emotion, they find that the prefrontal cortex engages moral emotions by actively "representing social knowledge" and with emotions guiding moral judgments by attaching values to behavior (5).

23. Cf. Kimball, on the "infanticidal logic" of Judeo-Christianity.

24. See also Decety and Chaminade, "Neural," on their experiments with shared representation networks in the brain, using videos of actors telling neutral or sad stories, while showing similar or different expressions, and brain scans of spectators sympathizing with them.

25. See also Blakeslee and Blakeslee, on "intuition cells" (Von Economo neurons) in the anterior cingulate and frontal insula, regarding shared bodily representations.

26. Cf. Clay 102 on Hesiod's Pandora, in the Prometheus myth, as representing the institution of marriage: "unlike promiscuous beasts ... and the similarly promiscuous gods, human beings uniquely regulate sexuality and reproduction through marriage."

27. Cf. John Cox, *The Devil* 25, on Herod in the Chester Cycle, as a parody of God, who, with hubris akin to Lucifer's, boasts that he's godlike and yet destroys his own son by mistake. See also Clark and Sponsler 71, on the stereotypes promoted by Herod's mask and the Jews' conical hats.

28. Later, in *The Resurrection*, Jesus makes clear that his torturers, who tied him to the cross, were Jews: "With cords enough and coarse ropes tough / My limbs outdrawn by fell Jews rough" (465).

29. See Mills for a comparison of *Everyman* to its source play, the Dutch *Elckerlijc*, and to the earlier English morality play, *The Castle of Perseverance*, regarding its theatres of world, memory, salvation, and body.

30. Cf. Greene, "Secret," on human reactions to "unfair offers," with increased activity in the anterior insula, "a brain region associated with anger, disgust, and autonomic arousal" (54). The desire for retributive punishment also involves the caudate nucleus, associated with emotional motivation and reward.

31. Emotional contagion in humans will be considered in the conclusion here.

32. Cf. Stark, especially 213–14. As a sociologist and historian, he details various aspects in the appeal of Christianity as it expanded in Europe. Yet he oversimplifies the case when he describes it as being "more rational" than pagan predecessors because it provided a clear distinction between good and evil for more cosmopolitan societies (201–02).

33. See Kobialka on the relationship between medieval Church debates about the status of the Eucharistic bread and wine, as God's body and blood, and the representations of God or Jesus onstage.

34. Cf. John Parker 98–104, on the "antichristian economics" at this point and elsewhere in *Everyman*. "In the end Everyman actually succeeds, from one perspective, in buying his way out of mortality" when he wills his money to charity (102).

35. Cf. Paulson 132, on the possible doubling of Everyman's friends and attributes, especially by traveling players, and on Good Deeds as being similar to the character of Virtue in other morality plays.

36. See Godfrey on the performability of *Everyman*, originally and today.

37. The right frontal insula also maps social emotions through visceral sensations, as polar opposites, such as "lust-disgust" (Blakeslee and Blakeslee 188).

38. Cf. Ladd 58. See also Harper and Mize.

39. Cf. Roth 79–80, on the language problems of patients with right-hemisphere damage who "lose the ability to grasp ... abstraction, context, and metaphor, as well as the nuances of semantics and tone, which often require interplay between the two hemispheres." In Lacanian terms, this might be called the Imaginary aspect of the Symbolic order.

40. See also Ramachandran and Hirstein.

41. Cf. Girard, *Scapegoat* 190, 202.

Chapter 4

1. On the evolutionary implications of the Great Chain of Being, as it persisted in scientific thought, see Marks, "Great."

2. See Kordela 71–80, on the history of God in secular reason, from Spinoza and Leibniz in the seventeenth century to Lacan in the twentieth, for whom God is unconscious. See also Lacan, *Four* 59.

3. See Allan on the "occult philosophy" of Faustus, in relation to medieval analogies and Renaissance representations.

4. Cf. Weimann 182–85, for comparisons of Faustus to Everyman and of Mephostophilis to the allegorical character of Vice, regarding the popular tradition of comical and serious devils in English theatre.

5. Cf. Gallagher on the "materiality of ethics" that appears in the staging of Faustus's blood in this play (1).

6. See Poole, "*Dr. Faustus*," on the play's vacillations (like its Elizabethan audience) between a Catholic theology of free will, with the possibility of divine forgiveness, via purgatory, and a Calvinist theology of being predestined for heaven or hell. She also considers the Good and Bad Angels in relation to *Everyman*— with Faustus as the self-made Renaissance Man (102–03).

7. Cf. Womack on clownish devil and vice characters, unleashing "violent horseplay, nonsense, parodic doubletalk and metatheatrical jokes," from medieval to Renaissance theatre (131).

8. Cf. Halpern who argues that this makes the play "a comedy, though an infernal one," with Faustus getting in the end what he desired all along (485). Yet, Halpern also finds tragicomic, Adornian heroism in Faustus's death-drive, as it negates the commodified artwork and reified subject (489–90).

9. See Wall-Randell on the "bookish aesthetic" of the play, in relation to Renaissance print culture (264).

10. Cf. Cless on the ecological aspects of *Doctor Faustus*.

11. Cf. Halpern on the question of whether Faustus "owns" his soul, through "possessive individualism" and the capitalist wage contract (akin to that for playwrights), or whether it is simply "a gift from God" (461). See also John Cox, "'To obtain'" 30, 40–44. Cox views the play as "deconstructing the traditional opposition between God and the devil" regarding Faustus's soul (41).

12. See Hammill on Faustus's Imaginary and Symbolic identifications regarding his books.

13. Cf. Shariff and Peterson, who critique Libet's theory of conscious free will as being merely a veto power over prior unconscious decisions. They propose "anticipatory consciousness," with the choice of goals favoring certain unconscious action schemas, as a further form of free will or "close-enough" to it.

14. Cf. Poole, "Devil's" 197–205, on Freud's theory of devils as projections of repressed instincts — in relation to *Doctor Faustus*.

15. Cf. Wegner 209, on experiments that demonstrate how the "unconscious priming of self" might influence subjects to "attribute ambiguous situations to their own will," or, when "subliminally primed" in the opposite way, "with the thought of an agent that was not the self— God," especially if they professed a belief in God.

16. See Bevington for a comparison of hell and heaven in the final Wakefield plays with their staging in *Doctor Faustus*. Bevington argues that the A-text's more restrained staging is closer to Calvinist theology and Marlowe's original script, while the B-text's more spectacular version makes the play a "tragic melodrama" akin to the Wakefield Cycle (308).

17. Cf. Halpern 477–78, on the ancient Christian sense, from Augustine onward, of evil as a "lack of being," shown in Faustus as that which "inhibits his *potencia*."

18. Cf. Beauregard 86–108, on the Catholic aspects of *Hamlet*. See also Pizzato, *Ghosts*, for more on the spirits in *Hamlet* and the human brain.

19. Cf. Ashworth on the graveyard scene as Hamlet's "descent to Hades" (165).

20. Cf. Hart on the combination of wild, Asian, mother-goddess and Greek, virgin Artemis in Shakespeare's Diana. See also Bicks.

21. See Marsalek on parallels between medieval resurrection dramas and *The Winter's Tale*.

22. Cf. Hunt on Shakespeare's Jupiter as "a deity of work," on the redemptive power of Posthumus's dream vision, and on the play's ending as solving the riddle of the tablet (111, 122, 128).

23. Cf. Halpern on Prospero as "the opposite of Marlowe's Faustus ... the negation of Marlovian negation ... not subjected to the forces of darkness ... [yet enjoying the] complete, though sometimes grumbling, submission of the spirits he commands" (475). Halpern sees this as reflecting Shakespeare's superior economic position, as an actor-shareholder, to Marlowe's as just a playwright. See also Robert A. Logan's comparison of Faustus and Prospero, including the "egotistical pride that blinds them," as they become like the author "in playing God the Creator" (213, 216).

24. Cf. Roberts on the kinship between human culture and "the animal Wild," shown by the hybridity of Caliban and Ariel (114).

25. See also Davidson, "Toward."

26. See also Newberg et al., "Neuropsychological," especially 103: "abstract concepts such as good and evil ... likely arise with the development of the inferior parietal lobe ... with its connections with the limbic system."

27. Cf. Hogan 174–79, for an application of neurological research on the quicker, limbic and the slower, cortical pathways of emotion to different film experiences.

28. Cf. Girard, *Theater* 311–52, on mimetic envy, sacrifice, and satire in *The Winter's Tale*, *Cymbeline*, and *The Tempest*.

29. The Spanish captain captured the "giants" with a trick, offering them iron chains as a gift, which his men then bolted to the natives' feet. "When they saw that they had been deceived, they roared like bulls. And they cried aloud for Setebos to help them" (Pigafetta 15). There are also parallels between details in *The Tempest* and the encounter between Sir Francis Drake's explorers and the Patagonians in the 1570s (Frey 35–37).

30. Cf. Skura for a summary and psychoanalytic critique of this New Historicist view, soon after it began to emerge in 1988.

31. See also Emery 7. Cf. Bermel on the task of translating this and other Gozzi plays — and the appeal of their "plasticity" for experimental directors today (40).

32. See Nordmann on Kleist's communal "death-wish" and eventual suicide, in relation to a recent staging of this play (25).

33. Kleist offended his own rulers with this play, which was banned from performance in Berlin and Vienna (x).

34. Hrdy mentions the golden hamster mother who may "cull" her litters to improve quality or size, and eat a few pups to "recoup some of her investment" (46). Abandoning an entire litter, if it numbers below a certain threshold, occurs in these hamsters and also in larger mammals, such as lions and bears. Infanticide by rival or dominant mothers (sometimes with the eating of infants) occurs in many social mammals: chimps, ground squirrels, prairie gods, wild dogs, marmosets, mongooses, and others — "some fifty species in all" (52, 92). In baboons and macaques, higher-ranking females will harass other mothers with daughters, which are more competition for them than sons, causing increased infant mortality (82). In a study of black-tailed prairie dogs, it was found that nearly a quarter of the litters of weaker mothers were destroyed by rival, lactating females (93). Also, in certain rodent species, pregnant females will re-absorb their embryos if potentially infanticidal males are in their territory (89).

35. See Nickisch on Büchner's scientific background, and yet his critique of science in *Woyzeck*.

36. See Fink, *Clinical* 93, on the psychotic structure, which lacks an "anchoring point" in social reality, given to perverts and neurotics through the "Father's No," according to Lacan. See also Fink, *Lacanian*.

37. For a critique of Buss's evolutionary model of human desire as too deterministic, and an argument for integrating it with the "standard social science model" (stressing cultural variation due to behavioral learning and social construction), see Levy, *What* 139–47.

38. For a feminist view of the play, see Martin (about Marie as a sacrificial scapegoat) and Schafer (about Marie in relation to the male gaze).

39. See Dean 54–55, regarding Lacan's explanation of a mysterious murder by the Papin sisters. "For the crime, in fact, marks what Lacan calls ... the *passage à l'acte* by which the criminal moves from pathology to ... the relief effected by the self-punishment the crime permits." See also Pizzato, *Edges* 124–25, regarding this diagnosis of the Papin murder and Jean Genet's play, *The Maids*, inspired by it.

40. "Gorgeous clouds, tinted with sunlight.— EVA, robed in white, is discovered on the back of a milk-white dove, with expanded wings, as if just soaring upward.— Her hands are extended in benediction over ST. CLARE and UNCLE TOM who are kneeling and gazing up to her.— Expressive music.— Slow curtain" (6.6).

41. Cf. Saxton 80, on a counter-effect with blackface minstrel shows: "Within a few months after the appearance of [the novel] *Uncle Tom's Cabin*, minstrels had coopted the title and main characters, while reversing the message."

42. Cf. Atkinson and Adolphs 159, on the neuro-anatomy of facial recognition involving fear, disgust, and anger. See also Gordon 167 on "facial empathy."

43. For related issues in cognitive philosophy, see Johnson 55–57 on "moral accounting" as a fundamental cognitive metaphor for revenge; Clark, "Connectionism" 109–11 on "prototypes" as figures for moral understanding that are different from stereotypes; and DesAutels 141–42 on potential "gestalt shifts" in frameworks and components of morality.

44. Cf. Pizzato, *Edges* 155–65.

45. See Pizzato, *Edges* 63–106, on various versions of God in Artaud's work (in relation to Derrida, psychoanalysis, and deconstruction).

46. See also Pizzato, "Nietzschean," on the Apollonian and Dionysian neuro-theatre of *Our Town*. Cf. the postmodern whimsy of a providential God as Puerto Rican janitor in an afterlife sauna in *Steambath*, by Bruce Jay Friedman (1970).

47. See Pizzato, "Soyinka's."

48. See Pizzato, "Brechtian."

49. See also Huerta, *Chicano* 40–42.

50. Recent plays about Greek myths include Cherrie Moraga's *The Hungry Woman: A Mexican Medea* (2001), Mary Zimmerman's *Metamorphoses* (2001), and Sarah Ruhl's *Eurydice* (2003). Another alternative may be seen in Constance Congdon's *Tales of the Lost Formicans* (1989), in which "Aliens" watch, direct, replay, and interpret fragments of various human lives. One of the humans, Evelyn, whose husband is becoming senile and thus communicating with the aliens, repeatedly says at one point: "Where is God?" (24).

51. Yet, in a different scene, Harper is told by the statue of a Mormon pioneer mother that God also causes change in humans: "God splits the skin with a jagged thumbnail from throat to belly and then plunges a huge filthy hand in, he grabs hold of your bloody tubes and they slip to evade his grasp but he squeezes hard, he *insists*, he pulls and pulls till all your innards are yanked out.... And then he stuffs them back, dirty, tangled and torn" (229–30).

52. My translation of: "Wer, wenn ich schriee hörte mich denn aus die Engel Ordnungen?" (Rilke 1: 441).

Chapter 5

1. Cf. Robert K. Logan. See also Levy, *Neuroethics* 29–63 on the "extended mind" of modern technology in relation to "neuroethics."

2. Cf. Rizzolatti and Sinigaglia 158–59, on the double evolution of human communication, with "emotive vocal expression" (linked to language prosody and music) involving deeper brain areas, as with primate calls, while human gestural and verbal language involves higher cortical areas.

3. By "our ancestors" I do not mean a direct lineage from Paleolithic to modern Europeans, but from humankind then to now.

4. See Dawkins on the extended phenotype of memes, expressing gene-like replications in human culture. See also Miller 271.

5. For an example, see my essay on Anthony Harvey's 1966 film, *Dutchman*, based on Amiri Baraka's play (Pizzato, "Skins").

6. Some insects and animals, such as ants, beavers, and various species of birds, do exhibit natural drives similar to human engineering and art. Also, elephants and apes have been taught to mimic human artistic expression, by making paintings while in captivity.

7. Baboon troops of up to 100 individuals are based on matrilineal kinship and a transitive rank hierarchy, with a dominant male who mates with many females, but only lasts as alpha for 7–8 months; females stay within the group but males enter or leave, and compete for dominance, often committing infanticide when taking power (Cheney and Seyfarth 57–58, 117). At other levels of the hierarchy, "Sexual consortships and male dominance ranks fluctuate over periods of days or weeks" (116). Cf. Miller 183.

8. See also Tomasello, "Two," on the human "understanding of others as intentional beings like the self"—as our major cognitive difference from other primates (180). For further details, see Tomasello, *Origins*. Cf. Meltzoff, who says that by 18 months human infants develop "an essential aspect of the adult theory of mind, namely that people (and not things) act in purposeful, intentional ways" (35). Cf. also Rochat 87, on human infants of that age, when they start to pass the self-recognition test (touching a dot of rouge on one's own face, as seen in the mirror)—a test which adult orangutans and chimpanzees pass as well (but not baboons). This corresponds to the Lacanian mirror stage, at 6–18 months of age.

9. Contrary to other primate researchers (such as Franz de Waal), Cheney and Seyfarth argue that baboons "have only limited sensitivity to grief and anxiety in others," even when mediating in others' aggressive disputes (196–97).

10. While not using the phrase, "theatre of consciousness" (as Bernard Baars does), Cheney and Seyfarth point out that the brain's medial prefrontal cortex is activated in both the assessment of others' mental states and "when people consider their *own* thoughts and beliefs" (201–02). Likewise, Tomasello considers "skills of recursive mindreading," with knowledge of the other's knowledge of one's own knowledge, as leading to joint goals and shared intentionality in humans, unlike in other great apes (324, 335).

11. Animal spirits sometimes appear as guides to humans in fantasy films. But the specific figures in prehistoric caves have not been used directly, as far as I know. The 2004 animated and live action French sci-fi film *Immortal* (directed by Enki Bilal) shows the Egyptian gods, Horus, Anubis, and Bastet. It features the falcon-headed Horus returning to a futuristic Manhattan, from a pyramid floating above it, in order to possess a human male's body and inseminate a special female. Also, a trend may be building to show ancient Greek gods and mythic warriors. As of this writing, several such films were released or are scheduled for release in 2010-11. Louis Leterrier's *Clash of the Titans* (a remake of the 1981 Desmond Davis and Ray Harryhausen film, which presented Laurence Olivier as Zeus, plus other Olympian gods, watching and toying with humans), Brett Ratner's *God of War* (with the hero from a current videogame going on a quest to destroy Ares, the god of war), Tarsem Singh's *Immortals* (with the mythic hero, Theseus, fighting the Titans and with other gods joining in the battle alongside humans), and Chris Columbus's *Percy Jackson and the Olympians* (with American teenagers and teachers as demigods, satyr, fury, and centaur). This popular interest in mythic heroes and gods has increased through fantasy novels and video games, as well as improved computer graphics in cinema. Perhaps it is symptomatic of a postmodern mix of neoclassicism and romanticism, moving beyond Christian mono-theism, through eclectic and nostalgic ideals.

12. Charlton Heston appears as a conventional white-bearded God in the clouds in the Paul Hogan comedy, *Angel and Me* (1990, John Cornell). He gives a crook the chance to believe he can act as an "angel of mercy." In the charming, magical realist film, *God is Brazilian* (2004, Carlos Diegues), God appears as a robust old man in white hair and a trim beard, dressed in light blue pants and a checkered shirt. He is compassionate about human suffering, though non-interventionist, and criticizes what humans have done with the world. After appearing to a fisherman, He travels with him and a Mary Magdalene-type girl, searching for a saint to leave in charge while He takes a vacation. But the saint insists on remaining an atheist, despite proof of God's existence. Godlike creators have also ap-

peared in various sci-fi films, such as *Blade Runner* and the *Matrix* series.

13. See Baird 333–38, on the "imaginary audience" in adolescent cognition, as abstract thought, self-awareness, and morality develop.

14. *It's a Wonderful Life* (1946, Frank Capra) is the classic example. But see Fowkes, 124–31, for more recent ones. She finds such angel-helpers to be masterful figures in comedies that involve "masochistic" human ghosts returning to certain parts of their lives, with the angel having a greater knowledge of destiny. This is also akin, I would add, to the Stage Manager in Thornton Wilder's play, *Our Town* (1938), who guides Emily's ghost in her visit to the past. On similarities between ghosts and angels in Hollywood films, see Edwards 84–86.

15. In a related study (Harris et al.), brain scans of believers and non-believers showed a greater activation of the ventromedial prefrontal cortex (involving emotional associations, self-representation, reward systems, and goals) with *beliefs* about both religious statements and ordinary facts, such as "angels exist" or "eagles exist." But religious *thinking* was associated more with brain areas involving pain, negative emotions, self-representation, planning, and cognitive conflicts (the anterior insula, ventral striatum, and anterior cingulate cortex), while judging ordinary facts relied more on memory retrieval networks.

16. Cf. Lakoff and Johnson on the Strict Father and Nurturing Parent as metaphorical schemas for competition and cooperation in evolution, regarding current ideas of morality and American political parties.

17. See, for example, *Date with an Angel* (1987, Tom McLoughlin), in which a beautiful female angel with a broken wing (Emmanuelle Béart) falls into an engaged man's swimming pool, the morning after his bachelor party. She brings joy to people with her glowing presence. Yet the hero's friends and his fiancée's father try to exploit the angel's sublime beauty through the advertising and news media. The hero hides her in the woods where she is surrounded by gentle animals. She bathes naked in a lake and her wing heals, restored by nature. But she can only speak in high-pitched chirps, squeals, and screams. She thus becomes the perfect object for the male fetishizing gaze: a dream woman with iconic beauty who never says a word. Eventually, the angel shows a wrathful nature against her antagonists — sending lightning and rain upon them. Yet she ultimately represents the nurturing aspect of God, whose voice is heard at the start of the film saying (about the hero), "Bring him home with love."

18. At one point in the film, the angels recall how they also evolved in witnessing early human evolution (according to the English subtitles): "Remember how one morning, out of the savannah, his forehead smeared with grass, the biped appeared, our long-awaited likeness? Its first word was a shout. Was it 'Ah' or 'Oh,' or merely a groan? We were at last able to laugh for the first time. And from this man's shout and that of his followers, we learned to speak." They also mention "the dancing in a circle, cave drawings, and writing" of early humans.

19. Heaven and hell, as realms beyond earth, are depicted more explicitly in other films, such as *Heaven Can Wait* (1978, by Warren Beatty and Buck Henry), *What Dreams May Come* (1998, by Vincent Ward, based on a Richard Matheson novel), *The Five People You Meet in Heaven* (2004, a TV movie based on a book by Mitch Albom), and *Dante's Inferno* (2007, an animated film by Sean Meredith).

20. An allegorical figure of Death is shown as a seductive, surreal female in *Orphée* (1950, Jean Cocteau), as a cloaked chess-player in *The Seventh Seal* (1957, Ingmar Bergman), and with postmodern, comical, identity twists in *Meet Joe Black* (1998, Martin Brest).

21. Megatron describes God as "funny" before She appears. And then again, after She presses Bethany's nose. A more traditional, yet alternative goddess is evoked as an idea and image, in *Woman on Top* (2000, Fina Torres). In this romantic comedy, a Brazilian chef in San Francisco (Penélope Cruz) gets famous on her own TV show, "Passion Food," yet misses her husband, who betrayed her. She then draws on the aid of the Afro-Brazilian oricha, Yemanja, goddess of the sea.

22. A very destructive Gabriel fights the rebellious but human-helping archangel Michael in *Legion* (2010, Scott Charles Stewart). Angels are also at war, with Gabriel (Christopher Walken) as destructive and rebellious in *The Prophecy* (1995, Gregory Widen) and *The Prophecy II* (1998, Greg Spence). The former includes "angelic script" on a cave wall, inspiring supernatural visions. Gabriel appears in a much friendlier form, played by Billy Connolly, helping an 11-year-old, British, working-class boy to do good deeds, in *Gabriel and Me* (2001, Udayan Prasad).

23. Constantine also says that God and the Devil made a wager, indirectly influencing human souls, toward good or evil, "maybe just for the fun of it" — with a rule of no direct contact. But that rule is being broken as full-fledged demons appear on earth, not just "half breeds." Along with such intriguing theological ideas, the film offers conventional mystery-thriller scenes, shoot outs, and an evil Mexican alien crossing the U.S. border with the "Spear of Destiny" — repeating melodramatic and racist stereotypes.

24. Two versions of an *Exorcist* "prequel" were made. Paul Schrader's horror film, which he initially scripted and directed, was remade by Warner Brothers and director Renny Harlin as an

action movie. But after the release of Hardin's film and its failure at the box office in 2004, Schrader's initial version was released in 2005.

25. For a feminist analysis of this aspect of the film, using the psychoanalytic theories of Julia Kristeva, see Creed.

26. I use the terms melodrama and tragedy as modes of drama across periods with different potential effects, rather than as narrowly defined genres like 1940s women's melodrama in film or 1950s Westerns. I argue that tragic edges can be found at certain moments in works that mostly follow the melodramatic formula (which does not begin with that term in eighteenth-century theatre, but goes back to the ancient Greek origins of tragedy). Cf. Lang 20; Singer 6–7, 57–58; Pizzato, *Theatres*; and Linda Williams.

27. Cf. LeDoux, *Emotional*, on links between the prefrontal cortex and limbic amygdala in the "extinction" of phobic fear responses through exposure therapy (170), in the ordinary planning and performing of emotional actions (177), and in behavioral and psychoanalytic treatments to control emotions through implicit learning or explicit awareness (265). Exposure therapy (as a type of behavioral treatment) is used to help veterans with post-traumatic stress disorder, like that shown by Jacob in the film. Yet, LeDoux points out the difficulty in gaining such control, due to evolved pathways from cortex to amygdala being "far weaker" than emotional networks from amygdala to cortex. See also 284–86, 303, on the direct and indirect influences of amygdala emotions on cortical systems, affecting working (short-term) memory, and the potential for "balance" between them.

28. In *Hellboy II* (2008, del Toro), the hero-devil fights similar demons, including insect-like Tooth Fairies and an octopus-like "elemental" Forest God.

29. Sympathy for the Joker increased due to the actor's death, just before the film's release. Heath Ledger also appears as the trickster Tony, along with the Devil (Tom Waits) in Terry Gilliam's *The Imaginarium of Dr. Parnassus* (2009), which was completed with multiple actors in the role due to Ledger's death.

30. Another black angel-cum-crook is featured in the tragic and mysterious film, *The Angel Levine* (1970, Jan Kadar). A street hustler (Harry Belafonte) comes back from the dead as a helpful, yet insecure angel, challenging the racist views of an old Jewish tailor (Zero Mostel), whose wife is dying. The film raises questions about poverty, prejudice, fate, and God's apparent disinterest or malevolence. Even the angel's identity is uncertain, as he is unable to regain the trust of his black girlfriend and may be just a thief who survived being hit by a car.

31. A 1967 film of the same title, directed by Stanley Donen, shows a short order cook (Dudley Moore) who sells his soul to a male devil (Peter Cook) for seven wishes and then, in another Faustian twist, meets the seven deadly sins, including Lust (Raquel Welch). Faustian deals are also made in the teen comedies, *Hunk* (1987, Lawrence Bassoff), with a she-devil, and *Deal of a Lifetime* (1999, Paul Levine), with a male Mephistophilis.

32. Elliot finds certain surprises as he is granted a series of wishes in exchange for his soul. He becomes rich and powerful as a Colombian drug-lord, but is then targeted for overthrow. He becomes a sensitive man to catch his dream girl, but then suffers as a freckle-faced softy on the beach. He becomes a basketball superstar, but with a tiny penis. He becomes a sophisticated intellectual who is gay and cannot have the dream girl. And he becomes President Lincoln being assassinated.

33. The Faustian contest of good versus evil meme-spirits becomes pornographic in the French film, *Exterminating Angels* (2006, Jean-Claude Brisseau). A filmmaker is observed and influenced by his grandmother's ghost and by unseen young women dressed in black, as he pursues the mystery of feminine *jouissance* by videotaping actresses as they masturbate and make love together.

34. Both *Angel Heart* and *The Devil's Advocate* involve Hollywood stereotypes of "voodoo" associated with Satan as pagan Other to Christianity. Yet they also show businessmen as satanic figures within modern consumer capitalism.

35. The biblical line, "His blood will be on our heads and the heads of our children," was cut from the subtitles to avoid a further appearance of anti–Semitism, but it remains spoken in Aramaic.

36. Cf. Winchell 226–29, about political attacks on the film in relation to Gibson's own anti–Semitic tirade when arrested for drunk driving in 2006.

37. According to Artaud, "A violent and concentrated action," onstage or perhaps onscreen, "is a kind of lyricism: it summons up supernatural images, a bloodstream of images, a bleeding spurt of images in the poet's head and in the spectator's as well" (*Theater* 82). He goes on to say: "Whatever the conflicts that haunt the mind of a given period, I defy any spectator to whom such violent scenes will have transferred their blood ... to give himself up, once outside the theater, to ideas of war, riot, and blatant murder." He then calls such tragic violence and its cathartic effect "a purification."

38. The melodramatic battle is given a tragic twist, however, near the end of the garden scene. While his apostles fight with their swords to defend him, Jesus shows compassion for a victim who lost his ear in the battle. He heals the wound, replacing the ear, and then tells Peter to

stop fighting, because "those who live by the sword shall die by the sword."

39. Such temptations are explored further in Martin Scorsese's 1988 film of Christ's passion, based on Kazantzakis's novel. See Pizzato, *Theatres* 138–49.

40. Or Symptoms, in a neuro–Lacanian sense, structuring the Symbolic, Imaginary, and Real orders of left cortices, right cortices, and subcortical networks.

Chapter 6

1. Episodes of *Millennium* were rebroadcast on the FX cable channel after the series ended in 1999 and were also released on DVD starting in 2004.

2. On the ritual ordering of spectators' lives in TV watching, see Dayan and Katz (especially chapters 4–5); Ferré 12–14, 21; Goethals; Kellner; and Liebes and Curran. See also Hill 608–09, for a survey of other theorists comparing film and television spectatorship. While some critics (Sylvia Harvey and Anne Friedberg) lament the loss of concentrated attention and aura in the shift from cinema to film on video, and others (Raymond Williams, John Ellis, and Timothy Corrigan) stress the flow experience, segmentalization, and glance aesthetics of TV, Hill remarks: "the viewing of film on television and video is not necessarily as inattentive — as the oppositions drawn between film and television viewing sometimes suggest" (609).

3. See Seltzer 14, 173–74, about murderers who were influenced by the media, actively studying their predecessors.

4. For various views on the cathartic effect of TV violence as purging aggression (using the terms "discharge," "blowing off steam," and "draining off aggressive energy"), see Crabb and Goldstein 366; Fowles, "Violence" 43 and *Case* 81; Hughes and Hasbrouck 148; Russell 168–71; and von Feilitzen. But I would distinguish the profound effects of tragic violence, involving clarification of emotions and purification of desires, from these views of a temporary purging. Cf. Fowles, *Case* 115–16, who argues the other way, in favor of a "redemptive violence" in TV melodrama, rather than the "nullifying violence of tragedy."

5. Žižek uses many specific films (but few television examples) to illustrate certain ideas in Lacan and their application to popular culture, with some references to Brecht as well. Žižek explores ideological issues and trends related to catharsis and mimesis, yet focuses on philosophical, rather than theatrical or therapeutic aspects of the Lacanian cure as the "ethics of the Real." See Žižek, *Art*; "Cyberspace"; *Enjoy*; *For They* 272; and *Looking*. See also Kunkle, "Žižek's," and Zupancic. Baudrillard has written more directly about the television medium as a simulacrum of violence, with the Loud family, for example, in a TV reality show of 1971 (49–55). Baudrillard's theories are based to some degree on Lacan's orders of the Symbolic, Imaginary, and Real — referring also to Artaud's theatre of cruelty (72). But Baudrillard dismisses the Real, as always already reproduced or "hyperreal," in contrast to Žižek's ethical concern with the Real through a more orthodox Lacanian view. See also Fink, "Ethics" 538.

6. *Millennium* received high praise and extreme criticism when it aired; it "divided critics into best and worst show of the season camps" ("'Millennium' Is"). Reviewers called it: "mournful realism," "unrelentingly downbeat," "grisly," "dark and uncompromising ... [plus] well-crafted, literate and suspenseful," "intelligently written, well acted, [yet] ... nauseatingly violent," "just old-fashioned stomach-turning tripe made to seem cynically hip," "the most dour hour on television," "the scariest series of the season," and "the best new show of the year." See, respectively, "'Millennium' Giving"; "First Hour"; Carman; Cox, "Creepy"; Levesque; Tom Shales, *Washington Post* (qtd. in "Critics"); Cox, "Chuckling"; Diane Holloway, *Austin American-Statesman* (qtd. in "Critics"); and *People Magazine* (also qtd. in "Critics"). Lance Henriksen, in the lead role of Frank Black, was nominated for a Golden Globe Award in each of *Millennium's* three seasons. The series received numerous nominations for other awards and won the People's Choice Award for "Favorite New Dramatic Series" in 1997. Its first season ratings were higher than those of its predecessor hit, *The X-Files*, in that show's premiere season ("Politics"). But *Millennium's* popularity dropped in its second and third seasons. It was canceled in 1999, prior to its own premise of the world's end in 2000. Yet this show was a milestone in TV history precisely because it challenged the mass audience so gravely, pushing the limits of violence and tragedy in the home theatre, while appealing to specific American anxieties on the eve of the new millennium (and 9/11). Its apocalyptic framework and prophetic references connected it with many Christian viewers, as well as those in a "mean-world syndrome" from the daily news media's fear-mongering (Gerbner et al.). But it eventually questioned the melodramatic righteousness and vengeance of those fundamentalist fantasies, with cathartic insights that are now even more prescient in our current decade.

7. The show's Millennium Group is based on an actual, forensic and behavioral science company called the Academy Group, which involves former FBI agents working as consultants and profilers for local law enforcement, but without the religious cult associations. Their logo is a red dragon, based on the Thomas Harris novel of that title. They discern the intense fantasies of

criminals from the evidence — somewhat like Frank's profiling visions.

8. See Alibek and Handelman 123–33, on the development of the Marburg virus as a biological weapon by the Soviet Union in the late 1980s. Marburg, a filovirus related to Ebola, produces "blood pooling underneath ... [and] oozing through" the skin, with blood streaming from the "nose, mouth, and genitals"—as the "body's internal organs literally begin to melt away" (125). It can spread through the inhalation of microscopic particles "with no direct physical contact" (22, 126).

9. See Murray 154 for neurological evidence that viewing TV violence activates certain areas of the brain's right hemisphere "similar to the memory storage of traumatic events by posttraumatic stress disorder patients." See also Ted Cox, "Creepy," on *Millennium* as a "metaphor for violence on TV. When Black imagines a murder, it flickers across the screen the way channels blip by when someone is surfing with the remote."

10. Cf. Zillman on the possibility for viewers to find a "cure" for anxieties through watching TV violence, as a "self-administered behavior-modification program" ("Anatomy" 160), especially with characters who are initially melodramatic, with clear-cut good or evil traits, but then develop a "tragic flaw" as the good protagonist, or "display a positive side" despite being the evil antagonist ("Mechanisms" 48–49). See also Zillmann, "Anatomy" 148–49, and "Psychology."

11. Many studies in the past quarter century have shown a strong correlation between watching high levels of TV violence and viewers' "desensitization" toward aggressive spectacle and the victim's suffering. This may "weaken some viewers' psychological restraints on violent behavior" (Hough and Erwin 413). Some researchers argue that children learn "aggressive scripts for behavior from observing media violence" (Huesmann 256). See also Centerwall; Hough and Erwin 412; Huesmann and Miller; Murray 152. This seems to affect boys more than girls, through the "catalyst" of identification with aggressive male characters onscreen (Hughes and Hasbrouck 143). But one study found that "viewing television violence can lead to the development of aggression for girls as well as for boys"—through similar identifications (Moise and Huesmann 383). Other reports offer many examples of violent acts, especially by children and teenagers, directly copying violence onscreen. As one scholar puts it: "the evidence indicates that the more people are exposed to violent television drama the more they are likely to be violent in their everyday lives. This may be especially true for children" (Condry 86). See also Bushman and Huesmann for a summary of evidence, from laboratory, cross-sectional field, and longitudinal field studies, about the effects of TV violence on children (up to the early teens), increasing both short-term and long-term aggressive behaviors.

12. Shrum also argues, against the critics of cultivation theory, that "people are often unaware of the source of the information they retrieve from memory" and thus viewers may increasingly believe their experience of TV violence as reality (264, 267–68). On different family contexts for children viewing TV violence as justified, see Krcmar.

13. According to the National Television Violence Study, 1994–95, "almost one-third of all programs (32 percent) can be characterized as showing no negative consequences of violence" (Wilson et al. 135). Only 16 percent depicted "long-term pain and suffering." See also "TV Violence and Kids" 24. According to Bushman and Huesmann 228, that study during the years 1996–98 demonstrates that TV violence is often sanitized and glamorized, with 73 percent of violent perpetrators showing no remorse, 55 percent of victims showing no pain or suffering, and only 15 percent of violent programs portraying long-term consequences to family, friends, and community.

14. Research over many decades by George Gerbner and his colleagues has shown that viewers of TV violence exhibit an increased fear of real-life aggression, by overestimating specific dangers outside their home theatres, through ritualized beliefs in violence onscreen. See Gerbner et al. 180; Philo 104; Shrum 261; Lyon 145; and Cantor. For a critique of Gerbner's research and theories, see Fowles, *Case* 40–43. Other researchers have used electro-dermal responses to show that when violence is "implied" offscreen, through various visual cues and the "background audio," it may elicit even more fear than when violence is shown directly onscreen (Kalamas and Gruber). Such physiological stress in the mass audience is market driven; so another scholar concludes that broadcasters should be forced to pay for the social cost of TV violence, as with any other risk assessment of "pollution" (Hamilton 3, 27–28, 39). However, some scholars have argued against the evidence of effects theory (that violence onscreen encourages imitation) and of cultivation theory (that screen violence increases spectators' fears). Instead of cultivating greater fear in viewers, TV violence may attract those who are already fearful, thus skewing the statistics (Chandler).

15. Cinema is more like theatre with its special, rather than ordinary, and public, instead of domestic, ritual space. It structures audience participation communally, through the "hypnotic relationship" between spectators and screen (Burch 124). Of course, DVDs have brought films into the home theatre, too.

16. In a study of violent TV characters in the 1994–95 season, only 13 percent were found to be "both good and bad," while 45 percent were just bad, "motivated primarily by self–interest,"

and 24 percent were purely good, "motivated by concern for others" (Wilson et al. 117).

17. See also Sharratt 285: "And while watching TV, the real home in which we watch operates as a norm which is curiously suspended but still present while another reality takes its place and takes place in it; the living-room becomes a palpable 'off-screen space' which continually reassures us against the lurking threats in the imaginary off-screen space of the horror movie."

18. Apter develops a different, non–psychoanalytic, "reversal theory" explanation. He defines a "safety-zone frame" in cinema and theatre, where the spectator is encouraged — with the "ritual" of taking a seat before the "enchanted zone" of the stage or screen — to enjoy the arousal of that "dangerous edge," reversing the threat of it, through a process of "self-substitution" (identification) with the characters in the drama (61–62). Here he also uses the term "detachment," although without reference to Brecht's distancing effect, nor to any theatre or film theories.

19. See Gould 8–9. He traces film and TV melodrama not only back to the eighteenth-century term for "drama heightened by music," and to nineteenth-century Grand-Guignol violence, but also to the ancient Greek "taste for dramas ending in the violent triumph of the good." He also mentions that violent melodrama today appeals to all classes of spectators. Cf. Eric Bentley's designation of Victorian stage melodrama as "the poor man's catharsis" (qtd. in Morse 17). See also Wilson and Goldfarb 327–28, on specific genres of nineteenth-century melodrama as leading to certain types of film and TV today: "Domestic melodrama became soap opera. Frontier melodrama became the western.... Crime melodrama became the popular mystery or detective show.... Nautical melodrama, which dealt with sailors and pirates, was the forerunner of swashbuckler films.... Equestrian melodrama — which featured horses performing spectacular tricks — and other popular melodramas which had animals as heroic stars were the predecessors of television and film melodramas featuring animals."

20. For specific steps in the early historical development from the theatrical proscenium to the cinema frame (and curtain), see Paul. See also Collier and Vardac, on the romantic and realist influences of eighteenth and nineteenth-century theatre in the film techniques of Murnau and Griffith, respectively.

21. Spectators may move through a movie, conceiving an entire diegetic realm, while the camera zooms, pans, angles, dollies, or cuts to new shots. But such suturing by the spectator, through the camera's eye, is still akin to the theatre viewer's co-creation of stage fiction through the mind's eye: focusing on various details and adding personal associations, traveling through space vicariously with the live actor (or physically with a mobile audience), and imagining an entire world offstage.

22. Cf. Doane 339, on film sound as it "*envelops* the spectator," unlike the visual frame of the screen.

23. For details and examples, see Paul 327–28.

24. Live actors onstage are affected in their performance by reactions in the audience. Likewise, although in a very different interactive medium, performances on millions of TV screens are edited by spectators switching between channels (or adjusting the volume and picture quality) with a touch of their remote-control buttons. Of course, TV can also be "live," unlike film, and sometimes involves its audience more directly in the action, as with onscreen voting by phone or the internet.

25. Cf. Raymond Williams 56 and Sharratt 283–88.

26. See Copjec's argument that film theory has misinterpreted the Lacanian "gaze" as Foucauldian, as coinciding with the spectator's panoptic look as meaningful and powerful. "In Lacan, on the other hand, the gaze is located 'behind' the image, as that which fails to appear in it and thus as that which makes all its meanings suspect" (36).

27. For a general view of imitation, mirror neurons, and emotional contagion regarding morality and compassion, see Bloom 114–18.

28. In February 1996, to avoid governmental regulation, the broadcast TV networks promised to institute an age-based ratings system, starting in January 1997. Due to continued public and political pressure, certain symbols were also added in October 1997: V for violence, S for sexual content, L for strong language, D for suggestive dialogue, and FV for fantasy violence in children's shows. See William Schneiderman and "Senate Committee" for further details. *Millennium* premiered in October 1996 and acquired the TV-14 rating during its first season, plus the V, S, and L ratings the following year.

29. See Sepinwall (of Newark's *Star-Ledger*) on the series premiere in 1996: "Literate, dark, and graphic, 'Millennium' may just be the most terrifying series in TV history. It's certainly the most disturbing." Bianco (of the *Pittsburgh Post-Gazette*) said its "essential violent ugliness goes beyond anything ever done on TV" (qtd. in "Critics"). See also Alyssa Katz's early review of the show's violence: "Its devotion to the twisted ways of the criminally insane raises interesting questions about the V-chip — among them, how many *Mortal Kombat*-nurtured kids will use it to protect their parents from *Millennium*'s severed heads and live burials."

30. See Alyssa Katz on the significance of the main character's name, for a certain segment of the mass audience: "Carter slyly borrowed his name from the former frontman for the Pixies,

an alternative rock cult figure whose songs move between fantasy and reality." See also Horsley 383–84, for a critique of Frank Black as the "central failing" of *Millennium*.

31. Frank's family became less of a focus in the show's third season (1998–99), after the death of his wife. The series plot then focused on the male and female detective pair, Frank Black and fellow FBI agent Emma Hollis (Klea Scott), like the Mulder and Scully paradigm in *The X-Files*, though with Frank as an older mentor.

32. The famous *X-Files* phrase, "The Truth is Out There," appeared on a poster behind Fox Mulder's office desk. The title "out there" also appeared in the banner at the beginning of each *Millennium* episode during its second season— replacing the first season phrase "who cares?" See Carter, concerning the *X-Files* motto: "it's double-edged, a joke. The truth is out there. It's also far out there. But, in fact, that's what we're all seeking" (84). Cf. Delasara 19, on this *X-Files* phrase as being "anti-postmodern"—although Carter's comment gives it a postmodern twist.

33. Cf. Carter's response, when asked if he himself is paranoid: "Very, very paranoid. I'm acutely aware of fear and betrayal. My father had a bad relationship with his mother. She had left his father at an early age, so he was keenly attuned to her betrayal of him. I think that's something that was passed down" (38). When pressed again about his own paranoia, Carter responds: "my mother, who I loved dearly, could never keep a secret. So if as a kid you go to your mother and you tell her something, and she can't keep it secret, it develops in you, you know, a sense that nothing is safe."

34. This plot device, introduced in the *Millennium* "Pilot," derives from the films *Manhunter* (1986) and *Silence of the Lambs* (1991), made from two novels by Thomas Harris. As Dyer points out (17), the real-life prototype for Frank Black and for similar detective "profilers" in these films, who catch the serial killer by learning how he thinks, is John Douglas—who was an FBI agent, author of *Mindhunter*, and technical advisor on the latter film. In actual practice, the profiling technique is psychological rather than psychic (Douglas and Olshaker 150–51). Yet Douglas's book, like the films, shows the profiler as tormented by his ability to get inside the killer's mind, which *Millennium* shares visually with the TV audience through Frank's flashes (or flashbacks) of the killing scene and killer's POV. Cf. the prologue of Douglas's book, entitled "I Must Be in Hell" (15–24). Frank's flashes of serial killing also recall the telepathic visions of the fashion photographer in *The Eyes of Laura Mars* (1978). Cf. Newitz 77 and Simpson 144n29. See also Joel Black.

35. Visually and acoustically, this recurrent special effect may have been influenced by a similar technique in Adrian Lyne's film, *Jacob's Ladder* (1990).

36. In Lacanian theory, human minds have either neurotic, perverse, or psychotic structures; yet neurotics may exhibit perverse and psychotic traits. In fact, psychotic and perverse structures are the logical, developmental foundations for the neurotic, Oedipal sacrifice of *jouissance*—and the further sacrifice of that sacrifice in the Lacanian cure. See Fink, *Clinical* 76–77, 165–66, 179, 194–95, 209–10, and *Lacanian* 53, 60, 69, 72.

37. The episode was entitled "The Beginning and the End." While four first season episodes, including the pilot, were written by the series creator, Chris Carter, none of the second season episodes were. Seven writers (or writing teams) were used for that season, but twelve of the twenty-three episodes, including this one, were written by Glen Morgan and James Wong. Morgan and Wong also wrote three episodes in the first season, along with Carter's four, although nine other writers or writing teams were used for the remaining fourteen shows. Cf. Carter on the lack of "good writers" for both *The X-Files* and *Millennium* (36).

38. A serial killer who teases the pursuing detective to become like him by killing him also forms the plot of the film *Se7en*. See Dyer for further parallels of plot and character between various films about serial killers with detective "profilers" and the *Millennium* series.

39. Earlier in the same episode, while Frank's wife is missing, he asks his Millennium Group colleague, Peter Watts (Terry O'Quinn), "What must I sacrifice to have her back?" Watts says he doubts "if you can sacrifice one thing to get another."

40. Cf. Fink, *Clinical* 172–74, on the pervert's refusal to sacrifice *jouissance* as the neurotic does, according to Freud, or in a more Lacanian sense, the failure of the paternal function to enforce the sacrifice. This leads to other types of symptomatic sacrifice (sadomasochism, voyeurism, fetishism, etc.) that try to make the law appear. The psychotic, however, does not even reach this secondary stage of Oedipal sacrifice that fails for perverts—"separation" from the mother through the naming function of the father's law—because he has not undergone the primary repression of "alienation" prior to separation. See also 178–79 for a summary of these three basic structures of the human mind regarding sacrifice: Lacanian alienation as the prohibition of infantile *jouissance* with the (m)Other and separation as the symbolization of the (m)Other's desire, related to the Freudian sense of primary and secondary repression.

41. Cf. Fink, *Clinical* 192. He points out that such presymbolic *jouissance* is always already lost, even to perverts. They do not return to the polymorphous perversity of childhood, with *jouis-*

sance everywhere on their bodies, but instead repeatedly enact an incomplete castration.

42. In "The Time Is Now," a continuation of "The Fourth Horseman," Frank receives one hypodermic syringe of the vaccine from fellow Group member, Lara Means, not directly from the Group. Lara leaves the syringe in an envelope with Frank's name on it, while becoming crazed with apocalyptic visions in a cabin where the Group has put her. She shoots a gun at Frank (seeing him as the devil) when he finds her there, and then becomes catatonic. Apparently, the Group gave her the vaccine syringe, but she chose to leave it for Frank as she went mad.

43. Cf. Lacan, *Seminar II*, 214, 229–33, on the "beyond of Oedipus" at Colonus, as depicted in Sophocles's play. See also Felman 128–59. And see Fink, *Lacanian* xii–xiii, 47, 62, 68, 79, on the subject's "assumption" of responsibility for the Other's desire as the fate of his or her existence. This is Lacan's interpretation of Freud's statement: "Wo Es war, soll Ich werden" (where it/id was, shall I/ego become)—as psychoanalytic cure.

44. Cf. Boothby, "Psychical Meaning," 357, on the two destinies of the death drive: "spectacular violence" on the Imaginary level or "sublimation of the self-deconstructive drive" in the Symbolic dimension. I would relate the former, Imaginary level of death-drive expression to the melodramatic violence of most television dramas. (Boothby uses the examples of his five-year-old son's "fascination with images of dismemberment" and the actual violence of Colombian bandits [339–42, 358].) But some TV exceptions, like certain episodes of *Millennium*, reach toward a more tragic sublimation as well. See also Boothby, *Death and Desire*.

45. In both theatre and psychoanalysis, the idea of "catharsis" has long been theorized as a potential cure for personal and social problems. Like Aristotle, Freud and Lacan focus on the problem of a tragic flaw (*hamartia*), a repeatedly "missed aim" in the neurotic's Oedipal sacrifices. In Lacanian terms, the persistence of the subject's unfulfilled and contradictory desires points to a deeper, erotic, death drive: the painful repetition of particular symptoms that express the knotted chains of signifiers in the unconscious. The subject gets a small degree of substitute pleasure through the sublimation of *jouissance*, in the continued Oedipal separation from the (m)Other, with the Name and No of the Father normalizing and repressing primal passions. Yet, the subject continues to be sacrificed by the weight of patriarchal prohibitions and by the irruption of maternal (or Other) desires—involving specific plot twists of unconscious signifiers and various character fantasies that mask the lack of being in being human. Cf. Žižek, *Metastases* 67–68, on the guilt paid to the superego through the Oedipal sacrifice of desire: "Superego is like the extortioner slowly bleeding us to death — the more he gets, the stronger his hold on us." See also Fink, *Clinical* 179–80, 184.

46. See Sconce 184–86, on the distinction between gaze and glance, suturing and surfing, in film and television theory. Sconce questions this distinction as essentializing both media: "there are, after all, disengaged cineastes just as there are transfixed video viewers" (185).

47. At the beginning of the *Millennium* series, Frank and his family return to Seattle, after living in Washington, DC, where he had been an FBI agent and developed his profiling gift, but also became a target for vengeful criminals. The violence of criminals, of the coming millennium, and within Frank follows him, however, and finds his family in their new home in Seattle.

48. For various ideas of the psychoanalytic death drive, throughout the history of Western culture, see Dollimore, although he favors Freud over Lacan (194–96).

49. Cf. Kaplan, *Women* 24, regarding the Oedipus complex and film spectatorship.

50. The desire for such an apocalypse continues to tempt many fundamentalists today, on both sides of the terrorist "holy war."

51. See Aristotle's *Poetics*, chapter 6: "A tragedy, then, is the imitation of an action ... with incidents arousing pity [sympathy] and fear, wherewith to accomplish its catharsis of such emotions" (1460). In book 8, chapter 7, of his *Politics*, Aristotle also describes how certain melodies, like that of sacred music, cathartically purge "feelings such as pity and fear, or, again, enthusiasm" (1315). Aristotle relates this to "music at the theatre" and describes two classes of "spectators," the educated and the vulgar, who deserve different types of music for catharsis, including "perverted modes" for the lower class (1315–16).

52. The yellow house was not included in the title images for *Millennium* in its third season (after the death of Catherine).

53. See Seltzer 22, on his description of America's current "wound culture, in which death is theater for the living." See also Kristeva, "Modern Theater," and *Powers* 26–29, for her discussion of theatre semiotics and catharsis, though without reference to the *chora*.

54. The apocalyptic fear of a "millennium bug" in computers was eventually used as a plot device in *Millennium*'s third season episode "Teotwawki."

55. See Earl's comparison of Aristotle and Freud. He insists that although Freud "abandoned the cathartic method, it ... remains as a residue in Freud's later work, and in fact still persists in various forms today ..." (83). This point is made, too, by Laplanche and Pontalis 61. See also Timothy Wiles 27, for a comparative

translation of Freud's cathartic method as purgation and Stanislavski's as purification.

56. "Psychopathic Characters on the Stage" was written in 1905 or 1906, but Freud then gave it to his friend Max Graf; it was not published until 1942. See Lacoue-Labarthe for a comparison of this essay to Nietzsche's *The Birth of Tragedy*, with both pointing to the tragic spectator's masochism, as a "link between catharsis and mimesis" (182).

57. See Palombo, for a model of therapeutic catharsis, using current complexity theory (involving emergence, complex adaptive systems, and phase transitions), which might also apply, in my view, to cathartic insights with theatre, film, and TV spectatorship.

58. See Fink, *Clinical* 165, where he defines neurosis as a "protest" against the sacrifice of *jouissance* imposed by one's parents, which eventually becomes desire in relation to the law.

59. Cf. Fink, *Clinical* 3–41, especially 14, on the analyst/analysand relationship, with "the analyst as an actor, a function, a placeholder, a blank screen, or a mirror."

60. Cf. Lonsdale 79, on Plato's "homeopathic model" of catharsis in relation to Aristotle's model, which involves "the use of external agitation to quell or purge internal disorder."

61. Cf. Fink, *Clinical* 97–98, 165. I would speculate that a pervert's violent acts, which try to make the law appear by transgressing it (as Fink explains), could be inspired by the law's appearance in, yet lack of power to stop, the repetitions of melodramatic violence on TV. Psychotic spectators might also be inspired by TV violence, since a common symptom of psychotics is to hear voices and see visions from the TV set as being directed at them personally (like hallucinatory spirits). Cf. Leader and Groves 111. As Fink points out, the psychotic, due to an absence of Oedipalization, "is more prone to immediate action, and plagued by little if any guilt after putting someone in the hospital, killing someone, raping someone, or carrying out some other criminal act" (98).

62. Cf. Žižek, *Metastases* 76: "The problem of contemporary media resides not in their enticing us to confound fiction with reality but, rather, in their 'hyperrealist' character by means of which they *saturate the void that keeps open the space for symbolic fiction*.... And it is here that violence comes on to the stage, in the guise of the psychotic *passage à l'acte*." Cf. also Kunkle, "Psychosis."

63. Regarding the third aspect, see Žižek, *Enjoy* 59: "This 'withdrawal' of the subject from the Other is what Lacan calls 'subjective destitution': not an act of sacrifice (which always implies the Other as its addressee) but an act of abandonment which sacrifices the very sacrifice." Cf. Dunand 248, 253–55, who says that the *jouis-sance* of the Other "has to be sacrificed as 'a thing of nothing,' calculated in terms of time and work" (254). Cf. also Fink, *Clinical* 69–71, on how the analysand "remains stuck" on the loss of *jouissance* to the Other, refusing "to allow the Other to enjoy ... the jouissance sacrificed" — until the sacrifice of that sacrifice (and reconfiguration of the fundamental fantasy) in the Lacanian cure. See also Lacan, *Écrits* 323–24.

64. Cf. Fink, *Clinical* 48, on the analyst's role in giving an "oracular" interpretation that plays on "two levels simultaneously (figurative and literal, affective and physical)" and thus hits the Real, unsticking the analysand's desire.

65. Cf. Lacan's myth of the "lamella," an amoeba-like phantom related to the death drive (and perhaps to Kristeva's sense of the maternal *chora*), which he describes as being like a monster in a horror movie or TV show: "something that would not feel good dripping down your face, noiselessly while you sleep, in order to brand it" ("Position" 273). See also Brousse 113–14.

66. Cf. Fink, *Clinical* 108–09, on the difference in the Lacanian treatment of a neurotic or psychotic: "the therapist tries to disrupt the neurotic's all too quick and convenient meaning-making activity, hoping to affect what is unconscious, not the ego. But with the psychotic ... the therapist must build up a sense of self ... that defines who the psychotic is and what his or her place is in the world." It is thus conceivable that the tragic episodes of *Millennium* could "disrupt" the view of some neurotic spectators and affect a cathartic awareness of unconscious truth, whereas its more melodramatic parts could build up a violent sense of self in some psychotics (or neurotics) and affect their active place in the real world. Similarly, the sadism or fetishism onscreen may not be enough for some perverse spectators and may inspire their own mimetic actions to make the law appear beyond the screen.

67. Cf. Timothy Wiles 4–8. He relates three different translations of *katharsis* (purgation, clarification, and purification) to the modern theories of Stanislavski, Artaud, Brecht, and Grotowski. Wiles himself, following Leon Golden and O.B. Hardison, prefers the translation of *katharsis* as "clarification" (like Nussbaum). On the translation of the term as "purification," see Davis. See also Salkever on Aristotle's specification of fear and pity as the two emotions involved in tragic catharsis, because they "inspire deliberation" (295). Catharsis uses these emotions to put "order into disorderly or incoherent souls" in the audience that might be tempted, like the tragic hero before his catastrophe, to believe in the hubris of "power and wealth" (297). Cf. Alford, *Psychoanalytic* 154–55, who disagrees with Salkever. See also Scheff for other emotions involved in his sociological theory of catharsis. However, in a Lacan-

ian theory of theatrical catharsis, the spectator would also experience the purgation, clarification, and purification of *desire*, beyond Aristotelian pity and fear (Lacan, *Seminar VII* 244–48, 323). The spectator would thus approach (as in analysis) a transformative cathartic awareness of the drive's painful, repeated *jouissance*, not just in the tragic hero onstage or onscreen, but also in oneself.

68. According to Lacanian theory, neurotics may have perverse or psychotic traits. They also pass through the psychotic and perverse structures in early childhood, thus forming the foundations for their neurotic symptoms.

69. See Kiehl 124, on the "acquired sociopathic personality" of patients with orbital frontal damage in the prefrontal cortex. Such problems can also be caused by lesions in the anterior cingulate and temporal lobes (125–27). Cf. Quartz and Sejnowski 200, 206–07, on diminished activity in the orbitofrontal cortex of an individual sociopathic killer and yet its possible hyperarousal in a group of killers through the "contagion of ideas," with obsessive ideation, compulsive repetition, rapid desensitization to violence, and blunting of emotional response. See also Damasio et al.; Kennett and Fine; and Roskies 192 — on aspects of moral decision making that involve the ventromedial area of the prefrontal cortex.

70. Cf. Kaplan, *Motherhood* 28, on the pleasures of cinematic regression, especially regarding maternal melodramas onscreen. Kaplan distinguishes "melodramatic elements" in certain novels and films, which make the reader/spectator "complicit" in the patriarchal unconscious, from "realistic elements" in other works, resisting oppressive institutions — and from deconstructive, postmodern texts that provide the possibility for the spectator/reader to experience "herself *otherwise*" (74).

71. Brecht's critique in the 1930s of Aristotelian catharsis speaks to the continued danger of mimetic violence today, through social rites and personal repetition compulsions, from stage and screen to real life. Brecht directly attacked "Aristotle's recipe for ... catharsis (the spiritual cleansing of the spectator)" (87). See also Brecht 57, 78–79, 135, 181. Cf. Augusto Boal's similar attack on Aristotelian catharsis, as a "purging of antisocial elements" (Carlson, *Theories* 475).

72. Brecht saw Aristotelian drama as "bundling together" the hero's tragic plot events as an "inexorable fate," shared by sympathetic, fearful spectators (87). Instead, Brecht's non-Aristotelian drama, as "epic theatre," would present plots and characters that spectators initially sympathize with, as familiar, but then feel defamiliarized, estranged, and distanced from — through specific performance techniques.

73. Brecht's theatre would provoke a different kind of catharsis (though he did not use that term) for critical thought and action in real life: a clarification of tragic fate and of the coercive mechanisms of theatre as changeable in society, because they are of "human contriving" (87).

74. Brecht critiqued the ancient ideal of the tragic plot, with events that reveal the "innermost being" of the hero, as being similar to "burlesque shows on Broadway, where the public, with yells of 'Take it off!,' forces the girls to expose their bodies more and more" (87).

75. Cf. Salkever 298–99.

76. After Glen Morgan and James Wong produced the second season (and ratings fell), Chris Carter returned as the active producer of *Millennium*'s third season, joined by Michael Duggan and Chip Johannessen. Carter reportedly wanted the show to focus "tragically" on Frank's second nervous breakdown, after the death of his wife, as it had on his unstable recovery from his first breakdown at the beginning of the series (Huff 128). But the show's melodramatic shift may have been inevitable given its falling ratings after the first season — and its competition with a similar series about a female detective, profiling serial killers with her psychic powers: NBC's less violent, yet more melodramatic *Profiler*. In its third and final season, *Millennium* also added a female detective, Emma Hollis, as Frank's young FBI protégé — while making the Group a melodramatically evil patriarchy like *The X-Files*'s dark-suited "Smoking Man" and his colleagues.

77. John Douglas calls these childhood characteristics of serial killers (bed-wetting, arson, and animal dissection) the "homicidal triad" (Douglas and Olshaker 144). He also states: "virtually all serial killers come from dysfunctional backgrounds of sexual or physical abuse, drugs or alcoholism, or any of the related problems" (356). In the *Millennium* episode, "The Beginning and the End," Catherine's kidnapper shows his (and the writers') knowledge of Douglas's term, while taunting her about Frank's efforts to rescue her: "With me his gift is useless. I don't fit. No homicidal triad in my childhood.... The profile will become about him, who he is, who he's being."

78. Cf. Bracher, *Writing* 128, on unconscious desires "perpetuating" victimization.

79. In Yeats's poem, "The Second Coming" (1921), this phrase also occurs in an apocalyptic context: "Things fall apart; the centre cannot hold; / The blood-dimmed tide is loosed, and everywhere / The ceremony of innocence is drowned" (184–85).

80. Cf. Douglas and Olshaker 149–50, where former FBI agent Douglas recounts his involvement in a particular case after he saw an article in his local paper. When he profiled the killer for the police, one of them asked: "Are you a psychic, Douglas?"

81. Cf. Douglas and Olshaker 359–64. See "After" for a 20-year timeline of the Green River case, including the DNA evidence against and confessions of Gary L. Ridgway in 2001 to 2003 (after $15 million was spent unsuccessfully from 1982 to 1991 to catch him). Ridgway, who passed a polygraph test while denying his guilt in 1984, confessed two decades later to 42 of the 49 murders that were attributed to the Green River serial killer, plus 6 more.

82. Cf. Douglas's description of his profiling technique: recreating the crime scene in his head, putting himself in the victim's place as the attacker threatens her, feeling her fear as he approaches her, and feeling "her pain as he rapes her or beats her or cuts her" (173). Douglas compares this to an actor looking for "subtext" (292) and says that he would continue to see the details of each case in his dreams (362).

83. Frank's simple diagnosis of the psychotic killer as bearing maternal guilt and sexual confusion relates in some ways to Fink's description of the source of psychosis. Although the psychotic does not "feel guilt" like neurotics, feminization often occurs in psychotic men, with an invasive Other *jouissance* (*Clinical* 98–99), which Frank may be referring to here as "anger."

84. Cf. Douglas and Olshaker 112: "In our research, we discovered that ... frequently serial offenders had failed in their efforts to join police departments and had taken jobs in related fields."

85. "Who are you to condemn me? They're the guilty. I took responsibility. You've seen them out there, where Satan has them.... It's the great plague.... This is prophecy, the final judgment and victory."

86. When Catherine tells him: "You can't ask me to pretend I don't know what you do"; Frank responds: "Everyone pretends. We all make believe. These men I help catch make us."

87. During this episode, "Paper Dove," Henry thanks the Polaroid Man for picking and giving him a picture of this victim prior to the killing. The Polaroid Man says he wanted the murder to be done while Frank Black was in the area (Arlington, Virginia). Frank is lured into a false sense of security after he catches Henry, which results in the kidnapping of Catherine by the Polaroid Man at the end of the episode.

88. While he is in the garage with his triptych of victims' photos (including a new, potential victim, Frank's wife, Catherine), Henry's mother calls him. "I hear my little man out there. Henri. Henri. Come to your darling mama." When he meets her in the kitchen, Henry promises that he will do his chores "later." But she becomes stern with him: "Oh, sure, you promise. But how you keep when you don't know what you promise? You got to read the list, my little man." Henry defies her (because of his duty to other ritual chores): "I'll do it when I get back." But she insists, "You do it now." Then she becomes cheerful with her power, treating Henry like a child to enforce her *jouissance*. "Don't look so glum, my little man. Heh? I know how to make you smile, cheer you up." She pinches Henry's cheeks with her yellow rubber gloves. "It has never failed to work," she says, laughing and hugging him. Then he plunges a knife into her back.

89. This corresponds to the final crime of California serial killer Ed Kemper (known as the "Coed Killer") in 1973, although in his case the garbage disposal spit the mother's voice box back at him. She had also treated him with more direct cruelty than shown with Henry's mother in the *Millennium* episode. See Douglas and Olshaker 109–15.

90. Henry knows Frank because the Polaroid Man, as primal, perverse father, lured Frank into this particular serial crime — and almost lured Henry into killing Frank's wife. This episode ends, in fact, with Catherine being kidnapped by the Polaroid Man, who in the next episode will capture Frank in the Polaroid image of becoming a killer himself. Like Frank, the TV viewer might come to a cathartic understanding of Henry's sacrificial catharsis through some kind of association with his or her own family. Indeed, Frank reverses Henry's matricidal rite by saving the mother (Catherine) while killing the perverse father (the Polaroid Man), in the next episode of the series. Yet this melodramatic triumph over evil is not a pure victory for the hero. Frank realizes his kinship to the killers, to Henry and the Polaroid Man, in catching the former and killing the latter. He also loses his wife as she witnesses his vengeful violence — which goes beyond saving her life, turning him into a similar monster. Later, their marriage breaks up because Catherine has seen her husband's violence.

91. The blonde women were killed in "The Innocents" and its sequel, "Exegesis," at the start of the third season.

92. As Žižek puts it, "Sacrifice conceals the abyss of the Other's desire ... the Other's lack, inconsistency, 'inexistence' ..." (*Enjoy* 56). A Lacanian cure, sacrificing the illusion of the Other's consistency (provided by the normative Oedipal sacrifice), would open up that abyss — like, but more positively than, the serial killer manifesting a hole in his own being and a lack in the Other's morality, through the signature of his apparently random violence. In a similar vein, Žižek calls the serial killer "a scapegoat embodying sacred violence" and refers to Dr. Hannibal Lecter (the cannibal psychiatrist in the Thomas Harris novel and Jonathan Demme film, *The Silence of the Lambs*), as "the closest mass culture can get to the figure of the Lacanian analyst" (*Enjoy* 57, 67n38). See also Badley 146, who describes Lecter (performed by Anthony Hopkins in the film) as "Freud with repressions removed,

transformed by Lacan and French feminism, ... who leads Starling to the Mother power — that of the lioness hunter — in herself."

93. Cf. Davidson, "Toward" 122, on plasticity in prefrontal cortex emotion circuits, especially during childhood.

94. The episode never explains why Helen and Tom do not have the baptism in their own parish church.

95. Tom tells Catherine, as he waits in her house for news on the case: "I'm trying to prepare myself for whatever happens. I try to anticipate the future so as not to dwell on what you can't control from the past ... [but] all I see are pictures of her suffering and I'm not there to save her." Later, when he, Frank, and Peter Watts find Helen's ring and much blood in a remote forest cabin, Tom says: "I feel nothing. There's nothing inside. None of this is real."

96. The Law's impotence is shown in this episode not only with Frank's inability to apprehend the kidnapper after he has located him, but also through Frank's visit to a mental hospital — where Green had been incarcerated and then prematurely released.

97. Cf. Grigg 64, on the psychotic's father as "inadequate or fraudulent with respect to the law and therefore found to be an ineffective vehicle for the Name-of-the-Father." According to Grigg, psychosis frequently occurs when the father "poses as the incarnation of high ideals." See also Fink, *Clinical* 99, on the psychotic's "domineering, monstrous," Imaginary father, due to the failure of the Symbolic paternal function.

98. Following that tragic irony, the episode offers a happy ending, with Helen, having recovered from her physical injuries (but still showing a scar on her face) being handed her baby again by Catherine. Yet Catherine's words also sound a tragic note, suggesting the psychological scars that will remain for Tom, Helen, and her child: "Here's the little man you've been waiting to see."

99. The episode's initial teaser showed the angel in Garry's hands as he finished making it, while his wife and children came home, carrying grocery bags and barely speaking to him, shortly before the murders occurred.

100. Cf. Turim 149, on the flashback in film melodrama (of the 1940s): "This single key to truth renders the narrative revelation orderly *and* psychoanalytically false. But another reading of the embedded flashback form, a more deconstructive look at these films, indicates another psychoanalytical analysis of these films on a less obvious level."

101. Both flashbacks of the crime are shown (in color) as narrative illustrations: first of Garry's false confession, and then of Frank's version of the crime. In this episode the viewer only sees Frank's gift of a profiling flash once (showing the older son standing on the steps, saying: "What are you doing?"), when he is in the Garry home with Didi, reconsidering the evidence.

102. According to Cheryl Meyer, mothers who purposely kill their offspring are "frequently described as devoted to their children" (E1).

103. On the Lacanian subject as a "breach" between signifiers in the unconscious, see Fink, *Lacanian* 72–79. Fink says that the goal of Lacanian treatment is the splitting of master signifiers, thus sacrificing the dominant sacrifice of castration, traversing the dead-ends of fantasy, and precipitating subjectivity in/as the breach (78–79).

104. Cf. Arsenio on the "happy victimizer" effect in young children who do not empathize with potential victims, especially when they themselves suffer from insecure attachment with adults due to emotional abandonment. "Some children may [be] so self-focused that they do not easily recognize their victims' pain and loss, whereas other children may learn to ignore victims' emotional reactions as irrelevant when those reactions conflict with the victimizers' own desires and goals." The latter may be "especially predictive of higher levels of 'happy victimization' and proactive/instrumental aggression" (59). See also Malti.

105. In a Lacanian sense, the phallic Serpent becomes another figure, like the angel, separating the child from the mother, or Adam and Eve from Eden, through the temptation of the Other's fruit as forbidden. The eating becomes sinful, the nakedness obscene, and the first humans guilty in their divine desire. Yet this scene of "original sin" is reenacted by every Oedipal child as it starts to desire lost objects of pleasure, as guilty knowledge. Thus, the primal sacrifice of *jouissance* is explained, in the symbolic castration of the Genesis myth, as resulting from human disobedience against God. This restores a partial sense of being through separation and frames the mirror-stage illusion of ego, after the child has lost its being in a prior alienation from the (m)Other — and from the mythical garden of Mother Nature, as culture prunes the child's brain. Cf. Fink, *Lacanian* 53. See also Kim Parker, who does not mention the angel in his Lacanian reading of Genesis, but does describe the tree as a "phallus," representing the Father's language and law, as well as "the cunning intervention of the serpent, to whom God gives language, the vehicle of both separation and articulation of desire" (27–28). Parker explains how all humans experience the loss of Eden in the Lacanian mirror (20). See also Piskorowski 314–15.

106. Some of the commercials broadcast during this episode included Cindy Crawford for Kay Jewelers asking, "What's it take to be her hero this Christmas?" (Diamond earrings and a matching bracelet were shown, for $99.95 each.) A Sears commercial then reminded the viewer:

"You've only got one week to go." Santa Claus was also shown buying hundreds of "gigga-pets" at Kentucky Fried Chicken (for $7.99 each). The gigga-pet was the first toy that Frank bought for Jordan at the beginning of this episode.

107. In another episode scripted by Maher and Reindl, "A Single Blade of Grass," Frank is drawn into a Native American ritual sacrifice, through his psychic gift. He also says that his grandmother was part Indian, suggesting that he inherited his gift, and Jordan hers, through that Other spirituality in American culture and in his genes.

108. See Lacan, *Seminar XX* 73–77, and Fink, *Lacanian* 112–22. See also Stuart Schneiderman 182: "The angel whose entire being is inflamed with love of God is a model for feminine enjoyment...."

109. The dog is also somewhat ghost-like. It barks and Frank's father speaks to it, but it is never shown inside the house after he opens the door.

110. There is also a discrepancy (or "continuity problem") between the angel drawings: Frank is shown making marks at the edges of the wings, but the one Frank has in the present does not have those marks.

111. Cf. Fink, *Lacanian* 54: "In separation, the subject attempts to fill the mOther's lack— demonstrated by the various manifestations of her desire for something else—with his or her own lack of being..., seeking out the precise boundaries of the Other's lack in order to fill it with him or her self."

112. Cf. Fink, *Lacanian* 78–79, 114–15.

113. Cf. Molly Ann Rothenberg's Lacanian view of angels and ghosts in recent screen drama: "Only by traversing the fantasy of angelic existence will we cease to live as ghosts and experience ourselves as fully human" (38).

114. Quotes were often shown at the start of *Millennium* episodes, somewhat like the Brechtian device of scene titles, used in his epic theatre to distance spectators and stimulate thought.

115. At the beginning of the first season, Frank has moved back to Seattle with his family after recovering from a nervous breakdown during his FBI work in Washington, DC. At the start of the third season, he is again recovering from a breakdown, which he suffered after his wife's death at the end of the second season, and has returned to working with the FBI (because he left the Millennium Group, feeling that it was responsible for the plague that killed Catherine).

116. Cf. Revelation 20:8.

117. Cf. Elin Diamond on her theory of gestic feminism in theatre.

118. Cf. Pizzato, "Jeffrey," on the ritual theatricality of the crimes, trial, and news reports of serial killer Jeffrey Dahmer.

119. Cf. Alford, *Psychology*, on the "natural evil" of the infant's aggressive hate toward the mother, even prior to love, according to Kleinian psychoanalytic theory (like Milton's Satan, with envious joy in destroying the good), as the basis for needing a "natural law" of reparation across human cultures. He also points out the difference here between Klein's version of the death drive and Freud's (68).

120. Cf. Žižek, *Parallax* 198, on the Lacanian inversion of the Freudian notion that fear focuses on an object which anxiety lacks. Instead, "fear blurs its object," whereas anxiety has a precise, primal object (*objet a* as cause of desire) and emerges "when we get to close to it." This may relate to the tragic dimension of some *Millennium* episodes. See also Fink, *Lacanian* 86, 91–94, 103.

121. Frank Black returned to the screen for one more show after the end of his series, battling the Millennium Group in an episode of *The X-Files*, broadcast on November 28, 1999. He helps Scully and Mulder stop the Group from causing an Armageddon on New Year's Eve of 2000 and is reunited with his daughter, Jordan.

Conclusion

1. Cf. Rotman on the emergence of God and Mind through the ancient Hebrew and Greek traditions: "the hypostatizations of the 'I'-effect that writing permits" (130). He also considers the postmodern potential of moving beyond God to new "ghosts" through the "ongoing dethronement of alphabetic culture" (137). And he explores bio-linguistic ghosts using the works of Terrence Deacon and Merlin Donald, which are also used here (113–16).

2. For example, 25 percent of the world's pollution, which leads to global warming and especially impacts poor countries, is produced by the United States, with 5 percent of the world's population, according to the Sierra Club's website. See also Neil Gaiman's novel, *American Gods*, where European gods of the past revive in America to battle current media gods. Cf. Surowiecki.

3. Cf. Ridge and Key for research on increased aggressive tendencies in readers of religious texts (both believers and non-believers, but more so for believers), when violence is justified as revenge, especially under divine authority.

4. Cf. Žižek, *Parallax* 184–85, on the paradox of the Holocaust, usually considered as the ultimate challenge to theology, yet demanding such a cosmic theatre to frame the scope of the catastrophe.

5. See Damasio, "Neural Basis" 18. "Although certain systems in the brain are clearly related to moral behavior, they are not set by genes to operate for the purposes of morality and ethics." He notes that brain damage can cause the inability to apply ethical rules, despite recall

of them, or (with damage at an earlier age) the inability to learn as well as apply such rules (18–19). For a critique of the "new atheism" of Dawkins, Dennett, Harris, and Hitchens, as involving a fundamentalist belief in scientific naturalism, see Armstrong, *The Case* 302–09.

6. Cf. Harrington 24–25, on the cruelty of the "cosmic process" in nature, both outside and within humans, according to early evolutionary theory, and the potential of reason and science to "fight" it. See also Davidson and Harrington 214–21, for a dialogue of Western scientists with the Dalai Lama on the problems of basing secular ethics in biology.

7. Cf. Kimball on the "infanticidal logic" of evolution as it appears in Western culture. Human offspring are also more likely to be abused by their stepfathers than by their biological fathers (de Waal, *Our* 108).

8. De Waal calls revenge the "flip side of reciprocity"—with examples from chimp behavior (*Our* 204–09).

9. See Eisenberg for research on various subjects' abilities to "maintain their vicarious emotional reaction to another's distress at a tolerable range ... [without becoming] self-focused and overwhelmed by their emotion" (138). See also Owen for evidence that film viewers may enjoy the challenge of a puzzling film, even if "not fully understanding the plot," when they identify with the hero and feel immersed in the world onscreen (304).

10. de Waal argues that humans are more like bonobos than other apes in our neoteny (*Our* 240–41).

11. Cf. Clark, *Natural-Born*, on the "cyborg" nature of the human brain and its "endless succession of designer environments" (197).

12. See Hurley 171, on the "chameleon effect" when "stereotypes automatically activate corresponding behavior" in the observer, according to psychological research. Hurley considers this in relation to media violence. See also Rizzolatti and Sinigaglia 151, on patients with echopraxia, who "have a compulsive tendency to imitate the acts of others immediately," due to neurological damage in the frontal lobes, which eliminates the "braking mechanism" that modifies such mirror-neuron, potential-action mimicry in the rest of us.

13. See Bloom 146–50, on positive "forces of moral change" in humans, including mutual interdependence, contact, persuasion through images and stories, and accretion of moral insight. See also Sopcak for research on how disgust and anxiety, found to be greater in US–American than in Brazilian and German readers, may block compassion while reading violent literary texts.

14. See Hatfield et al. See also Sullins.

15. Cf. Papousek et al. for a study of emotional contagion involving female subjects and a film of an actress giving sad, cheerful, or neutral expressions. The subjects showed varying degrees of emotion perception and "regulation." See also Atkinson and Adolphs 168–72.

16. Cf. Staub 171–72, on how utopian ideologies with moral righteousness may produce scapegoats and genocide.

17. Cf. the Dalai Lama's argument, in dialogue with Western scientists, that compassion may be more fundamental to human nature than competitive emotions (in Davidson and Harrington 85–102).

18. Cf. McGeer 229, on three dimensions of "moral nature" (with evidence from human psychological development, cross-cultural studies, primatology, and autism): compassion for others, concern with social position, and "concern with 'cosmic' structure and position, growing out of the need to bring order and meaning to our lives and fostered by our capacity to view ourselves in intertemporal terms." She also relates this cosmic concern to the human sense of "awe," even at the beauty science reveals, and to an evolved engagement in long-term planning (250–51), as well as certain periods in Western culture (287).

19. See Goodale and Milner 78, on ordinary human pantomimed movements, as "slower, less accurate, and somewhat stylized" (with the object absent), unlike a mime artist's gestures, which exaggerate certain aspects to be convincing with imaginary objects.

20. See Rizzolatti and Sinigaglia, who cite theatre director Peter Brook's assertion that mirror neurons demonstrate a biological foundation for the actor-spectator relationship—a sharing on which, these neuroscientists say, "the theatre evolves and revolves" (ix). See also Pineda, for various articles on mirror neuron systems in relation to "social cognition."

21. See also Rizzolatti and Sinigaglia 85–91, on communicative mirror neurons, which fire with monkey and human facial expressions (lip smacking, tongue protrusion, and lip extension), as intransitive gestures related to transitive grooming and feeding actions. "Lip smacking without grooming therefore appears to be a *form of ritualized motor act* that transforms object-related functions into communicative functions ..." (91). These researchers also describe an emotional mirror neuron system, centered in the insula, with the triggering of disgust and pain in oneself, when perceived in the other person (183–89). But they caution that compassion, while based on this system, depends on "many factors" that extend or inhibit such a "sharing [of] someone's emotive state at a visceromotor level" (191).

22. Iacoboni relates his theory of a mirror neuron "chain" to Maurice Merleau-Ponty's phenomenological idea that the other's intentions inhabit one's body (77–78). Cf. Lacan, *Four* 71–76, 114–15, for his psychoanalytic theory that one's desire is the desire of the Other, in relation

to the work of Merleau-Ponty. See also Gallese et al. 397; Rizzolatti and Sinigaglia 130–31.

23. Fink explains how the usual overwriting of the Imaginary order by the Symbolic, starting with the mirror stage, does not occur in psychosis. The Imaginary continues to predominate, also causing language disturbances and "confusion between self and other" due to the failure of the paternal function. Cf. Brugger 210, on neurological evidence that "periods of acute psychosis are associated with an absence of a clear pattern of hemispheric dominance for language" (in the left cortex). Brugger also speculates that this is why schizophrenia was preserved in human evolution, with the "selective advantage of being able to associate across stereotyped borders."

24. Cf. Fink, *Clinical* 97–98: "even slight provocation can lead the psychotic to engage in seriously punishing behavior," with lack of control over primal drives and with "little if any guilt" after murder, rape, or other criminal acts.

25. See also Fink, *Clinical* 176–78, on the pervert becoming the phallus for the (m)other, the object of her desire, filling in her lack, stuck at the level of serving her, and thus being overwhelmed by her demand. "Only the mother's demand exists; she is lacking in nothing ... that is symbolizable for the child." With the neurotic, however, as a "full-fledged subject," the naming of her desire "forces the child out of his position as object, and propels him into the quest for the elusive key to her desire." Fink also relates perversion to the current crisis in patriarchy: "the all-too-common contemporary father who never worked out his own problems with authority ..." (180).

26. See Kennett and Fine 178, 185, on the loss of moral motivation, or at least social decision making, caused by ventromedial PFC damage. Cf. Roskies 192, 98, on the VMPFC as a link between moral cognition and moral affect, with damage to it resulting in the inability to form new moral motives, though longer term moral knowledge is retained, as with damage to the hippocampus producing short-term amnesia. See Prinz 266–67, on "social exchange" problems (impulsiveness, abnormal emotions, and social faux pas) in patients with VMPFC injuries, although he disagrees with others who define it as a cheater-detection module. See also Damasio et al.

27. See Gazzaniga, *Ethical* 148–50, 172, on intuitive judgments, which are then rationalized by the left-brain "interpreter" into moral ideas and beliefs. Cf. Davidson, "Affective," on the left PFC participating more in positive emotions and the right in negative, with the ventromedial PFC involved in forming emotional goals (1199).

28. Cf. Lehrer on the emotional brain's unconscious computational wisdom, experienced as the right "feeling," while it interacts with the prefrontal lobes, though it may also override, or be suppressed by their metacognitive controls (47–48, 76–81, 107). Lehrer points to various studies that show that the prefrontal cortex is like the "conductor of an orchestra," in relation to the rest of the brain, when producing a placebo or pricey taste effect, making beliefs into real sensations (117, 146–48). PFC consciousness, with its slower, limited processing, often tries to minimize complexity by chunking information and projecting simplistic expectations that alter the emotional brain's computations with "misleading shortcuts" and "shoddy top-down thinking" (155, 204). Lehrer also mentions research showing that people with more activity in the superior temporal sulcus, while sympathizing with others' emotions, "are more likely to exhibit altruistic behavior" (183).

29. *Homo homini lupus*, man is a wolf to man, appears in a play by Plautus, *Asinaria*, in a slightly different phrase.

30. Based on these studies, MindSign, a company in San Diego, California, now offers filmmakers and advertisers a brain scanning method for increasing their impact on audiences.

31. See Moll et al. 6, on research associating the ventral prefrontal cortex with "implicit stereotypes," which may be unconscious, as negative or positive attributes, but are "powerful determinants of behaviors that can be justified by convincing logico-verbal arguments."

32. In summarizing their studies, Hasson, Landesman, et al. make a distinction, not between mind-control (or escapist) entertainment and Artaudian or Brechtian catharsis, but between filmmakers who control or distort reality and others who "remain faithful" to it, enabling multiple interpretations, as in André Bazin's preference for realistic over imagistic film effects (16). They also acknowledge that a film becomes propaganda when it controls spectator's brains with "a tight grip" (18). And they admit that a low degree of brain area correlations could mean not just disengagement but rather "an intensely engaged but variable (across individuals) processing of a movie sequence," as with an "art film."

33. Iacoboni mentions that the correlation between media violence watching and aggressive behavior is closer than with passive smoke or asbestos exposure and cancer (208–09). See Bushman and Huesmann for more details.

34. Cf. Hurley 168–69, on the difference between human "imitation" and chimp "emulation" (copying goals, not detailed performances), in relation to mirror neurons and media violence. Hurley considers neurological evidence that imitation becomes impulsive when not inhibited by the prefrontal cortex (169–70) and that media violence creates "cognitive scripts" for real-life

behaviors (181). See also the various articles in the collection edited by Hurley and Chater, exploring human imitation drives through neuroscience, animal behavior, child development, and culture.

35. See Subrahmanyam et al. 85–86, for a review of experimental evidence that playing violent videogames, even for brief periods, increases children's aggressive play, aggressive responses to questions, and aggressive ideation. It is also significant, regarding the themes of this book, that some videogames, with massive online connections across the Web, involve the player in a godlike role creating new virtual life forms ("Spore") or communities of virtual people ("Sims" and "Civilization") or environments for being worshipped by them ("Black and White").

36. This corresponds to the Lacanian theory of the psychoanalytic cure, as a "new configuration of the analysand's fundamental fantasy, and thus a new relation ... to the Other's desire and the Other's jouissance," in which "the lost object is finally given up" (Fink, *Clinical* 70–71).

37. Cf. de Waal's evolutionary argument that "self-recognition and higher forms of empathy emerged together in the branch leading to humans and apes" (*Our* 194). See also Wrangham and Peterson's point that only two animals, humans and chimpanzees, display "intense, male-initiated territorial aggression, including lethal raiding into neighboring communities in search of vulnerable enemies to attack and kill" (24).

38. Cf. Wright, who explores evolving images of god in Western and non–Western cultures. He concludes with a scientific idea of God as a "special creative explanation" for natural and cultural evolution toward a moral order in the human species (450), or as a cooperative Logos involving family love and social sympathy (455–58). He also compares religious belief in God to scientific belief in subatomic particles and sees both as being in dialogue with nature (452–53).

39. Cf. neuroscientist Francisco Varela on the cognitive self as virtual, regarding ethics, Lacan, and Buddhism (60–66).

40. Cf. Richerson and Boyd, an evolutionary psychologist and anthropologist, who argue that modern "institutions of complex societies are manifestly built on ancient and tribal [social] instincts ... deriving from cultural evolutionary processes" (235).

41. Cf. Hood 236–48, who uses scientific research to argue for a "biology of belief" in the supernatural, involving an intuitive "supersense" in all human brains, with the dopamine pleasure system enabling greater pattern perception by some people, although controlled, in varying degrees, by the dorsal lateral prefrontal cortex.

42. Cf. Teilhard de Chardin; Fiddes; and Spong. See also Kosslyn, "Science," for a neuroscientist's theory of God as the "Ultimate Superset," involving emergent properties and downward causality.

Bibliography

Adam [The Mystery of Adam]. Trans. from Latin and French by Edward Noble Stone. *World Drama*. Vol. 1. Ed. Barrett H. Clark. New York: Dover, 1933. 304–21.

Aeschylus. *The Orestes Plays of Aeschylus*. Trans. Paul Roche. New York: NAL, 1962.

———. *Prometheus Bound*. Trans. James Scully and C. J. Herington. Oxford: Oxford University Press, 1989.

"After Many Years of Pursuit, DNA Testing Helped Lead Investigators to Green River Killer." *The Seattle Times* 6 Nov. 2003: A16.

Agamben, Giorgio. *Homo Sacer*. Stanford: Stanford University Press, 1998.

Aiken, George L. *Uncle Tom's Cabin*. New York: Samuel French, 1858. Online. http://www.iath.virginia.edu/utc/onstage/scripts/aikenhp.html

Albright, Carol Rausch, and James B. Ashbrook. *Where God Lives in the Human Brain*. Naperville, Illinois: Sourcebooks, 2001.

Alford, C. Fred. *The Psychoanalytic Theory of Greek Tragedy*. New Haven: Yale University Press, 1992.

———. *Psychology and the Natural Law of Reparation*. Cambridge: Cambridge University Press, 2006.

Alibek, Ken, and Stephen Handelman. *Biohazard*. New York: Random House, 1999.

Allan, Neil. "An Age in Love with Wonders." *Literature Compass* 2.1 (2005): 1–16.

Alper, Matthew. *The "God" Part of the Brain*. New York: Rogue Press, 2001.

"Anamnesis." *Millennium*. Writ. Erin Mahler and Kay Rindel. Dir. John P. Kousakis. Fox. 17 Apr. 1998.

Apollodorus. *The Library of Greek Mythology*. Trans. Robin Hard. Oxford: Oxford University Press, 1997.

Apter, Michael J. *The Dangerous Edge: The Psychology of Excitement*. New York: Macmillan, 1992.

Aristophanes. *Birds*. Trans. Peter Meineck. *Aristophanes I*. Indianapolis: Hackett, 1999. 262–81.

Aristotle. *The Basic Works of Aristotle*. Ed. Richard McKeon. New York: Random House, 1941.

Armstrong, Karen. *The Case for God*. New York: Knopf, 2009.

———. *The Great Transformation: The Beginning of Our Religious Traditions*. New York: Knopf, 2006.

———. *A History of God*. New York: Knopf, 1994.

Arsenio, William F. "Happy Victimizers and Moral Responsibility." *Talking to Children About Responsibility and Control of Emotions*. Ed. Michael Schleifer and Cynthia Martiny. Calgary: Detselig, 2006. 49–63.

Artaud, Antonin. *Collected Works*. Volume 1. London: Calder and Boyers: 1968.

———. *The Theater and Its Double*. New York: Grove, 1958.

Asendorpf, Jens B. "Self-awareness, Other-awareness, and Secondary Representation." *The Imitative Mind: Development, Evolution, and Brain Bases*. Ed. Andrew N. Meltzoff and Wolfgang Prinz. Cambridge: Cambridge University Press, 2002. 63–73.

Ashby, Clifford. *Classical Greek Theatre: New Views of an Old Subject*. Iowa City: University of Iowa Press, 1999.

Ashworth, Ann. "Hamlet's Epic Descent into Lucian's Hades." *Journal of Evolutionary Psychology* 12 (1991): 165–73.

Atkinson, Anthony P. and Ralph Adolphs. "Visual Emotion Perception." In *Emotion and Consciousness*. Lisa Feldman Barrett, Paul M. Niedenthal, and Piotr Winkielman, eds. New York: Guilford, 2005. 150–84.

Atran, Scott. *In Gods We Trust: The Evolutionary Landscape of Religion*. Oxford: Oxford University Press, 2002.

Aunger, Robert. *The Electric Meme*. New York: Free Press, 2002.

Aziz-Zadeh, Lisa, Stephen M. Wilson, Giacomo Rizzolatti, and Marco Iacoboni. "Congruent Embodied Representations for Visually Presented Actions and Linguistic Phrases Describing Actions." *Current Biology* 16.18 (Sept. 2006): 1818–1823.

Baars, Bernard J. *In the Theater of Consciousness: The Workspace of the Mind*. Oxford: Oxford University Press, 1997.

Baars, Bernard J., Michael R. Fehling, Mark LaPolla, and Katherine McGovern. "Consciousness Creates Access." *Scientific Approaches to Consciousness*. Ed. Jonathan D. Cohen and Jonathan W. Schooler. Mahwah, NJ: Lawrence Erlbaum, 1997. 423–44.

Bacon, Helen H. "The Furies' Homecoming." *Classical Philology* 96.1 (2001): 48–59.

Badley, Linda. *Film, Horror, and the Body Fantastic*. Westport, Connecticut: Greenwood, 1995.

Bahn, Paul. *Cave Art*. London: Francis Lincoln, 2007.

Baird, Abigail A. "Adolescent Moral Reasoning." *Moral Psychology*. Vol. 3. Ed. Walter Sinnott-Armstrong. Cambridge: MIT Press, 2008. 323–41.

Balsera, Viviana Díaz. "Cleansing Mexican Antiquity." *A Companion to the Literature of Colonial America*. Ed. Susan Castillo and Ivy Schweitzer. Oxford: Blackwell, 2005. 292–305.

Banyas, Carol A. "Evolution and Phylogenetic History of the Frontal Lobes." Miller and Cummings 83–106.

Barash, David P. "Targets of Aggression." *The Chronicle of Higher Education* 54.6 (2007): B6.

Barash, David P. and Nanelle R. Barash. *Madame Bovary's Ovaries*. New York: Delacorte, 2005.

Barlow, Connie, ed. *From Gaia to Selfish Genes*. Cambridge: MIT Press, 1991.

Barrett, James. "Pentheus and the Spectator in Euripides' *Bacchae*." *American Journal of Philology* 119.3 (1998): 337–360.

Barrett, Lisa Feldman, Paul M. Niedenthal, and Piotr Winkielman, eds. *Emotion and Consciousness*. New York: Guilford, 2005.

Baudrillard, Jean. *Simulations*. New York: Semiotext(e), 1983.

Beauregard, David N. *Catholic Theology in Shakespeare's Plays*. Cranbury, NJ: Associated University Presses, 2008.

Beckwith, Sarah. "Ritual, Church and Theatre." *Culture and History 1350–1600*. Ed. David Aers. Detroit: Wayne State University Press, 1992. 65–90.

———. *Signifying God: Social Relation and Symbolic Act in the York Corpus Christi Plays*. Chicago: University of Chicago Press, 2001.

Bedazzled. Dir. Harold Ramis. Twentieth Century–Fox, 2000.

"The Beginning and the End." *Millennium*. Writ. Glen Morgan and James Wong. Dir. Thomas Wright. Fox. 9 Sept. 1997.

Belfiore, Eleanora, and Oliver Bennett. *The Social Impact of the Arts*. New York: Palgrave, 2008.

Bergland, Renée. "Toltec Mirrors." *A Companion to the Literature of Colonial America*. Ed. Susan Castillo and Ivy Schweitzer. Oxford: Blackwell, 2005. 141–58.

Bermel, Albert. "Gozzi in the Mainstream." *Theater* 20.1 (1988): 38–41.

Bernal, Martin. *Black Athena: The Afroasiatic Roots of Classical Civilization*. Vol. 1. New Brunswick: Rutgers University Press, 1987.

Bevington, David. "One Hell of an Ending." *"Bring Furth the Pagants."* Edited by David N. Klausner and Karen Sawyer Marsalek. Toronto: University of Toronto Press, 2007. 292–310.

Bicks, Caroline. "Backsliding at Ephesus." *Pericles: Critical Essays*. Ed. David Skeele. New York: Garland, 2000. 205–27.

Bjorklund, David F. and Katherine Kipp. "Social Cognition, Inhibition, and Theory of Mind: The Evolution of Human Intelligence." *The Evolution of Intelligence*. Ed. Robert J. Sternberg and James C. Kaufman. Mahwah, New Jersey: Lawrence Erlbaum, 2002. 27–54.

Black, Joel. *The Aesthetics of Murder*. Baltimore: Johns Hopkins University Press, 1991.

Blackmore, Susan. *The Meme Machine*. Oxford: Oxford University Press, 1999.

Blakeslee, Sandra, and Matthew Blakeslee. *The Body Has a Mind of Its Own*. New York: Random House, 2007.

Bloom, Paul. *Descartes' Baby*. New York: Perseus, 2004.

Boehm, Christopher. "Conflict and the Evolution of Social Control." In *Evolutionary Origins of Morality*. Edited by Leonard Katz. Bowling Green: Bowling Green State University, 2000. 79–101.

———. "Egalitarian Behaviour and the Evolution of Political Intelligence." In *Machiavellian Intelligence II*. Edited by Andrew Whiten and Richard W. Byrne. Cambridge: Cambridge University Press, 1997. 341–64.

———. *Hierarchy in the Forest*. Cambridge: Harvard University Press, 1999.

Bolte Taylor, Jill. *My Stroke of Insight*. New York: Viking, 2008.

Bonda, Eva, Michael Petrides, David Ostry, and Alan Evans. "Specific Involvement of Human Parietal Systems and the Amygdala in the Perception of Biological Motion." *Journal of Neuroscience* 16.11 (June 1996): 3737–44.

Boothby, Richard. *Death and Desire: Psychoanalytic Theory in Lacan's Return to Freud*. New York: Routledge, 1991.

———. "The Psychical Meaning of Life and Death." *Disseminating Lacan*. Ed. David Pettigrew and François Raffoul. Albany: State University of New York Press, 1996. 337–63.

Bordwell, David, and Noël Carroll, eds. *Post-Theory: Reconstructing Film Studies*. Madison: University of Wisconsin Press, 1996.

"Borrowed Time." *Millennium*. Writ. Chip Johannessen. Dir. Dwight Little. Fox. 15 Jan. 1999.

Boyer, Pascal. *Religion Explained*. New York: Basic, 2001.

Bracher, Mark. *The Writing Cure: Psychoanalysis, Composition, and the Aims of Education*. Carbondale: Southern Illinois University Press, 1999.

Bradshaw, John L. "The Evolution of Intellect: Cognitive, Neurological, and Primatological Aspects and Hominid Culture." *The Evolution of Intelligence*. Ed. Robert J. Sternberg and James C. Kaufman. Mahwah, New Jersey: Lawrence Erlbaum: 2002. 55–78.

Brecht, Bertolt. *Brecht on Theatre*. Trans. John Willett. New York: Hill, 1964.

Bremmer, Jan N. "Scapegoat Rituals in Ancient Greece." *Greek Religion*. Ed. Richard Buxton. Oxford: Oxford University Press, 2000. 272–93.

Breuil, Abbé H. *Four Hundred Centuries of Cave Art*. Trans. Mary E. Boyle. New York: Hacker Art, 1979.

Briller, Bert. "Are There Answers to Television Violence?" *Television Quarterly* 29.4 (1995): 37–47.

Brockett, Oscar G., and Franklin J. Hildy. *History of the Theatre*. Boston: Allyn and Bacon, 2003.

Brodie, Richard. *Virus of the Mind: The New Science of the Meme*. Seattle: Integral, 1996.

Brousse, Marie-Hélène. "The Drive (II)." *Reading Seminar XI*. Ed. Richard Feldstein, Bruce Fink, and Maire Jaanus. Albany: State University of New York Press, 1995. 109–17.

Brown, A. L. "The Erinyes in the *Oresteia*: Real Life, the Supernatural, and the Stage." *Journal of Hellenic Studies* 53 (1983): 13–34.

Bruce Almighty. Dir. Tom Shadyac. Universal, 2003.

Brugger, Peter. "From Haunted Brain to Haunted Science." *Hauntings and Poltergeists*. Ed. James Houran and Rense Lange. Jefferson, North Carolina: McFarland, 2001. 195–213.

Büchner, Georg. *Woyzeck*. Trans. Carl Richard Mueller. *The Modern Theatre*. Ed. Robert W. Corrigan. New York: Macmillan, 1964. 7–19.

Buckman, Robert. *Can We Be Good Without God? Biology, Behavior, and the Need to Believe*. Amherst, New York: Prometheus, 2002.

Buckner, Randy L., Jessica R. Andrews-Hanna, and Daniel L. Schacter. "The Brain's Default Network." *Annals of the New York Academy of Sciences* 1124 (2008): 1–38.

Budge, E. A. Wallis. *The Gods of the Egyptians or Studies in Egyptian Mythology*. Vol. 2. New York: Dover, 1969.

Burch, Noël. *Theory of Film Practice*. 1973. Trans. Helen R. Lane. Princeton: Princeton University Press, 1981.

Burkert, Walter. *Creation of the Sacred: Tracks of Biology in Early Religions*. Cambridge: Harvard University Press, 1996.

———. *Greek Religion*. Cambridge: Harvard University Press, 1985.

Burns, Edward. *The Chester Mystery Cycle: A New Staging Text*. Liverpool: Liverpool University Press, 1987.

Bushman, Brad J., and L. Rowell Huesmann. "Effects of Televised Violence on Aggression." *Handbook of Children and the Media*. Ed. Dorothy G. Singer and Jerome L. Singer. London: Sage, 2001. 223–54.

Buss, David M. *The Evolution of Desire*. New York: Basic, 1994.

Butler, Ann B., and William Hodos. *Comparative Vertebrate Neuroanatomy.* 2nd edition. NY: Wiley-Interscience, 2005.

Byrne, Richard. "The Primate Origins of Human Intelligence." *The Evolution of Intelligence.* Ed. Robert J. Sternberg and James C. Kaufman. Mahwah, New Jersey: Lawrence Erlbaum, 2002. 79–96.

_____. *The Thinking Ape: Evolutionary Origins of Intelligence.* Oxford: Oxford University Press, 1995.

Caldwell, Richard. *The Origin of the Gods.* Oxford: Oxford University Press, 1982.

Calvo-Merino, B., C. Jola, D. E. Glaser, and P. Haggard. "Towards a Sensorimotor Aesthetics of Performing Art." *Consciousness and Cognition* 17 (2008): 911–22.

Cantor, Joanne. "The Media and Children's Fears, Anxieties, and Perceptions of Danger." *Handbook of Children and the Media.* Ed. Dorothy G. Singer and Jerome L. Singer. London: Sage, 2001. 207–21.

Carlson, Marvin. *Theories of the Theatre.* Ithaca: Cornell University Press, 1984.

Carman, John. "'Millennium' Drips With All But Suspense." *San Francisco Chronicle,* 24 Oct. 1996: D1.

Carter, Chris. Interview with David Lipsky. *Rolling Stone* 20 Feb. 1997: 36–84.

Cawley, A. C., ed. *The Wakefield Pageants in the Towneley Cycle.* Manchester: University Press, 1958.

Cela-Conde, Camilo J., Francisco J. Ayala, Enric Munar, Fernando Maestú, Marcos Nadal, Miguel A. Capó, David del Río, Juan J. López, Tomás Ortiz, Claudio Mirasso, and Gisèle Marty. "Sex-related Similarities and Differences in the Neural Correlates of Beauty." *PNAS* 106.10 (2009): 3847–52.

Centerwall, Brandon S. "Television and Violent Crime." *The Public Interest* 111 (spring 1993): 56–71.

Césaire, Aimé. *A Tempest.* 1969. Trans. Richard Miller. New York: Ubu Repertory Theater, 1992.

Chandler, Daniel. "Cultivation Theory." Sept. 1995. Internet. Online: http://www.aber.ac.uk/~dgc/cultiv.html.

Changeux, Jean-Pierre. "Creation, Art, and the Brain." Changeux et al. 1–10.

Changeux, Jean-Pierre, Antonio R. Damasio, Wolf Singer, and Yves Christen, eds. *Neurobiology of Human Values.* Heidelberg: Springer-Verlag, 2005.

Chemers, Michael Mark. "Anti-Semitism, Surrogacy, and the Invocation of Mohammed in the *Play of the Sacrament.*" *Comparative Drama* 41.1 (2007): 25–55.

Cheney, Dorothy L., and Robert M. Seyfarth. *Baboon Metaphysics: The Evolution of a Social Mind.* Chicago: University of Chicago Press, 2007.

Churchill, Caryl. *Cloud 9.* New York: Routledge, 1991.

Churchland, Patricia Smith. *Brain-Wise: Studies in Neurophilosophy.* Cambridge: MIT Press, 2002.

Ciompi, Luc, and Jaak Panksepp. "Energetic Effects of Emotions on Cognitions." *Consciousness and Emotion.* Ed. Ralph D. Ellis and Natika Newton. Amsterdam: John Benjamins, 2005. 23–56.

City of Angels. Dir. Brad Silberling. Warner Brothers, 1998.

Clark, Andy. "Connectionism, Moral Cognition, and Collaborative Problem Solving." *Mind and Morals.* Eds. Larry May, Marilyn Friedman, and Andy Clark. Cambridge: MIT Press, 1996. 109–28.

_____. *Natural-Born Cyborgs.* Oxford: Oxford University Press, 2003.

Clark, Robert L. A., and Claire Sponsler. "Othered Bodies: Racial Cross-Dressing in the *Mistere de la Sainte Hostie* and the Croxton *Play of the Sacrament.*" *Journal of Medieval and Early Modern Studies* 29.1 (1999): 61–88.

Clay, Jenny Strauss. *Hesiod's Cosmos.* Cambridge: Cambridge University Press, 2003.

Clements, A.M., S.L. Rimrodt, J.R. Abel, J.G. Blankner, S.H. Mostofsky, J.J. Pekar, M.B. Denckla, and L.E. Cutting. "Sex Differences in Cerebral Laterality of Language and Visuospatial Processing." *Brain and Language* 98.2 (2006): 150–58.

Cless, Downing. "Ecologically Conjuring Doctor Faustus." *Journal of Dramatic Theory and Criticism* 20 (Spring 2006): 145–67.

Clottes, Jean. *Cave Art.* London: Phaidon, 2008.

_____. "Sticking Bones into Cracks in the Upper Paleolithic." *Becoming Human.* Ed. Colin Renfrew and Iain Morley. Cambridge: Cambridge University Press, 2009. 195–211.

Clottes, Jean, and David Lewis-Williams. *The Shamans of Prehistory: Trance and Magic in the Painted Caves.* Trans. Sophie Hawkes. New York: Abrams, 1998.

Cole, Susan Guettel. "Procession and Celebration at the Dionysia." *Theater and Society in the Classical World.* Ed. Ruth Scodel.

Ann Arbor: University of Michigan Press, 1993. 25–38.

Collier, Jo Leslie. *From Wagner to Murnau: The Transposition of Romanticism from Stage to Screen.* Ann Arbor: UMI, 1988.

Collins, Francis S. *The Language of God.* New York: Free Press, 2006.

Condry, John. *The Psychology of Television.* Hillsdale, New Jersey: Lawrence Erlbaum Associates, 1989.

Congdon, Constance. *Tales of the Lost Formicans and Other Plays.* New York: Theatre Communications Group, 1994.

Constantine. Dir. Francis Lawrence. Warner Brothers, 2005.

Copjec, Joan. *Read My Desire: Lacan Against the Historicists.* Cambridge: MIT Press, 1994.

Corballis, Michael C. "Evolution of the Generative Mind." *The Evolution of Intelligence.* Ed. Robert J. Sternberg and James C. Kaufman. Mahwah, New Jersey: Lawrence Erlbaum, 2002. 117–44.

_____. *The Lopsided Ape.* Oxford: University of Oxford Press, 1991.

Cornford, F. M. *Greek Religious Thought.* New York: AMS, 1969.

Corrigall, Jenny, and Heward Wilkinson, eds. *Revolutionary Connections.* London: Karnac, 2003.

"Covenant." *Millennium.* Writ. Robert Moresco. Dir. Roderick J. Pridy. Fox. 21 Mar. 1997.

Coveney, Peter, and Roger Highfield. *Frontiers of Complexity.* New York: Fawcett Columbine, 1995.

Cox, John D. *The Devil and the Sacred in English Drama, 1350–1642.* Cambridge: Cambridge University Press, 2000.

_____. "'To obtain his soul.'" *Early Theatre* 5.2 (2002): 29–46.

Cox, Ted. "Chuckling on the Dark Side: Fox's Self-Important 'Millennium' Gets a Delightful Dose of Humor." *Daily Herald* (Arlington Heights, IL) 21 Nov. 1997: 28.

_____. "Creepy, Yes, But Also Uncompromising: Fox's 'Millennium' is Nothing Less Than an Investigation into the Nature of Evil." *Daily Herald* (Arlington Heights, IL) 25 Oct. 1996: 24.

Cozolino, Louis J. *The Neuroscience of Human Relationships.* New York: Norton, 2006.

_____. *The Neuroscience of Psychotherapy.* New York: Norton, 2002.

Crabb, Peter B., and Jeffrey H. Goldstein. "The Social Psychology of Watching Sports: From Ilium to Living Room." *Responding to the Screen: Reception and Reaction Processes.* Ed. Jennings Bryant and Dolf Zillmann. Hillsdale, New Jersey: Lawrence Erlbaum Associates, 1991. 355–71.

Creed, Barbara. *The Monstrous Feminine.* London: Routledge, 1993.

"Critics Have Trouble With 'Millennium's Graphic Nature." *Seattle Post-Intelligencer* 25 Oct. 1996: 39.

"The Curse of Frank Black." *Millennium.* Writ. Glen Morgan and James Wong. Dir. Ralph Hemmeker. Fox. 31 Oct. 1997.

Curtis, Gregory. *The Cave Painters.* New York: Knopf, 2006.

Damasio, Antonio. *Descartes' Error: Emotion, Reason, and the Human Brain.* New York: Putnam, 1994.

_____. *The Feeling of What Happens: Body and Emotion in the Making of Consciousness.* New York: Harcourt, 1999.

_____. *Looking for Spinoza: Joy, Sorrow, and the Feeling Brain.* Orlando: Harcourt, 2003.

_____. "The Neural Basis of Social Behavior." *Neuroethics.* Ed. Steven J. Marcus. New York: Dana, 2002. 14–19.

_____. "The Neurobiological Grounding of Human Values." *Neurobiology of Human Values.* Eds. Jean-Pierre Changeux, Antonio R. Damasio, Wolf Singer, and Yves Christen. Heidelberg: Spring, 2005. 47–56.

Damasio, Antonio R., Daniel Tranel, and Hanna Damasio. "Individuals with Sociopathic Behavior Caused by Frontal Damage Fail to Respond Autonomically to Social Stimuli." *Behavioural Brain Research* 41 (1990): 81–94.

Damasio, Hanna. "Disorders of Social Conduct Following Damage to Prefrontal Cortices." Changeux et al. 37–46.

d'Aquli, Eugene G., Charles D. Laughlin, Jr., John McManus, et al. *The Spectrum of Ritual.* New York: Columbia University Press, 1979.

d'Aquili, Eugene G., and Andrew B. Newberg. "The Neuropsychology of Aesthetic, Spiritual, and Mystical States." *NeuroTheology.* Ed. Rhawn Joseph. San Jose: University Press, 2002. 243–50.

_____. *The Mystical Mind: Probing the Biology of Religious Experience.* Minneapolis: Fortress, 1999.

The Dark Knight. Dir. Christopher Nolin. Warner Brothers, 2008.

Davidson, Richard. "Affective Style, Psychopathology, and Resilience." *American Psychologist.* 55.11 (2000): 1196–1214.

_____. "Neural Substrates of Affective Style and Value." Changeux et al. 67–90.

———. "Toward a Biology of Positive Affect and Compassion." Davidson and Harrington 107–30.
Davidson, Richard J., and Anne Harrington, eds. *Visions of Compassion*. Oxford: Oxford University Press, 2002.
Davies, Paul. *The Cosmic Jackpot: Why Our Universe Is Just Right for Life*. Boston: Houghton Mifflin, 2007.
Davis, Michael. *Aristotle's Poetics: the Poetry of Philosophy*. Lanham, Maryland: Rowman and Littlefield, 1992.
Dawkins, Richard. *The Extended Phenotype*. Oxford: Oxford University Press, 1999.
———. *The God Delusion*. New York: Houghton Mifflin, 2006.
———. *The Selfish Gene*. Oxford: Oxford University Press, 1976.
Day, Sean. "Some Demographic and Sociocultural Aspects of Synesthesia." *Synesthesia*. Ed. Lynn C. Robertson and Noam Sagiv. Oxford: Oxford University Press, 2005. 11–33.
Dayan, Daniel, and Elihu Katz. *Media Events: The Live Broadcasting of History*. Cambridge: Harvard University Press, 1992.
Deacon, Terence W. *The Symbolic Species: The Co-Evolution of Language and the Brain*. New York: Norton, 1997.
Dean, Carolyn. "Law and Sacrifice." *Representations* 13 (1986): 42–62.
Decety, Jean. "Neurophysiological Evidence for Simulation of Action." *Simulation and Knowledge of Action*. Ed. Jerome Dokic and Joëlle Proust. Amsterdam: John Benjamins, 2002. 53–72.
Decety, Jean, and Thierry Chaminade. "Neural Correlates of Feeling Sympathy." *Neuropsychologia* 41 (2003): 127–38.
———. "When the Self Represents the Other: A New Cognitive Neuroscience View on Psychological Identification." *Consciousness and Cognition* 12 (2003): 577–96.
Decety, Jean, and Philip L. Jackson. "A Social Neuroscience Perspective on Empathy." *Current Directions in Psychological Science* 15.2 (2006): 54–58.
Decety, Jean, and Jessica A. Sommerville. "Shared Representations between Self and Other." *Trends in Cognitive Sciences* 7.12 (2003): 527–33.
Dehaene, Stanislas. "How a Primate Brain Comes to Know Some Mathematical Truths." *Neurobiology of Human Values*. In Changeux et al. 143–56.
Delasara, Jan. *PopLit, PopCult, and* The X-Files*: A Critical Exploration*. Jefferson, NC: McFarland, 2000.

Dennett, Daniel. *Breaking the Spell: Religion as a Natural Phenomenon*. New York: Penguin, 2006.
———. *Consciousness Explained*. Boston: Little, Brown and Co., 1991.
———. *Darwin's Dangerous Idea*. New York: Simon & Schuster, 1995.
DesAutels, Peggy. "Gestalt Shifts in Moral Perception." In May et al. 129–44.
Detienne, Marcel. "Forgetting Delphi between Apollo and Dionysus." *Classical Philology* 96.2 (2001): 147–58.
The Devil and Max Devlin. Dir. Steven Hilliard Stem. Buena Vista, 1981.
The Devil's Advocate. Dir. Taylor Hackford. Warner Brothers, 1997.
de Waal, Frans. *Good Natured*. Cambridge: Harvard University Press, 1996.
———. *Our Inner Ape*. New York: Penguin, 2005.
———. *Primates and Philosophers: How Morality Evolved*. Princeton: Princeton University Press, 2006.
Diamond, Elin. *Unmaking Mimesis: Essays on Feminism and Theater*. London: Routledge, 1997.
Diamond, Jared. *Guns, Germs, and Steel*. New York: Norton, 1997.
———. *Why Is Sex Fun?* New York: Basic, 1997.
Dissanayake, Ellen. *Art and Intimacy*. Seattle: University of Washington Press, 2000.
Doane, Mary Ann. "The Voice in the Cinema: The Articulation of Body and Space." *Narrative, Apparatus, Ideology*. Ed. Philip Rosen. New York: Columbia University Press, 1986. 335–48.
Dodds, E. R. *The Greeks and the Irrational*. Berkeley: University of California Press, 1951.
Dogma. Dir. Kevin Smith. Lions Gate, 1999.
Doidge, Norman. *The Brain That Changes Itself*. New York: Viking, 2007.
Dollimore, Jonathan. *Death, Desire and Loss in Western Culture*. New York: Routledge, 1998.
Donald, Merlin. *A Mind So Rare: The Evolution of Human Consciousness*. New York: Norton, 2001.
———. *Origins of the Modern Mind: Three Stages in the Evolution of Culture and Cognition*. Cambridge: Harvard University Press, 1991.
———. "The Roots of Art and Religion in Ancient Material." *Becoming Human*. Ed. Colin Renfrew and Iain Morley. Cambridge: Cambridge University Press, 2009. 95–103.

Dooley, John A. "Common Motifs and Effects in Shamanic Passage Rites and No Theatre." *Shaman* 8.2 (2000): 99–129.

Douglas, John, and Mark Olshaker. *Mindhunter.* New York: Scribner, 1995.

Dox, Donnalee. "Theatrical Space, Mutable Space, and the Space of Imagination: Three Readings of the *Croxton Play of the Sacrament*." *Medieval Practices of Space.* Ed. Barbara A. Hanawalt and Michal Kobialka. Minneapolis: University of Minnesota Press, 2000. 167–98.

Dunand, Anne. "The End of Analysis (I)" and "The End of Analysis (II)." *Reading Seminar XI.* Ed. Richard Feldstein, Bruce Fink, and Maire Jaanus. Albany: State University of New York Press, 1995. 243–49 and 251–56.

Dunand, Francoise, and Christiane Zivie-Coche. *Gods and Men in Ancient Egypt.* Ithaca: Cornell University Press, 2004.

Dunbar, Robin. *Grooming, Gossip, and the Evolution of Language.* Cambridge: Harvard University Press, 1996.

———. "The Social Brain: Mind, Language, and Society in Evolutionary Perspective." *Annual Review of Anthropology* 2003 (32): 163–81.

Dyer, Richard. "Kill and Kill Again." *Sight and Sound* 7.9 (Sept. 1997): 14–17.

Dyson, Freeman J. *Infinite in All Directions.* New York: Harper & Row, 1988.

Earl, James W. "Identification and Catharsis." *Pragmatism's Freud: The Moral Disposition of Psychoanalysis.* Baltimore: Johns Hopkins University Press, 1986. 79–92.

Edelman, Gerald M. *Bright Air, Brilliant Fire: On the Matter of the Mind.* New York: Basic, 1992.

———. *Wider Than the Sky.* New Haven: Yale University Press, 2004.

Edelman, Gerald M., and Giulio Tononi. *A Universe of Consciousness: How Matter Becomes Mind.* New York: Basic, 2000.

Edwards, Emily D. "A House That Tries to be Haunted." *Hauntings and Poltergeists.* Ed. James Houran and Rense Lange. Jefferson, North Carolina: McFarland, 2001. 82–119.

Edwards-Lee, Terri A., and Ronald E. Saul. "Neuropsychiatry and the Right Frontal Lobe." In Miller and Cummings. 304–20.

Ehret, Günter. "Hemisphere Dominance of Brain Function—Which Functions are Lateralized and Why?" *23 Problems in Systems Neuroscience.* Ed. J. Leo van Hemmen and Terrence J. Sejnowski. Oxford: Oxford University Press, 2006. 44–61.

Eisenberg, Nancy. "Empathy-Related Emotional Responses, Altruism, and Their Socialization." Davidson and Harrington 131–64.

Elliott, John R., Jr. *Playing God: Medieval Mysteries on the Modern Stage.* Toronto: University of Toronto Press, 1989.

Elliott, Rebecca, Raymond J. Dolan, and Chris D. Frith. "Dissociable Functions in the Medial and Lateral Orbitofrontal Cortex." *Cerebral Cortex* 10 (2000): 308–17.

Emery, Ted. "Introduction." Carlo Gozzi. *Five Tales for the Theatre.* Ed. and trans. Albert Bermel and Ted Emery. Chicago: University of Chicago Press, 1989. 1–19.

Enders, Jody. *Death by Drama and Other Medieval Urban Legends.* Chicago: University of Chicago Press, 2002.

———. "Dramatic Memories and Tortured Spaces in the *Mistere de la Sainte Hostie*." *Medieval Practices of Space.* Ed. Barbara A. Hanawalt and Michal Kobialka. Minneapolis: University of Minnesota Press, 2000. 199–222.

———. *The Medieval Theater of Cruelty.* Ithaca: Cornell University Press, 1999.

———. "Theater Makes History." *Speculum* 79.4 (2004): 991–1016.

Eshleman, Clayton. *Juniper Fuse: Upper Paleolithic Imagination and the Construction of the Underworld.* Middletown, CT: Wesleyan University, 2003.

Euripides. *The Bacchae.* Trans. Paul Roche. *Three Plays of Euripides.* New York: Norton, 1974. 78–126.

———. *Electra.* Trans. John Davie. *Electra and Other Plays.* London: Penguin, 1998. 135–74.

———. *Hippolytus.* Trans. David Grene. *Greek Tragedies.* Vol. 1. Ed. David Grene and Richmond Lattimore. Chicago: University of Chicago Press, 1960. 231–91.

———. *Medea.* Trans. Paul Roche. *Three Plays of Euripides.* New York: Norton, 1974. 33–77.

———. *Orestes.* Trans. John Peck and Frank Nisetich. Oxford: Oxford University Press, 1995.

———. *The Trojan Women.* Trans. Paul Roche. *Euripides: Ten Plays.* New York: Penguin, 1998. 457–512.

Evan Almighty. Dir. Tom Shadyac. Universal, 2006.

Everett, Daniel L. *Don't Sleep, There are Snakes.* New York: Pantheon, 2008.

Everyman and Medieval Miracle Plays. Ed. A. C. Cawley. New York: Dutton, 1959.

"Exegesis." *Millennium*. Writ. Chip Johannessen. Dir. Ralph Hemecker. Fox. 9 Oct. 1998.

Fauconnier, Gilles, and Mark Turner. *The Way We Think*. New York: Basic, 2002.

Feinberg, Todd E. *Altered Egos: How the Brain Creates the Self*. Oxford: Oxford University Press, 2001.

Felman, Shoshana. *Jacques Lacan and the Adventure of Insight*. Cambridge: Harvard University Press, 1987.

Ferré, John P. *Channels of Belief: Religion and American Commercial Television*. Ames: Iowa State University Press, 1990.

Fiddes, Paul S. *The Creative Suffering of God*. Oxford: Clarendon, 1988.

Fink, Bruce. *A Clinical Introduction to Lacanian Psychoanalysis*. Cambridge: Harvard University Press, 1997.

———. "The Ethics of Psychoanalysis: A Lacanian Perspective." *Psychoanalytic Review* 86.4 (Aug. 1999): 529–45.

———. *The Lacanian Subject*. Princeton: Princeton University Press, 1995.

"First Hour of Fox Series Seemed Like a Millennium." *The Seattle Times* 31 Oct 1996: E6.

"The Fourth Horseman." *Millennium*. Writ. Glen Morgan and James Wong. Dir. Dwight Little. Fox. 8 May 1998.

Fowkes, Katherine A. *Giving Up the Ghost: Spirits, Ghosts, and Angels in Mainstream Comedy Films*. Detroit: Wayne State University Press, 1998.

Fowles, Jib. *The Case for Television Violence*. Thousand Oaks, California: Sage, 1999.

———. "The Violence Against Television Violence." *Television Quarterly* 28.1 (1996): 41–45.

Freeman, Anthony. "God as an Emergent Property." *Journal of Consciousness Studies* 8.9–10 (2001): 147–59.

Freeman, Walter J. *Societies of Brains*. Hillsdale, NJ: Lawrence Erlbaum, 1995.

Freud, Sigmund. "Psychopathic Characters on the Stage." *Psychoanalytic Quarterly* 11 (1942): 459–64.

———. *The Standard Edition of the Complete Psychological Works of Sigmund Freud*. London: Hogarth, 1953–74.

Frey, Charles. "*The Tempest* and the New World." *Shakespeare Quarterly* 30 (1979): 29–41.

Fuster, Joaquín M. *Cortex and Mind*. Oxford: Oxford University Press, 2003.

Gaiman, Neil. *American Gods*. New York: Harper, 2003.

Gallagher, Lowell. "Faustus's Blood and the (Messianic) Question of Ethics." *ELH* 73 (2006): 1–29.

Gallese, Vittorio, and Alvin Goldman. "Mirror Neurons and the Simulation Theory of Mind-Reading." *Trends in Cognitive Neuroscience* 2.12 (1998): 493–502.

Gallese, Vittorio, Christian Keysers, and Giacomo Rizzolatti. "A Unifying View of the Basis of Social Cognition." *Trends in Cognitive Neuroscience* 8.9 (2004): 396–401.

Gaster, Theodor H., ed. *Thespis: Ritual, Myth, and Drama in the Ancient Near East*. Rev. Ed. Garden City, NY: Doubleday, 1961.

Gazzaniga, Michael S. *The Ethical Brain*. New York: Dana, 2005.

———. *The Mind's Past*. Berkeley: University of California Press, 1998.

Gerbner, George, Larry Gross, Nancy Signorielli, Michael Morgan, and Marilyn Jackson-Beeck. "The Demonstration of Power: Violence Profile No. 10." *Journal of Communications* 29.3 (summer 1979): 177–96.

Geschwind, Daniel H., and Marco Iacoboni. "Structural and Functional Asymmetries of the Human Frontal Lobes." In Miller and Cummings. 45–70.

Gibbons, Reginald, trans. *Bakkhai*. By Euripides. Oxford: Oxford University Press, 2001.

Gibbs, Raymond W., Jr. *Embodiment and Cognitive Science*. Cambridge: Cambridge University Press, 2006.

Girard, René. *The Scapegoat*. Baltimore: Johns Hopkins University Press, 1986.

———. *A Theater of Envy*. Oxford: Oxford University Press, 1991.

———. *Violence and the Sacred*. Baltimore: Johns Hopkins University Press, 1977.

Glenberg, Arthur M. "What Memory Is For." *Behavioral and Brain Sciences* 20 (1997): 1–19.

Godfrey, Bob. "Everyman (Re)Considered." *European Medieval Drama* 4 (2000): 155–68.

Goethals, Gregor T. *The TV Ritual: Worship at the Video Altar*. Boston: Beacon, 1981.

Goldman, Alvin I. "Simulation Theory and Mental Concepts." *Simulation and Knowledge of Action*. Ed. Jerome Dokic and Joëlle Proust. Amsterdam: John Benjamins, 2002. 1–19.

Goleman, Daniel. *Destructive Emotions*. New York: Bantam, 2003.

Goodale, Melvyn A. and A. David Milner. *Sight Unseen*. Oxford: Oxford University Press, 2004.

Gordon, Robert M. "Sympathy, Simulation, and the Impartial Spectator." May et al. 165–80.

Gough, William C., and Robert L. Shacklett. "What Science Can and Can't Say About Spirits, Part 3." *Journal of Religion and Psychical Research* 24.1 (2001): 48–57.

Gould, Thomas. "The Uses of Violence in Drama." *Violence in Drama*. Ed. James Redmond. Cambridge: Cambridge University Press, 1991. 1–14.

Gozzi, Carlo. *Five Tales for the Theatre*. Ed. and trans. Albert Bermel and Ted Emery. Chicago: University of Chicago Press, 1989.

Green, Viviane, ed. *Emotional Development in Psychoanalysis, Attachment Theory and Neuroscience*. New York: Brunner-Routledge, 2003.

Greenblatt, Stephen. *Hamlet in Purgatory*. Princeton: Princeton University Press, 2001.

Greene, Joshua. "Emotion and Cognition in Moral Judgment." In Changeux et al. 57–66.

———. "The Secret Joke of Kant's Soul." *Moral Psychology*. Vol. 3. Ed. Walter Sinnott-Armstrong. Cambridge: MIT Press, 2008. 35–79.

Greene, Joshua, and Jonathan Haidt. "How (and Where) Does Moral Judgment Work?" *Trends in Cognitive Science* 6.12 (Dec. 2002): 517–23.

Greenspan, Stanley I., and Stuart G. Shanker. *The First Idea: How Symbols, Language, and Intelligence Evolved From Our Primate Ancestors to Modern Humans*. Cambridge, MA: Perseus, 2004.

Gregory, Richard. Interview. *Conversations on Consciousness*. By Susan Blackmore. Oxford University Press, 2006. 104–14.

Grigg, Russell. "From the Mechanism of Psychosis to the Universal Condition of the Symptom: On Foreclosure." *Key Concepts of Lacanian Psychoanalysis*. Ed. Dany Nobus. London: Rebus, 1998. 48–74.

Groopman, Jerome. "God on the Brain." *New Yorker* 17 Sept. 2001: 165–68.

Guthrie, R. Dale. *The Nature of Paleolithic Art*. Chicago: University of Chicago Press, 2005.

Guthrie, Stewart Elliott. *Faces in the Clouds: A New Theory of Religion*. New York: Oxford University Press, 1993.

Haidt, Jonathan. "The Emotional Dog and Its Rational Tail: A Social Intuitionist Approach to Moral Judgment." *Psychological Review* 108.4 (2001): 814–34.

———. *The Happiness Hypothesis*. New York: Basic, 2006.

Halifax, Joan. *Shaman: The Wounded Healer*. London: Thames and Hudson, 1982.

Halpern, Richard. "Marlowe's Theater of Night." *ELH* 71 (2004): 455–95.

Hamer, Dean. *The God Gene*. New York: Doubleday, 2004.

Hamilton, James T. *Channeling Violence: The Economic Market for Violent Television Programming*. Princeton: Princeton University Press, 1998.

Hammill, Graham. "Faustus's Fortunes." *ELH* 63.2 (1996): 309–36.

Harper, Elizabeth, and Britt Mize. "Material Economy, Spiritual Economy, and Social Critique in *Everyman*." *Comparative Drama* (Sept. 2006): 236–311.

Harrington, Anne. "A Science of Compassion or a Compassionate Science?" In Davidson and Harrington. 18–30.

Harris, Sam. *The End of Faith*. New York: Norton, 2004.

Harris, Sam, Jonas T. Kaplan, Ashley Curiel, Susan Y. Bookheimer, Marco Iacoboni, and Mark S. Cohen. "The Neural Correlates of Religious and Nonreligious Belief." *PLoS ONE* (2009) 4.10: e7272.

Hart, F. Elizabeth. "'Great is Diana' of Shakespeare's Ephesus." *SEL* 43.2 (Spring 2003): 347–74.

Harth, Erich. *The Creative Loop: How the Brain Makes a Mind*. New York: Addison-Wesley, 1993.

Hasson, Uri, Ohad Landesman, Barbara Knappmeyer, Ignacio Vallines, Nava Rubin, and David J. Heeger. "Neurocinematics." *Projections* 2.1 (2008): 1–26.

Hasson, Uri, Yuval Nir, Ifat Levy, Galit Fuhrmann, and Rafael Malach. "Intersubject Synchronization of Cortical Activity During Natural Vision." *Science* 303 (2004): 1634–40.

Hatfield, Elaine, John T. Cacioppo, and Richard L. Rapson. *Emotional Contagion*. Cambridge: Cambridge University Press, 1994.

Haught, John F. *Is Nature Enough?* Cambridge: Cambridge University Press, 2006.

Hayden, Brian. *Shamans, Sorcerers, and Saints*. Washington: Smithsonian, 2003.

Hecht, Jennifer Michael. *Doubt: A History*. New York: Harper, 2003.

Heinrichs, Albert. "'He Has a God in Him': Human and Divine in the Modern Perception of Dionysus." *Masks of Dionysus*. Ed. Thomas H. Carpenter and Christopher A.

Faraone. Ithaca: Cornell University Press, 1993. 13–43.
Hellboy. Dir. Guillermo del Toro. Columbia, 2004.
Helm, James J. "Aeschylus' Genealogy of Morals." *Transactions of the American Philological Association* 134.1 (2004): 23–54.
Hill, John. "Film and Television." *The Oxford Guide to Film Studies.* Ed. John Hill and Pamela Church Gibson. Oxford: Oxford University Press, 1998. 605–10.
Hinde, Robert A. *Why Gods Persist.* London: Routledge, 1999.
Hobson, J. Allan. "Consciousness as a State-Dependent Phenomenon." *Scientific Approaches to Consciousness.* Ed. Jonathan D. Cohen and Jonathan W. Schooler. Mahwah, NJ: Lawrence Erlbaum, 1997. 379–96.
Hogan, Patrick Colm. *Cognitive Science, Literature, and the Arts.* New York: Routledge, 2003.
Holland, Norman. "Literary Creativity: A Neuropsychoanalytic View." *Evolutionary and Neurocognitive Approaches to Aesthetics, Creativity, and the Arts.* Ed. Colin Martindale, Paul Locher, and Vladimir M. Petrov. Amityville, New York: Baywood, 2007. 165–80.
Hood, Bruce M. *Supersense.* New York: Harper, 2009.
Hoppal, Mihaly, and Keith D. Howard, eds. *Shamans and Cultures.* Los Angeles: International Society for Trans-Oceanic Research, 1993.
Horgan, John. *Rational Mysticism.* New York: Houghton Mifflin, 2003.
Horsley, Jake. *The Blood Poets: A Cinema of Savagery 1958–1999.* Vol. 2. London: Scarecrow Press, 1999.
Houdé, Oliver. "Cerebral Basis of Human Errors." In Changeux et al. 137–42.
Hough, Kirstin J., and Philip G. Erwin. "Children's Attitudes Toward Violence on Television." *The Journal of Psychology* 131.4 (1997): 411–15.
Hrdy, Sarah Blaffer. *Mother Nature.* New York: Ballantine, 1999.
Huerta, Jorge. *Chicano Drama.* Cambridge: Cambridge University Press, 2000.
———. "A Representation of Death in an Anti-Vietnam War Play by Luis Valdez: *Dark Root of a Scream.*" *Death in American Texts and Performances.* Ed. Lisa K. Perdigao and Mark Pizzato. Surrey: Ashgate, 2010.
Huesmann, L. Rowell. "Cross-National Communalities in the Learning of Aggression from Media Violence." *Television and the Aggressive Child.* Ed. L. Rowell Huesmann and Leonard D. Eron. Hillsdale, New Jersey: Lawrence Erlbaum Associates, 1986. 239–57.
Huesmann, L. Rowell, and Laurie S. Miller. "Long-Term Effects of Media Violence." *Aggressive Behavior: Current Perspectives.* Ed. L. Rowell Huesmann. New York: Plenum Press, 1994. 153–86.
Huff, Richard. "Now That Fox is in a Fix, What Will Future Hold? Carter's Science Project." *New York Daily News* 2 Oct. 1998, home ed.: 128.
Hughes, Jan N., and Jan E. Hasbrouck. "Television Violence: Implications for Violence Prevention." *School Psychology Review* 25.2 (1996): 134–51.
Hunt, Maurice. *Shakespeare's Labored Art.* New York: Peter Lang, 1995.
Hurley, Susan. "Imitation, Media Violence, and Freedom of Speech." *Philosophical Studies* 117 (2004): 165–218.
Hurley, Susan, and Nick Chater, eds. *Perspectives on Imitation.* Vols. 1 and 2. Cambridge: MIT Press, 2005.
Iacoboni, Marco. *Mirroring People.* New York: Farrar, Straus and Giroux, 2008.
"The Innocents." *Millennium.* Writ. Michael Duggan. Dir. Thomas Wright. Fox. 2 Oct. 1998.
Jablonka, Eva, and Marion Lamb. *Evolution in Four Dimensions.* Cambridge: MIT, 2005.
Jacob, Pierre, and Marc Jeannerod. *Ways of Seeing.* Oxford: Oxford University Press, 2003.
Jacob's Ladder. Dir. Adrian Lyne. TriStar, 1990.
Jameson, Michael. "The Asexuality of Dionysus." *Masks of Dionysus.* Ed. Thomas H. Carpenter and Christopher A. Faraone. Ithaca: Cornell University Press, 1993. 44–64.
Jaynes, Julian. *The Origin of Consciousness in the Breakdown of the Bicameral Mind.* Boston: Houghton Mifflin, 1976.
Johnson, Mark L. "How Moral Psychology Changes Moral Theory." May et al. 45–68.
Joseph, Rhawn. *The Transmitter to God: The Limbic System, the Soul, and Spirituality.* San Jose: University Press California, 2001.
Jourdan, Robert. *Music, the Brain, and Ecstasy.* New York: Avon, 1997.
Kagan, Jerome. "Morality and Its Development." *Moral Psychology.* Vol. 3. Ed. Walter Sinnott-Armstrong. Cambridge: MIT Press, 2008. 297–312.

Kahneman, Daniel, and Cass R. Sunstein. "Cognitive Psychology of Moral Intuitions." In Changeux et al. 91–106.

Kalamas, Alicia D. and Mandy L. Gruber. "Electrodermal Responses to Implied Versus Actual Violence on Television." *Journal of General Psychology* 125.1 (1998): 31.

Kalke, Christine M. "The Making of a Thyrsus: The Transformation of Pentheus in Euripides' *Bacchae*." *The American Journal of Philology* 106.4 (1985): 409–26.

Kandel, Eric R. "Biology and the Future of Psychoanalysis: A New Intellectual Framework for Psychiatry Revisited." *American Journal of Psychiatry* (Apr. 1999) 156.4: 505–24.

_____. *In Search of Memory*. New York: Norton, 2006.

_____. *Psychiatry, Psychoanalysis, and the New Biology of Mind*. New York: American Psychiatric, 2005.

Kaplan, E. Ann. *Motherhood and Representation: The Mother in Popular Culture and Melodrama*. London: Routledge, 1992.

_____. *Women and Film: Both Sides of the Camera*. New York: Methuen, 1983.

Kaplan-Solms, Karen and Mark Solms. *Clinical Studies in Neuro-Psychoanalysis*. London: Karnac, 2000.

Kapogiannis, Dimitrios, Aron K. Barbey, Michael Su, Giovanna Zamboni, Frank Krueger, and Jordan Grafman. "Cog-nitive and Neural Foundations of Religious Belief." PNAS 106.12 (2009): 4876–81.

Katz, Alyssa. Rev. of *Millennium*. *The Nation* 25 Nov. 1996: 35–36.

Katz, Leonard D., ed. *Evolutionary Origins of Morality*. Bowling Green: Bowling Green State University, 2000.

Kellner, Douglas. "Television, Mythology and Ritual." *Praxis* 6 (1982): 133–55.

Kennett, Jeanette, and Cordelia Fine. "Internalism and the Evidence from Psychopaths and 'Acquired Sociopaths.'" *Moral Psychology*. Vol. 3. Ed. Walter Sinnott-Armstrong. Cambridge: MIT Press, 2008. 173–90.

Kernodle, George R. *The Theatre in History*. Fayetteville: University of Arkansas Press, 1989.

Kiehl, Kent A. "Without Morals: The Cognitive Neuroscience of Criminal Psychopaths." *Moral Psychology*. Vol. 3. Ed. Walter Sinnott-Armstrong. Cambridge: MIT Press, 2008. 119–50.

Kihlstrom, John F. "Consciousness and Me-ness." *Scientific Approaches to Consciousness*. Ed. Jonathan D. Cohen and Jonathan W. Schooler. Mahwah, NJ: Lawrence Erlbaum, 1997. 451–68.

Kimball, A. Samuel. *The Infanticidal Logic of Evolution and Culture*. Newark: University of Delaware Press, 2007.

King, Barbara J. *Evolving God: A Provocative View of the Origins of Religion*. New York: Doubleday, 2007.

Kirby, E. T. *Ur-Drama: The Origins of Theatre*. New York: New York University Press, 1975.

Kister, Daniel A. "Comic Play in a Korean Shaman Rite." In Hoppal and Howard 40–46.

Kleist, Heinrich von. *Prince Friedrich of Homburg*. Trans. Diana Stone Peters and Frederick G. Peters. New York: New Directions, 1978.

Kobialka, Michal. *This Is My Body*. Ann Arbor: University of Michigan Press, 1999.

Koechlin, Etienne, Chrystèle Ody, and Frédérique Kouneiher. "The Architecture of Cognitive Control in the Human Prefrontal Cortex." *Science* 302 (2003): 1181–85.

Kordela, A. Kiarina. *$urplus*. Albany: State University of New York Press, 2007.

Kosslyn, Stephen M. "A Science of the Divine?" *What is Your Dangerous Idea?* Ed. John Brockman. London: Simon & Schuster, 2006. 164–74.

Krcmar, Marina. "The Contribution of Family Communication Patterns to Children's Interpretations of Television Violence." *Journal of Broadcasting and Electronic Media* 42.2 (1998): 250–64.

Kristeva, Julia. "Modern Theater Does Not Take (A) Place." *Sub-Stance* 18/19 (1977): 131–34.

_____. *Powers of Horror: An Essay on Abjection*. New York: Columbia University Press, 1982.

_____. *Revolution in Poetic Language*. New York: Columbia University Press, 1984.

Kubiak, Anthony, and Bryan Reynolds. "The Delusion of Critique." *Transversal Enterprises in the Drama of Shakespeare and His Contemporaries*. Ed. Bryan Reynolds. New York: Palgrave, 2006. 64–84.

Kunkle, Sheila. "Psychosis in a Cyberspace Age." *Other Voices* 1.3 (October 1998).

_____. "Žižek's Paradox." *Journal for Lacanian Studies* 1.2 (2003): 224–42.

Kushner, Tony. *Angels in America*. New York: Theatre Communications Group, 1995.

Lacan, Jacques. "Desire and the Interpretation of Desire in *Hamlet*." *Yale French Studies* 55/56 (1977): 11–52.

———. *Écrits: A Selection*. Trans. Bruce Fink. New York: Norton, 2002.
———. *The Four Fundamental Concepts of Psycho-Analysis*. Trans. Alan Sheridan. New York: Norton, 1978.
———. *The Seminar of Jacques Lacan: Book II, The Ego in Freud's Theory and in the Technique of Psychoanalysis, 1954–1955*. Trans. Sylvana Tomaselli. Cambridge: Cambridge University Press, 1988.
———. *The Seminar of Jacques Lacan: Book VII, The Ethics of Psychoanalysis, 1959–1960*. Trans. Dennis Porter. New York: Norton, 1997.
———. *The Seminar of Jacques Lacan: Book XX, On Feminine Sexuality, The Limits of Love and Knowledge (Encore), 1972–73*. Trans. Bruce Fink. New York: Norton, 1998.
Lacoue-Labarthe, Philippe. "Theatrum Analyticum." *Mimesis, Masochism, and Mime*. Ed. Timothy Murray. Ann Arbor: University of Michigan Press, 1997. 175–96.
Ladd, Roger A. "'My condicion is mannes soule to kill'—Everyman's Mercantile Salvation." *Comparative Drama* (2007) 41.1: 57–78.
Lakoff, George, and Mark Johnson. *Philosophy in the Flesh: The Embodied Mind and Its Challenge to Western Thought*. New York: Basic, 1999.
Lang, Robert. *American Film Melodrama: Griffith, Vidor, Minnelli*. Princeton: Princeton University Press, 1989.
Laplanche, J., and J.-B. Pontalis. *The Language of Psycho-Analysis*. New York: Norton, 1974.
Laughlin, Charles D., Jr., John McManus, and Eugene G. d'Aquili. *Brain, Symbol and Experience*. Boston: Shambhala, 1990.
Leader, Darian, and Judy Groves. *Lacan For Beginners*. London: Penguin, 1995.
Leary, Mark R. *The Curse of the Self*. Oxford: Oxford University Press, 2004.
LeDoux, Joseph. *The Emotional Brain*. New York: Simon & Schuster, 1996.
———. *Synaptic Self*. New York: Penguin, 2003.
Lee, Meewon. "Shamanistic Elements of Korean Folk Theatre, *Kamyon guk*." In Hoppal and Howard. 33–39.
Lefkowitz, Mary. *Greek Gods, Human Lives*. New Haven: Yale University Press, 2003.
Lehrer, Jonah. *How We Decide*. Boston: Houghton Mifflin, 2009.
Lemoine, Jacques. "The Diagnosis of Disease as a 'Shamanic Equation' among the Hmong of Laos and Thailand." Hoppal and Howard 111–19.

Lesko, Barbara S. *The Great Goddesses of Egypt*. Norman: University of Oklahoma Press, 1999.
Levesque, John. "'Millennium' Crosses the Line in Gruesomeness." *Seattle Post-Intelligencer* 23 Oct. 1996: C1.
Levitin, Daniel J. *This Is Your Brain on Music*. New York: Penguin, 2007.
Levy, Neil. *Neuroethics*. Cambridge: Cambridge University Press, 2007.
———. *What Makes Us Moral?* Oxford: One World, 2004.
Lewis-Williams, David. "Constructing a Cosmos: Architecture, Power and Domestication at Çatalhöyük." *Journal of Social Archaeology* 4.1 (2004): 28–59.
———. "Harnessing the Brain: Vision and Shamanism in Upper Paleolithic Western Europe." *Beyond Art: Pleistocene Image and Symbol*. Ed. Alan E. Leviton. San Francisco: California Academy of Sciences, 1997. 321–42.
———. "'Meaning' in Southern African San Rock Art: Another Impasse?" *South African Archaeological Bulletin* 54 (1999): 141–45.
———. *The Mind in the Cave*. London: Thames and Hudson, 2002.
———. "Of People and Pictures." *Becoming Human*. Ed. Colin Renfrew and Iain Morley. Cambridge: Cambridge University Press, 2009. 135–58.
Lewis-Williams, David J., and David G. Pearce. "Southern African San Rock Paintings as Social Intervention: A Study of Rain-Control Images." *African Archaeological Review* 21.4 (Dec. 2004): 199–228.
Liebes, Tamar, and James Curran, eds. *Media, Ritual and Identity*. London: Routledge, 1998.
Lima, Robert. *Stages of Evil: Occultism in Western Theater and Drama*. Lexington: University Press of Kentucky, 2005.
Lipton, Bruce H. *The Biology of Belief*. Santa Rosa: Mountain of Love, 2005.
Logan, Robert A. *Shakespeare's Marlowe*. Burlington, VT: Ashgate, 2007.
Logan, Robert K. *The Extended Mind*. Toronto: University of Toronto Press, 2007.
Lonsdale, Steven H. *Dance and Ritual Play in Greek Religion*. Baltimore: Johns Hopkins University Press, 1993.
Lorblanchet, Michel, and Francis Jach. *Cougnac*. Gourdon, France: Imagis, 2004.
Lovelock, James. *The Ages of Gaia*. New York: Norton, 1988.
Lynch, Aaron. *Thought Contagion: How Belief*

Lyon, Eleanor. "Media Murder and Mayhem: Violence on Network Television." *Marginal Conventions: Popular Culture, Mass Media and Social Deviance*. Ed. Clinton R. Sanders. Bowling Green: Bowling Green State University Popular Press, 1990. 144–54.

MacIntyre, Alasdair. *Dependent Rational Animals*. Chicago: Open Court, 2001.

Malti, Tina. "Moral Emotions and Aggressive Behavior in Childhood." *Emotions and Aggressive Behavior*. Ed. Georges Steffgen and Mario Gollwitzer. Cambridge: Hogrefe and Huber, 2007. 185–200.

Margulis, Lynn. *Symbiotic Planet*. New York: Basic, 1998.

Marks, Jonathan. "Great Chain of Being." *Encyclopedia of Race and Racism*. 68–73.

———. Review of *Demonic Males*, by Richard Wrangham and Dale Peterson. *Human Biology* 70 (1998): 143–46.

———. *What It Means To Be 98% Chimpanzee*. Berkeley: University of California Press, 2002.

———. *Why I am Not a Scientist*. Berkeley: University of California Press, 2009.

Marlowe, Christopher. *Doctor Faustus*. *The Plays of Christopher Marlowe*. Ed. Roma Gill. Oxford: Oxford University Press, 1971. 331–400.

Marsalek, Karen Sawyer. "'Awake Your Faith.'" "*Bring Furth the Pagants.*" Edited by David N. Klausner and Karen Sawyer Marsalek. Toronto: University of Toronto Press, 2007. 271–91.

Martin, Laura. "'Schlechtes Mensch/Gutes Opfer': The Role of Marie in Geog Büchner's *Woyzeck*." *Gendering German Studies*. Ed. Margaret Littler. Oxford: Blackwell, 1997. 51–66.

Matthews, Caitlin, and John Matthews. *Walkers Between the Worlds*. Rochester, VT: 2003.

Maurer, Daphne, and Catherine J. Mondloch. "Neonatal Synesthesia." *Synesthesia*. Oxford: Oxford University Press, 2006. 193–213.

McConachie, Bruce. *Engaging Audiences*. New York: Palgrave, 2008.

———. "Falsifiable Theories for Theatre and Performance Studies." *Theatre Journal* 59.4 (Dec. 2007): 553–77.

McConachie, Bruce, and F. Elizabeth Hart, eds. *Performance and Cognition*. New York: Routledge: 2006.

McCullough, Malcolm E. *Beyond Revenge*. San Francisco: Jossey-Bass, 2008.

McGeer, Victoria. "Varieties of Moral Agency." *Moral Psychology*. Vol. 3. Ed. Walter Sinnott-Armstrong. Cambridge: MIT Press, 2008. 227–57.

McGrath, Alister. *Dawkin's God*. Oxford: Blackwell, 2005.

———. *Science and Religion*. Oxford: Blackwell, 1999.

McNamara, Patrick. *Mind and Variability: Mental Darwinism, Memory, and Self*. Westport, Connecticut: Praeger, 1999.

———. "Religion and the Frontal Lobes." *Religion in Mind: Cognitive Perspectives on Religious Belief, Ritual, and Experience*. Ed. Jensine Andresen. Cambridge: Cambridge University Press, 2001. 237–56.

———. *Where God and Science Meet*. Vols. 1–3. Westport, Connecticut: Praeger, 2006.

Meeks, Dimitri, and Christine Favard-Meeks. *Daily Life of the Egyptian Gods*. Trans. G. M. Goshgarian. 1993. Ithaca: Cornell University Press, 1996.

Mellars, Paul. "Cognition and Climate." *Becoming Human*. Ed. Colin Renfrew and Iain Morley. Cambridge: Cambridge University Press, 2009. 212–31.

Meltzoff, Andrew N. "Elements of a Developmental Theory of Imitation." *The Imitative Mind: Development, Evolution, and Brain Bases*. Ed. Andrew N. Meltzoff and Wolfgang Prinz. Cambridge: Cambridge University Press, 2002. 19–41.

Metz, Christian. *The Imaginary Signifier*. Bloomington: Indiana University Press, 1982.

Meyer, Cheryl L. "Andrea Yates, the Mother Next Door." *Pittsburgh Post-Gazette* 1 July 2001, two star edition: E1.

Meyer, Marvin W. ed. *The Ancient Mysteries*. New York: Harper, 1987.

Michael. Dir. Nora Ephron. New Line, 1996.

"Midnight of the Century." *Millennium*. Writ. Erin Maher and Kay Reindl. Dir. Dwight Little. Fox. 19 Dec. 1997.

"'Millennium' Giving Star Gray Hairs: Show's Frank Black Lightens Up Off-Camera." *The Denver Post* 6 Dec 1998: I-01.

"'Millennium' Is Intense and Scary; 'EZ Streets' Goes Down Same Tense Road." *The Buffalo News* 25 Oct. 1996: C11.

Miller, Bruce L., and Jeffrey L. Cummings, eds. *The Human Frontal Lobes*. New York: Guilford, 1999.

Miller, Geoffrey. *The Mating Mind*. New York: Doubleday, 2000.

Mills, David. "Theatres of *Everyman*." *From Page to Performance*. Ed. John A. Alford. East Lansing: Michigan State University Press, 1995. 127–49.

Minsky, Marvin. *The Emotion Machine*. New York: Simon & Schuster, 2006.

———. *The Society of Mind*. New York: Simon & Schuster, 1986.

Mithen, Steven. "The Evolution of Imagination: An Archaeological Perspective." *SubStance* 94/95 (2001): 28–54.

———. "Out of the Mind: Material Culture and the Supernatural." *Becoming Human*. Ed. Colin Renfrew and Iain Morley. Cambridge: Cambridge University Press, 2009. 123–34.

———. *The Prehistory of the Mind*. London: Thames and Hudson, 1996.

Moise, Jessica F., and L. Rowell Huesmann. "Television Violence Viewing and Aggression in Females." *Annals of the New York Academy of Sciences* 794 (1996): 382.

Moll, Jorge, Ricardo de Oliveira-Souza, Roland Zahn, and Jordan Grafman. "The Cognitive Neuroscience of Moral Emotions." *Moral Psychology*. Vol. 3. Ed. Walter Sinnott-Armstrong. Cambridge: MIT Press, 2008. 1–17.

"Monster." *Millennium*. Writ. Glen Morgan and James Wong. Dir. Perry Lang. Fox. 17 Oct. 1997.

Montelle, Yann-Pierre. "Mimicry, Deception, Mimesis." *Rock Art Research* 23.1 (2006): 23–24.

———. *Palaeoperformance*. London: Seagull, 2009.

———. "Paleoperformance: Investigating the Human Use of Caves in the Upper Paleolithic." *New Perspectives on Prehistoric Art*. Ed. Günter Berghaus. Westport, Connecticut: Praeger, 2004. 131–52.

Moody, Raymond A., Jr. *Life After Life*. Atlanta: Mockingbird: 1975.

Morse, William R. "Desire and the Limits of Melodrama." *Melodrama*. Ed. James Redmond. Cambridge: Cambridge University Press, 1992. 17–30.

Mossman, J. M. "Chains of Imagery in *Prometheus Bound*." *Classical Quarterly* 46.1 (1996): 58–67.

Murphy, Nancey, and Warren S. Brown. *Did My Neurons Make Me Do It?* Oxford: Oxford University Press, 2007.

Murray, John P. "The Violent Face of Television: Research and Discussion." *The Faces of Televisual Media*. 2nd ed. Ed. Edward L. Palmer and Brian M. Young. Mahwah, NJ: Lawrence Erlbaum, 2003. 143–60.

Nadel, Jacqueline. "Imitation and Imitation Recognition." *The Imitative Mind: Development, Evolution, and Brain Bases*. Ed. Andrew N. Meltzoff and Wolfgang Prinz. Cambridge: Cambridge University Press, 2002. 42–62.

Nelson, Katherine. *Language in Cognitive Development*. Cambridge: Cambridge University Press, 1996.

Nesse, Randolph M., and Alan T. Lloyd. "The Evolution of Psychodynamic Mechanisms." *The Adapted Mind*. Ed. Jerome H. Barkow, Leda Cosmides, and John Tooby. Oxford: Oxford University Press, 1992. 601–24.

Nettle, Daniel. "What Happens in Hamlet? Exploring the Psychological Foundations of Drama." *The Literary Animal*. Ed. Jonathan Gottschall and David Sloan Wilson. Evanston: Northwestern University Press, 2005.

Neumann, Erich. *The Great Mother*. Princeton: Princeton University Press, 1955.

Newberg, Andrew B., Eugene G. d'Aquili, Stephanie K. Newberg, and Verushka deMarici. "The Neuropsychological Correlates of Forgiveness." *Forgiveness*. Ed. Michael E. McCullough, Kenneth I. Pargament, and Carl E. Thoresen. New York: Guilford, 2000. 91–110.

Newberg, Andrew, Eugene d'Aquili, and Vince Rause. *Why God Won't Go Away: Brain Science and the Biology of Belief*. New York: Ballantine, 2002.

Newberg, Andrew, and Mark Robert Waldman. *Why We Believe What We Believe*. New York: Free Press, 2006.

Newitz, Annalee. "Serial Killers, True Crime, and Economic Performance Anxiety." *Mythologies of Violence in Postmodern Media*. Ed. Christopher Sharrett. Detroit: Wayne State University Press, 1999. 65–83.

Nickisch, Curt Wendell. "Georg Büchner's Philosophy of Science." *Selecta* 18 (1997): 37–45.

Niedenthal, Paula M., Lawrence W. Barsalou, Francois Ric, and Silvia Krauth-Gruber. "Embodiment in the Acquisition and Use of Emotion Knowledge." Barrett et al. 21–50.

Nietzsche, Friedrich. *The Birth of Tragedy*. Trans. Walter Kaufmann. New York: Random House, 1967.

———. *The Gay Science*. *The Portable Nietzsche*. Ed. and trans. Walter Kaufmann. New York: Viking, 1959. 93–102.

Nisse, Ruth. *Defining Acts: Drama and the Politics of Interpretation in Late Medieval*

England. Notre Dame: University of Notre Dame Press, 2004.

Nordmann, Alfred. "Political Theater as Experimental Anthropology." *New German Critique* 66 (Fall 1995): 17–34.

Nussbaum, Martha C. *The Fragility of Goodness: Luck and Ethics in Greek Tragedy and Philosophy*. Cambridge: Cambridge University Press, 1986.

Obbink, Dirk. "Dionysus Poured Out: Ancient and Modern Theories of Sacrifice and Cultural Formation." *Masks of Dionysus*. Ed. Thomas H. Carpenter and Christopher A. Faraone. Ithaca: Cornell University Press, 1993. 65–86.

Oh, God!. Dir. Carl Reiner. Writ. Larry Gelbart. Warner Brothers, 1977.

Oh, God! Book II. Dir. Gilbert Cates. Writ. Josh Greenfeld. Warner Brothers, 1980.

Oh God, You Devil. Dir. Paul Bogart. Writ. Andrew Bergman. Warner Brothers, 1984.

Ornstein, Robert. *The Right Mind*. New York: Harcourt, 1997.

Owen, Bradford. "Narrative Comprehension and Enjoyment of Feature Films." *New Beginnings in Literary Studies*. Ed. Jan Auracher and Willie van Peer. Newcastle, UK: Cambridge Scholars, 2008. 286–307.

Padel, Ruth. *In and Out of the Mind: Greek Images of the Tragic Self*. Princeton University Press, 1992.

_____. *Whom Gods Destroy: Elements of Greek and Tragic Madness*. Princeton University Press, 1995.

Pagels, Elaine. *The Origin of Satan*. New York: Random House, 1995.

Palombo, Stanley R. *The Emergent Ego: Complexity and Coevolution in the Psychoanalytic Process*. Madison: International, 1999.

Panksepp, Jaak. *Affective Neuroscience: The Foundations of Human and Animal Emotions*. Oxford: Oxford University Press, 1998.

_____. "Emotions as Viewed by Psychoanalysis and Neuroscience: An Exercise in Consilience." *NeuroPsychoanalysis* 1 (1999): 15–38.

"Paper Dove." *Millennium*. Writ. Walon Green and Ted Mann. Dir. Thomas Wright. Fox. 16 May 1997.

Papousek, Ilona, H. Harald Freudenthaler, and Günter Schulter. "The Interplay of Perceiving and Regulating Emotions in Becoming Infected with Positive and Negative Moods." *Personality and Individual Differences* 45 (2008): 463–67.

Parker, Kim Ian. "Mirror, Mirror on the Wall, Must We Leave Eden, Once and For All? A Lacanian Pleasure Trip Through the Garden." *Journal for the Study of the Old Testament* 83 (1999): 19–29.

Parker, John. *The Aesthetics of Antichrist*. Ithaca: Cornell University Press, 2007.

The Passion of the Christ. Dir. Mel Gibson. Newmarket, 2004.

Paul, William. "Uncanny Theater: The Twin Inheritances of the Movies." *Paradoxa* 3.3–4 (1997): 321–47.

Paulson, Julie. "Death's Arrival and *Everyman*'s Separation." *Theatre Survey* 48.1 (May 2007): 121–41.

Penchansky, David. *What Rough Beast? Images of God in the Hebrew Bible*. Louisville, KY: Westminster John Knox Press, 1999.

Perry, Marvin, and Frederick M. Schweitzer. *Antisemitism*. New York: Palgrave, 2002.

Persinger, Michael A. "Experimental Simulation of the God Experience." *NeuroTheology*. Ed. Rhawn Joseph. San Jose: University Press, 2002. 279–92.

_____. "The Neuropsychiatry of Paranormal Experiences." *Journal of Neuropsychiatry and Clinical Neurosciences* 13.4 (2001): 515–23.

_____. *Neuropsychological Bases of God Beliefs*. New York: Praeger, 1987.

Pessoa, Luiz. "Seeing the World in the Same Way." *Science* 303 (2004): 1617–18.

Pfeiffer, John E. *The Creative Explosion*. New York: Harper & Row, 1982.

Philo, Greg. "The Media and Public Belief." *Media and Mental Distress*. Ed. Greg Philo. London: Longman, 1996. 82–104.

Pigafetta, Antonio. *The Voyage of Magellan*. Englewood Cliffs, NJ: Prentice-Hall, 1969.

"Pilot." *Millennium*. Writ. Chris Carter. Dir. David Nutter. Fox. 25 Oct. 1996.

Pineda, Jaime A., ed. *Mirror Neuron Systems*. New York: Springer, 2009.

Piskorowski, Anna. "In Search of Her Father: A Lacanian Approach to Genesis 2–3." *A Walk in the Garden: Biblical, Iconographical and Literary Images of Eden*. Ed. Paul Morris and Deborah Sawyer. 310–18.

Pizzato, Mark. "Brechtian and Aztec Violence in Valdez's *Zoot Suit*." *Journal of Popular Film and Television* 26.2 (Summer 1998): 52–61.

_____. *Edges of Loss: From Modern Drama to Postmodern Theory*. Ann Arbor: University of Michigan Press, 1998.

_____. *Ghosts of Theatre and Cinema in the Brain*. New York: Palgrave, 2006.

_____. "Jeffrey Dahmer and Media Cannibalism: The Lure and Failure of Ritual." *Mythologies of Violence in Postmodern Media*. Ed. Christopher Sharrett. Detroit: Wayne State University Press, 1999. 85–118.

———. "Nietzschean Neurotheatre: Apollinian and Dionysian Spirits in the Brain Matters of *Our Town*." *Nietzsche and the Rebirth of the Tragic*. Ed. Mary Ann Frese Witt. Madison, NJ: Fairleigh Dickinson University Press, 2007. 186–218.

———. "Skins of Desire in Evolution." *Death in American Texts and Performances*. Ed. Lisa K. Perdigao and Mark Pizzato. Surrey, England: Ashgate, 2010. 29–44.

———. "Soyinka's *Bacchae*, African Gods, and Postmodern Mirrors." *Journal of Religion and Theatre* 2.1 (Fall 2003). http://apollo.fa.mtu.edu/~dlbruch/rtjournal

———. *Theatres of Human Sacrifice: From Ancient Ritual to Screen Violence*. Albany: State University of New York Press, 2005.

The Play of the Sacrament. Specimens of the Pre-Shakespearean Drama. Ed. John M. Manly. New York: Dover, 1967. 239–76.

"The Politics of Programming: Sometimes, Creative Clout and Star Power are More Important than Ratings." *The Star-Ledger* (Newark, NJ) 19 Sept. 1997: 61.

Poole, Kristen. "The Devil's in the Archive." *Renaissance Drama* 35 (2006): 191–219.

———. "*Dr. Faustus* and Reformation Theology." *Early Modern English Drama*. Ed. Garrett A. Sullivan, Jr., Patrick Cheney, and Andrew Hadfield. New York: Oxford University Press, 2006. 96–107.

Price, Simon. *Reigions of the Ancient Greeks*. Cambridge: Cambridge University Press, 1999.

Prinz, Jesse J. *The Emotional Construction of Morals*. Oxford: Oxford University Press, 2007.

Quartz, Steven R., and Terrence J. Sejnowski. *Liars, Lovers, and Heroes*. New York: Harper, 2002.

Ragland-Sullivan, Ellie. *Jacques Lacan and the Philosophy of Psychoanalysis*. Urbana: University of Illinois Press, 1986.

Ramachandran, V. S. *A Brief Tour of Human Consciousness*. New York: Pi Press, 2004.

———. *The Emerging Mind*. London: Profile, 2003.

———. "Mirror Neurons and Imitation Learning as the Driving Force Behind 'the Great Leap Forward' in Human Evolution." *Edge* 69 (1 June 2000). Internet.

Ramachandran, V. S., and Sandra Blakeslee. *Phantoms in the Brain: Probing the Mysteries of the Human Mind*. New York: William Morrow, 1998.

Ramachandran, V. S., and William Hirstein. "The Science of Art: A Neurological Theory of Aesthetic Experience." *Journal of Consciousness Studies* 6.6–7 (1999): 15–51.

Ramachandran, V.S., and Edward M. Hubbard. "The Emergence of the Human Mind." *Synesthesia*. Ed. Lynn C. Robertson and Noam Sagiv. Oxford: Oxford University Press, 2005. 147–92.

———. "Synesthesia." *23 Problems in Systems Neuroscience*. Ed. J. Leo van Hemmen and Terrence J. Sejnowski. Oxford: Oxford University Press, 2006. 432–73.

Raphael, Max. *Prehistoric Cave Paintings*. Washington, DC: Old Dominion Foundation, 1945.

Read, Leslie Du S. "Beginnings of Theatre in Africa and the Americas." *The Oxford Illustrated History of Theatre*. Ed. John Russell Brown. Oxford: Oxford University Press, 1995. 93–105.

Rehm, Rush. *Greek Tragic Theatre*. London: Routledge, 1992.

———. *The Play of Space: Spatial Transformation in Greek Tragedy*. Princeton: Princeton University Press, 2002.

Reid, Robert G. B. *Biological Emergences*. Cambridge: MIT Press, 2007.

Renfrew, Jane M. "Neanderthal Symbolic Behavior?" *Becoming Human*. Ed. Colin Renfrew and Iain Morley. Cambridge: Cambridge University Press, 2009. 50–60.

Renfrew, Colin. "Situating the Creative Explosion: Universal or Local?" *Becoming Human*. Ed. Colin Renfrew and Iain Morley. Cambridge: Cambridge University Press, 2009. 74–92.

Revonsuo, Antti. "The Reinterpretation of Dreams: An Evolutionary Hypothesis of the Function of Dreaming." *Sleeping and Dreaming*. Ed. Edward F. Pace-Schott, Mark Solms, Mark Blagrove, and Stevan Harnad. Cambridge: Cambridge University Press, 2003. 85–111.

Richerson, Peter J., and Robert Boyd. *Not by Genes Alone: How Culture Transformed Human Evolution*. Chicago: University of Chicago Press, 2005.

Ridge, Robert D., and Colin W. Key. "From Sacred to Profane: The Effects of Reading Violent Religious Literature on Subsequent Human Aggression." *New Beginnings in Literary Studies*. Ed. Jan Auracher and Willie van Peer. New-castle, UK: Cambridge Scholars, 2008. 267–85.

Ridley, Matt. *The Red Queen: Sex and the Evolution of Human Nature*. New York: Macmillan, 1993.

Rifkin, Jeremy. *The Empathic Civilization*. New York: Penguin, 2009.

Rilke, Rainer Maria. *Werke*. Frankfurt: Insel, 1982.

Rivera, José. *Marisol and Other Plays*. New York: Theatre Communications Group, 1997.

Rizzolatti, Giacomo, and Michael A. Arbib. "Language Within Our Grasp." *Trends in Neuroscience* 21.5 (1998): 188–94.

Rizzolatti, Giacomo, and Laila Craighero. "Mirror Neuron." Changeux et al. 107–24.

Rizzolatti, Giacomo, and Corrado Sinigaglia. *Mirrors in the Brain—How Our Minds Share Our Actions*. Oxford: Oxford University Press, 2008.

Roberts, Jeanne Addison. *The Shakespearean Wild*. Lincoln: University of Nebraska Press, 1991.

Rochat, Philippe. "Ego Function of Early Imitation." *The Imitative Mind: Development, Evolution, and Brain Bases*. Ed. Andrew N. Meltzoff and Wolfgang Prinz. Cambridge: Cambridge University Press, 2002. 85–97.

Rosemary's Baby. Dir. Roman Polanski. Paramount, 1968.

Roskies, Adina L. "Internalism and the Evidence from Pathology." *Moral Psychology*. Vol. 3. Ed. Walter Sinnott-Armstrong. Cambridge: MIT Press, 2008. 191–206.

Roth, Melissa. *The Left Stuff*. New York: M. Evans, 2005.

Rothenberg, Jerome, ed. *Technicians of the Sacred*. Garden City, NY: 1968.

Rothenberg, Molly Anne. "The 'Newer Angels' and the Living Dead: The Ethics of Screening Obsessional Desire." *Camera Obscura* 40–41 (1997): 17–41.

Rotman, Brian. *Becoming Beside Ourselves*. Durham: Duke University Press, 2008.

Russell, Gordon W. "Psychological Issues in Sports Aggression." *Sports Violence*. Ed. Jeffrey H. Goldstein. New York: Springer, 1983. 157–81.

Ryan, Robert E. *The Strong Eye of Shamanism*. Rochester, Vermont: Inner Traditions, 1999.

Ryle, Gilbert. *The Concept of Mind*. London: Hutchinson, 1949.

Sabom, Michael B. *Recollections of Death: A Medical Investigation*. New York: Harper & Row, 1982.

Sacks, Oliver W. *Musicophilia*. New York: Knopf, 2007.

———. "Sigmund Freud: The Other Road." *Freud and the Neurosciences*. Ed. Giselher Guttman and Inge Scholz-Strasser. Vienna: Austrian Academy of Sciences, 1998. 11–22.

"Sacrament." *Millennium*. Writ. Frank Spotnitz. Dir. Michael Watkins. Fox. 21 Feb. 1997.

Salkever, Stephen G. "Tragedy and the Education of the *Demos*: Aristotle's Response to Plato." *Greek Tragedy and Political Theory*. Ed. J. Peter Euben. Berkeley: University of California Press, 1986. 274–303.

Satinover, Jeffrey. *The Quantum Brain*. New York: John Wiley and Sons, 2001.

Saver, Jeffrey L., and John Rabin. "The Neural Substrates of Religious Experience." *Journal of Neuropsychiatry and Clinical Neurosciences* 9 (1997): 498–510.

Saxton, Alexander. "Blackface Minstrelsy." *Inside the Minstrel Mask*. Ed. Annemarie Bean, James V. Hatch, and Brooks McNamara. Hanover, NH: Wesleyan University Press, 1996. 67–85.

Schacter, Daniel L. *Searching for Memory*. New York: Basic, 1996.

Schafer, Elizabeth. "The Male Gaze in *Woyzeck*." *Madness in Drama*. Ed. James Redmond. Cambridge: Cambridge University Press, 1993. 55–64.

Schechner, Richard. *Environmental Theater*. New York: Applause, 1994.

———. *The Future of Ritual*. London: Routledge, 1993.

———. *Performance Theory*. London: Routledge, 1988.

Scheff, Thomas J. *Catharsis in Healing, Ritual, and Drama*. Berkeley: University of California Press, 1979.

Scherb, Victor I. "Violence and the Social Body in the Croxton *Play of the Sacrament*." *Violence in Drama*. Ed. James Redmond. Cambridge: Cambridge University Press, 1991. 69–78.

Schneiderman, Stuart. *An Angel Passes: How the Sexes Became Undivided*. New York: New York University Press, 1988.

Schneiderman, William. "Voluntary or Not, Is it Censorship?" *National Journal* 29.29 (19 July 1997): 1490.

Schore, Allan N. *Affect Regulation and the Origin of the Self*. Hillsdale, New Jersey: Lawrence Erlbaum, 1994.

———. "The Human Unconscious: The Development of the Right Brain and its Role in Early Emotional Life." *Emotional Development in Psychoanalysis, Attachment Theory and Neuroscience*. Ed. Viviane Green.

New York: Brunner-Routledge, 2003. 23–54.

Schroeder, Gerald L. *The Hidden Face of God: Science Reveals the Ultimate Truth.* New York: Simon & Schuster, 2001.

Sconce, Jeffrey. *Haunted Media: Electronic Presence from Telegraphy to Television.* Durham, NC: Duke University Press, 2000.

Scott, Robert A. "Making Relics Work." *The Artful Mind.* Ed. Mark Turner. Oxford: Oxford University Press, 2006. 211–23.

Scott, William C. "Two Suns over Thebes: Imagery ad Stage Effects in the *Bacchae*." *Transactions of the American Philological Association* 105 (1975): 333–46.

Segal, Charles. *Dionysiac Poetics and Euripides' Bacchae.* 1982. Princeton: Princeton University Press, 1997.

Seltzer, Mark. *Serial Killers: Death and Life in America's Wound Culture.* New York: Routledge, 1998.

"Senate Committee Hears Complaints About TV Ratings System." *The News Media and the Law* 21.2 (Spring 1997): 22–23.

Sepinwall, Alan. "Another Dark, Disturbing Vision from the Creator of 'The X-Files': 'MillenniuM.'" *The Star-Ledger* (Newark, NJ), 25 Oct. 1996: 49.

Shaffer, Peter. *Amadeus.* New York: Harper & Row, 1981.

———. *Equus.* New York: Penguin 1977.

Shakespeare, William. *The Tempest.* Baltimore: Penguin, 1970.

———. *The Tragedy of Hamlet, Prince of Denmark.* New York: NAL, 1963.

———. *The Winter's Tale.* New York: Penguin, 1971.

Shapiro, Alan, and Peter Burian, trans. *The Oresteia.* By Aeschylus. Oxford: Oxford University Press, 2003.

Shariff, Azim F., and Jordan B. Peterson. "Anticipatory Consciousness, Libet's Veto and a Close-Enough Theory of Free Will." *Consciousness and Emotion.* Ed. Ralph D. Ellis and Natika Newton. Amsterdam: John Benjamins, 2005. 197–216.

Sharratt, Bernard. "The Politics of the Popular?—From Melodrama to Tel-evision." *Performance and Politics in Popular Drama.* Ed. David Bradby, Louis James, and Bernard Sharratt. Cambridge: Cambridge University Press, 1980. 275–95.

Shermer, Michael. *How We Believe: The Search for God in the Age of Science.* New York: Freeman, 2000.

Shim, Jung-Soon. "The Shaman and the Epic Theatre: The Nature of Han in the Korean Theatre." *New Theatre Quarterly* 20.3 (2004): 216–24.

Shlain, Leonard. *Sex, Time and Powe: How Women's Sexuality Shaped Evolution.* New York: Penguin, 2003.

Shrum, L. J. "Effects of Television Portrayals of Crime and Violence on Viewers' Perceptions of Reality: A Psychological Process Perspective." *Legal Studies Forum* 22.1–3 (1998): 257–68.

Siegel, Daniel J. *The Developing Mind.* New York: Guilford, 1999.

Simpson, Philip L. "The Politics of Apocalypse in the Cinema of Serial Murder." *Mythologies of Violence in Postmodern Media.* Ed. Christopher Sharrett. Detroit: Wayne State University Press, 1999. 119–44.

Singer, Ben. *Melodrama and Modernity: Early Sensational Cinema and Its Contexts.* New York: Columbia University Press, 2001.

"A Single Blade of Grass." *Millennium.* Writ. Erin Maher and Kay Reindl. Dir. Rodman Flender. Fox. 24 Oct. 1997.

Sin Noticias de Dios. Dir. Augustín Díaz Yanes. First Look International, 2001.

Sissa, Giulia, and Marcel Detienne. *The Daily Life of the Greeks Gods.* Stanford: Stanford University Press, 2000.

Skura, Meredith Anne. "The Case of Colonialism in *The Tempest*." *Caliban.* Ed Harold Bloom. New York: Chelsea, 1992. 221–48.

Smail, Daniel Lord. *On Deep History and the Brain.* Berkeley: University of California Press, 2008.

Smith, Joseph H. "Original Evil and the Time of the Image." *Journal for the Psychoanalysis of Culture and Society* 1.2 (Fall 1996): 35–46.

Sober, Elliott. "Kindness and Cruelty in Evolution." Davidson and Harrington 46–65.

Solms, Mark, and Oliver Turnbull. *The Brain and the Inner World.* New York: Other Press, 2002.

"Somehow, Satan Got Behind Me." *Millennium.* Writ. Darin Morgan. Dir. Darin Morgan. Fox. 1 May 1998.

Sopcak, Paul. "Compassion and Disgust as Markers of Cultural Differences in Reading Violence in Literary Texts." *New Beginnings in Literary Studies.* Ed. Jan Auracher and Willie van Peer. Newcastle, UK: Cambridge Scholars, 2008. 198–217.

Sophocles. *Aias [Ajax].* Trans. Herbert Golder and Richard Pevear. Oxford: Oxford University Press, 1999.

———. *Oedipus, the King. The Oedipus Cycle.*

Trans. Dudley Fitts and Robert Fitzgerald. New York: Harcourt, 1969. 1–78.
_____. *Philoctetes*. Trans. Carl Phillips. Oxford: Oxford University Press, 2003.
Sor Juana Inés de la Cruz. *Loa* to *The Divine Narcissus*. *Wadsworth Anthology of Drama*. Ed. W. B. Worthen. Fourth Edition. Boston: Wadsworth, 2004. 503–7.
Soule, Lesley Wade. *Actor as Anti-Character: Dionysus, the Devil, and the Boy Rosalind*. Westport, Connecticut: Greenwood Press, 2000.
Soyinka, Wole. *Collected Plays 1*. Oxford University Press, 1973.
_____. *Myth, Literature and the African World*. Cambridge: Cambridge UP, 1976.
Spong, John Shelby. *Why Christianity Must Change or Die*. San Francisco: Harper, 1998.
Stark, Rodney. *The Rise of Christianity*. Princeton: Princeton University Press, 1996.
Staub, August W., and Michael J. Hussey. "Saints and Cyborgs." *Journal of Dramatic Theory and Criticism* 13.1 (1998): 177–82.
Staub, Ervin. "Emergency Helping, Genocidal Violence, and the Evolution of Responsibility and Altruism in Children." Davidson and Harrington 165–81.
Subrahmanyam, Kaveri, Robert Kraut, Patricia Greenfield, and Elisheva Gross. "New Forms of Electronic Media." *Handbook of Children and the Media*. Ed. Dorothy G. Singer and Jerome L. Singer. London: Sage, 2001. 73–99.
Sullins, Ellen S. "Emotional Contagion Revisited." *Personality and Social Psychology Bulletin* 17.2 (1991): 166–74.
Surowiecki, James. *The Wisdom of Crowds*. New York: Anchor, 2004.
Taçon, Paul S. C. "Identifying Ancient Religious Thought and Iconography." *Becoming Human*. Ed. Colin Renfrew and Iain Morley. Cambridge: Cambridge University Press, 2009. 61–73.
Taylor, Shelley E. *The Tending Instinct*. New York: Holt, 2002.
Teilhard de Chardin, Pierre. *Christianity and Evolution*. New York: Harcourt, 1971.
"Teotwawki." *Millennium*. Writ. Chris Carter and Frank Spotnitz. Dir. Thomas Wright. Fox. 16 Oct. 1998.
"The Time is Now." *Millennium*. Writ. Glen Morgan and James Wong. Dir. Thomas Wright. Fox. 15 May 1998.
Tomasello, Michael. "How Are Humans Unique?" *New York Times Magazine* (25 May 2008): 15.

_____. *Origins of Human Communication*. Cambridge: MIT Press, 2008.
_____. "Two Hypotheses About Primate Cognition." *The Evolution of Cognition*. Ed. Cecilia Heyes and Ludwig Huber. Cambridge: MIT Press, 2000. 165–83.
Traunecker, Claude. *The Gods of Egypt*. Trans. David Lorton. 1992. Ithaca: Cornell University Press, 2001.
Treisman, Anne. "Synesthesia." *Synesthesia*. Ed. Lynn C. Robertson and Noam Sagiv. Oxford: Oxford University Press, 2005. 239–54.
Trevarthen, Colwyn. "Lateral Asymmetries in Infancy: Implications for the Development of the Two Hemispheres." *Neuroscience and Biobehavioral Reviews* 20.4 (1996): 571–86.
Trivers, Robert. *Social Evolution*. Menlo Park, California: Benjamin/Cummings, 1985.
Turim, Maureen. *Flashbacks in Film: Memory and History*. New York: Routledge, 1989.
Turnbull, Oliver, and Mark Solms. "Memory, Amnesia, and Intuition: A Neuro-Psychoanalytic Perspective." *Emotional Development in Psychoanalysis, Attachment Theory and Neuroscience*. Ed. Viviane Green. New York: Brunner-Routledge, 2003. 55–85.
Turner, Mark. *The Literary Mind*. Oxford: Oxford University Press, 1996.
Turner, Victor. "Body, Brain, and Culture." *Zygon* 18.3 (1983): 221–45.
Valdez, Luis. *Early Works*. Houston: Arte Publico, 1990.
Van Biema, David. "God vs. Science." *Time* 168.20 (13 Nov. 2006): 48–55.
Vardac, A. Nicholas. *Stage to Screen: Theatrical Method from Garrick to Griffith*. New York: Benjamin Blom, 1968.
Varela, Francisco J. *Ethical Know-How*. Stanford: Stanford University Press, 1999.
Varela, Francisco J., Evan Thompson, and Eleanor Rosch. *The Embodied Mind: Cognitive Science and Human Experience*. Cambridge: MIT Press, 1991.
Vaughan, Susan C. *The Talking Cure: The Science Behind Psychotherapy*. New York: Henry Holt, 1998.
Volkan, Vamik D. *The Need to Have Enemies and Allies*. North Vale, New Jersey: Jacob Aronson, 1988.
von Feilitzen, Cecilia. "Media Violence — Four Research Perspectives." *Approaches to Audiences*. Ed. Roger Dickinson. New York: Oxford University Press, 1998. 88–103.
Wade, Nicholas. *The Faith Instinct*. New York: Penguin, 2009.

The Wakefield Mystery Plays. Ed. Martial Rose. Garden City, New York: Doubleday, 1962.
Wall-Randell, Sarah. "*Doctor Faustus* and the Printer's Devil." *Studies in English Literature, 1500–1900* 48.2 (Spring 2008): 259–281.
"Weeds." *Millennium.* Writ. Frank Spotnitz. Dir. Michael Pattinson. Fox. 24 Jan. 1997.
Wegner, Daniel M. *The Illusion of Conscious Will.* Cambridge: MIT Press, 2002.
Weimann, Robert. *Shakespeare and the Popular Tradition in the Theater.* Baltimore: Johns Hopkins University Press, 1978.
Weingarten, S. M., D. G. Charlow, and E. Holmgren. "The Relationship of Hallucinations to the Depth Structures of the Temporal Lobes." *Acta Neurochirurgica* 24 (1977): 199–216.
White, Stephen. "Io's World." *The Journal of Hellenic Studies* 121 (2001): 107–40.
Whiten, Andrew, and Richard W. Byrne, eds. *Machiavellian Intelligence II.* Cambridge: Cambridge University Press, 1997.
Wilder, Thornton. *Three Plays.* New York: Avon, 1957.
Wiles, David. *Tragedy in Athens.* Cambridge: Cambridge University Press, 1997.
Wiles, Timothy J. *The Theater Event: Modern Theories of Performance.* Chicago: University of Chicago Press, 1980.
Wilkinson, Richard H. *The Complete Gods and Goddesses of Ancient Egypt.* New York: Thames, 2003.
Williams, Linda. *Playing the Race Card: Melodramas of Black and White From Uncle Tom to O.J. Simpson.* Princeton: Princeton University Press, 2001.
Williams, Raymond. *Television: Technology and Cultural Form.* New York: Schocken, 1975.
Wills, Christopher. *The Runaway Brain.* New York: Basic, 1993.
Wilson, Barbara J., et al. "Content Analysis of Entertainment Television: The 1994–95 Results." *Television Violence and Public Policy.* Ed. James T. Hamilton. Ann Arbor: University of Michigan Press, 1998. 105–48.
Wilson, Edward O. *Consilience: The Unity of Knowledge.* New York: Knopf, 1998.
Wilson, Edwin, and Alvin Goldfarb. *Living Theater: A History.* 3rd ed. Boston: McGraw, 2000.
Winchell, Mark Royden. *God, Man and Hollywood.* Wilmington: Intercollegiate Studies Institute, 2008.
Wings of Desire. Dir. Wim Wenders. Orion Classics, 1987.
Wink, Walter. *Unmasking the Powers.* Philadelphia: Fortress, 1986.
Winkielman, Michael. *Shamanism: The Neural Ecology of Consciousness and Healing.* Westport, Connecticut: Bergin and Garvey, 2000.
———. "Shamanism and Innate Brain Structures: The Original Neurotheology." *NeuroTheology.* Ed. Rhawn Joseph. San Jose: University Press, 2002. 387–96.
Winkielman, Piotr, Kent C. Berridge, and Julia L. Wilbarger. "Emotion, Behavior, and Conscious Experience." Barrett et al. 335–62.
Winkler, John J. "The Ephebes' Song: *Tragoidoi* and *Polis.*" *Nothing to Do with Dionysos?* Ed. John J. Winkler and Froma I. Zeitlin. Princeton: Princeton University Press, 1990. 20–62.
Wise, Jennifer. *Dionysus Writes: The Invention of Theatre in Ancient Greece.* Ithaca: Cornell University Press, 1998.
The Witches of Eastwick. Dir. George Miller. Warner Brothers, 1987.
Wolf, Maryanne. *Proust and the Squid: The Story and Science of the Reading Brain.* New York: Harper, 2007.
Womack, Peter. "Imagining Communities." *Culture and History 1350–1600.* Ed. David Aers. Detroit: Wayne State University Press, 1992. 91–146.
Wrangham, Richard, and Dale Peterson. *Demonic Males: Apes and the Origins of Human Violence.* Boston: Houghton Mifflin, 1996.
Wright, Robert. *The Evolution of God.* New York: Little, Brown and Co., 2009.
Yeats, William Butler. *The Collected Poems of W. B. Yeats.* New York: Macmillan, 1956.
York Plays. Ed. Lucy Toulmin Smith. New York: Russell and Russell, 1963.
Zeki, Semir. *Inner Vision.* Oxford: Oxford University Press, 1999.
———. "Neural Concept Formation and Art: Dante, Michelangelo, Wagner." *Journal of Consciousness Studies* 9.3 (2002): 53–76.
Zhang, Yehong. "Culture and Reading: The Influence of Western and Eastern Thought Systems on the Understanding of Fairy Tales." *New Beginnings in Literary Studies.* Ed. Jan Auracher and Willie van Peer. Newcastle, UK: Cambridge Scholars, 2008. 218–37.
Zillmann, Dolf. "Anatomy of Suspense." *The Entertainment Functions of Television.* Ed. Percy H. Tannenbaum. Hillsdale, New Jersey: Lawrence Erlbaum Associates, 1980. 133–63.

———. "Mechanisms of Emotional Involvement with Drama." *Poetics* 23 (1994): 33–51.

———. "The Psychology of the Appeal of Portrayals of Violence." *Why We Watch: The Attractions of Violent Entertainment*. Ed. Jeffrey Goldstein. Oxford: Oxford University Press, 1998. 179–211.

Žižek, Slavoj. *The Art of the Ridiculous Sublime: On David Lynch's* Lost Highway. Seattle: University of Washington Press, 2000.

———. "Cyberspace, or the Unbearable Closure of Being." *Endless Night: Cinema and Psychoanalysis, Parallel Histories*. Ed. Janet Bergstrom. Berkeley: University of California Press, 1999. 96–125.

———. *Enjoy Your Symptom! Jacques Lacan in Hollywood and Out*. New York: Routledge, 1992.

———. *For They Know Not What They Do: Enjoyment as a Political Factor*. London: Verso, 1991.

———. *Looking Awry: An Introduction to Jacques Lacan through Popular Culture*. Cambridge: MIT Press, 1992.

———. *The Metastases of Enjoyment: Six Essays on Woman and Causality*. London: Verso, 1994.

———. *The Parallax View*. Cambridge: MIT Press, 2006.

Zupancic, Alenka. *Ethics of the Real*. London: Verso, 2000.

Index

Absolute Universal Being (AUB) 25, 28, 30, 32–33
abstractive operator (left hemisphere) 25–26, 79, 85, 89, 301n67
Adam (The Mystery of Adam) 91
The Adventurer (Chaplin) 284
Aeschylus 14, 65–80, 166, 297n48, 300n47, 300n59
Agamben, Giorgio 289n1
Aiken, George L. *see Uncle Tom's Cabin*
Albright, Carol Rausch 19, 32
Alford, C. Fred 314n67, 318n119
Alibek, Ken 310n8
alienation 6, 28, 35, 44, 88, 106, 110, 111, 113, 116, 135, 137, 159, 161, 165–66, 173, 176, 180, 182, 184, 187, 210, 228, 231, 241, 259, 262, 269, 272, 276–77, 283, 285–86, 290n19, 296n45, 300n52, 304n50, 312n40, 317n105
alienation effect (Brecht) 165, 175, 241, 277
Allan, Neil 303n3
allegory 121, 151–52, 187
Alper, Matthew 19, 26, 33
alpha leaders (or lead actors) 1, 13–14, 25, 48, 50, 55, 59, 61, 76, 90, 103–5, 110, 115, 125, 137, 143, 146, 152, 156–57, 160, 164, 177, 179–85, 1887, 190, 192–95, 197, 207, 220–21, 236, 268, 281, 287, 306n7
altered state of consciousness 34–36, 43, 45–47, 50–52, 296n43, 297n65–66
altruism 68–69, 116, 148–49, 151–52, 173, 177, 280, 286, 292n41, 299n28, 299n35–37, 320n28
amygdala 22, 31, 36, 148, 291n29, 292n34, 294n20, 308n27
anachronism 97, 106, 109, 110, 113
Angel and Me 306n12
Angel Heart 217, 308n34
The Angel Levine 308n30
animal drives 5–6, 9. 14–16, 49–50, 54–55, 58, 62, 70–72, 77, 91–94, 100, 104–5, 115–16, 122–23, 127–28, 130, 139, 180, 185–86, 195, 221–22, 280, 286; *see also* territoriality
anti-Semitism 84–85, 93, 113, 125, 218–19, 301n2, 302n7, 308n35–36; *see also* genocide; holocaust
apes 64, 67, 94, 100, 115, 177–79, 216, 280, 290n18, 292n46, 299n26, 302n21, 306n6, 306n10, 319n10, 321n37; *see also* hominids; primates
Aphrodite 75, 80, 131
apocalypse 16, 87, 172, 218, 227–29, 231–33, 235–42, 244–46, 250, 262–63, 268, 271–72, 276–77, 309n6, 313n42, 313n50, 313n54, 315n79
Apollinaire, Guillaume 166
Apollo 71, 73–79, 140, 181–82, 185, 189, 300n47, 300n53, 300n59, 300n62, 301n68
Apollodorus 71
Apollonian 21–22, 52, 76, 80, 140, 169, 212, 294n16, 298n71, 305n46
Apter, Michael J. 230, 311n18
Aristophanes 80, 289n6, 301n70
Aristotle 70, 99, 220, 236–38, 241–42, 289n2, 291n32, 292n33, 295n25, 299n42, 313n45, 313n51, 313n55, 314n60, 314n67, 315n71–72
Armstrong, Karen 92, 298n7, 302n15–16, 319n5
Arsenio, William F. 317n104
Artaud, Antonin 167–68, 173, 175, 220, 225, 228, 240–42, 254, 256, 268–74, 277, 284, 287, 305n45, 308n37, 309n5, 314n67, 320n32
Artemis 71, 75, 80, 304n20
As You Like It 140
Asendorpf, Jens B. 290n19
Ashbrook, James B. 19, 32
Ashby, Clifford 62
Ashworth, Ann 304n19

Index

atheism 24, 68, 168, 279, 306n12, 319n5
Athena 181–82, 185, 189, 289n8, 300n63, 301n71
Atkinson, Anthony P. 164
Atran, Scott 295n26, 296n32
attention association area (in the brain) 28–29, 47
attunement 7, 77, 290n19, 300n61
audience effects 225–26, 229–30, 232, 235, 239, 242, 277, 282–85, 308n26, 308n37, 309n4, 310n9–14, 311n18, 314n61–62, 314n66, 320n30, 320n32–34, 321n35; see also Artaud, Antonin; Brecht, Bertolt; catharsis; copycat violence
Aunger, Robert 290n16
autonomic (arousal and quiescent) nervous system 29–33, 37, 46, 51, 62–63, 80, 285, 296n39–40, 297n66, 303n30
Aziz-Zadeh, Lisa 291n29
Aztecs 150–52, 170

Baars, Bernard J. 12, 77, 293n51–52, 306n10
baboons 5, 56, 178–80, 197, 279, 290n14, 305n34, 306n7–9; see also Machiavellian Intelligence; social metaphysics
The Bacchae 60–65, 73, 75–76, 79–80, 82, 168, 171, 191, 298n12, 298n16, 300n56, 301n73; see also Dionysus; Euripides
Bacon, Helen H. 301n64
Badley, Linda 316n92
Bahn, Paul 34, 39–40, 297n60
Baird, Abigail A. 307n13
Balsera, Viviana Díaz 151
Banyas, Carol A. 296n41
Baraka, Amiri 306n5
Barash, David P. 97–98, 290n12
Barlow, Connie 299n37
Barrett, James 298n18
Batman 210
Baudrillard, Jean 228, 309n5
Beauregard, David N. 304n18
Beckett, Samuel 167
Beckwith, Sarah 302n9
Bedazzled 213, 308n31
Belfiore, Eleanora 291n32
Belzebub 128, 132
Bergland, Renée 151
Bermel, Albert 305n31
Bernal, Martin 289n8
Bevington, David 304n16
Bicks, Caroline 304n20
binary operator (left hemisphere) 26–27, 32, 49, 52, 72, 79, 85, 89, 99, 128, 184, 189–90, 231, 284, 286
Bjorklund, David F. 94
Black, Joel 312n34
Blackmore, Susan 24, 290n16, 292n35, 294n21, 295n25, 295n28
Blade Runner 306n12
Blakeslee, Sandra 18, 20, 23, 52, 61–62, 72–73, 91, 132–33, 136, 291n25, 293n50,
293n7, 294n22–23, 297n57, 298n73, 301n72, 303n25, 303n27, 303n37
Bloom, Paul 296n42, 311n27, 319n13
Boal, Augusto 315n71
body image 22, 105, 112, 132–34, 293n7; see also sparagmos
Boehm, Christopher 13, 293n 57, 299n36
Bolte Taylor, Jill 291n30
Bonda, Eva 291n29
bonobos 64, 74, 319n10; see also chimpanzees
The Book of Life 217
Boothby, Richard 313n44
Bordwell, David 4
bottom-up brain networks see top-down brain networks
Boyer, Pascal 103, 293n2
Bracher, Mark 315n78
Bradshaw, John L. 94
brainstem 6–7, 9–10 , 18, 19, 29, 31, 32, 36, 50, 55, 73, 74, 77, 84, 105, 128, 187, 208, 281, 291n23, 299n27
Brave Heart 221
Brecht, Bertolt 165, 168, 173, 175, 220, 225, 228, 240–42, 254, 256, 267–70, 272, 277, 284, 287, 305n48, 309n5, 311n18, 314n67, 315n71–74, 318n114, 320n32
Bremmer, Jan N. 98–99
Breuil, Abbé H. 34, 50
Briller, Bert 230
Brockett, Oscar G. 58, 155
Brodie, Richard 295n26
Brousse, Marie-Hélène 314n65
Brown, A.L. 71
Bruce Almighty 190–94
Brugger, Peter 294n13, 320n23
Büchner, Georg 157, 160, 165, 305n35
Buckman, Robert 19
Buckner, Randy L. 292n49
Buddhism 24, 28–30, 47, 142, 295n28, 296n36, 319n6, 319n17, 321n39
Budge, E.A. Wallis 56
Burch, Noël 310n15
Burkert, Walter 17–19, 51, 115, 293n4, 298n1, 298n8, 299n31, 300n46, 300n55, 302n20
Burns, Edward 302n11–12
Burns, George 181–91, 194
Bush, George W. 227–28
Bushman, Brad J. 310n11, 310n13, 320n33
Buss, David M. 159
Butler, Ann B. 291n23
Byrne, Richard 51, 94, 299n24

Cage, Nicholas 200–01
Calderón, Pedro 153, 156
Caldwell, Richard 301n64
Calvo-Merino, B. 293n59
Cantor, Joanne 310n14
capuchin monkeys 100–01, 116–17, 119, 153, 188, 203–04, 207, 280
care system 112
Carlson, Marvin 155, 166, 315n71

Index 347

Carman, John 309*n*6
Carroll, Noel 4
Carter, Chris 231–32, 312*n*30, 312*n*32–33, 312*n*37, 315*n*76
Cartesian *cogito* 131, 135, 160
Cartesian theatre (of the mind) 11–12, 293*n*50
catharsis 8, 23, 27, 70, 73–74, 76, 89, 98, 113, 123, 125, 135, 138, 141, 145–47, 149, 168, 207–9, 220, 224, 228–31, 235, 237–42, 246–47, 249–50, 253, 257, 260, 266, 268, 270, 276–77, 281, 283–86, 291*n*32, 292*n*33, 297*n*48, 297*n*67, 300*n*49, 308*n*37, 309*n*4–6, 311*n*19, 313*n*45, 313*n*51, 313*n*53, 313*n*55–57, 314*n*60, 314*n*66–67, 314*n*71, 315*n*73, 316*n*90, 320*n*32; *see also* tragic flaw
Catholicism 29, 47, 84, 138, 151, 153, 156, 165, 171, 187, 203, 205, 218–19, 304*n*6, 304*n*18
causal operator (left hemisphere) 25–27, 49, 52, 72, 79, 85
Cawley, A.C. 302*n*11
Cela-Conde, Camilo J. 291*n*28
Centerwall, Brandon S. 310*n*11
cerebellum 55, 132
Ceres 146
Césaire, Aimé 169–70
Chaikin, Joseph 168
Chaminade, Thierry 107–8, 303*n*24
Chandler, Daniel 310*n*14
Changeux, Jean-Pierre 147
Chemers, Michael Mark 301*n*3
Cheney, Dorothy L. 178–80, 290*n*14, 306*n*7, 306*n*9–10
Chester Cycle (or Mystery) plays 302*n*11–12
child abuse 257–62
chimpanzees 56, 64, 67, 94, 100, 179, 279–80, 292*n*36, 298*n*20, 299*n*23–24, 305*n*34, 306*n*8, 319*n*8, 320*n*34, 321*n*37
chora (Kristeva) 9, 23, 61, 63, 70, 76, 78–79, 82, 91, 144, 237, 240, 253, 256, 264, 273, 275, 283*n*42, 313*n*53, 314*n*65
Christianity 14–15, 27, 84–89, 93, 97, 100, 105–8, 114–16, 121, 125–29, 141, 149–55, 162–72, 179, 197, 217–25, 228, 231, 252, 296*n*5, 296*n*45, 301*n*1, 301*n*5–6, 303*n*32, 304*n*17, 308*n*34; *see also* Judeo-Christianity
Christians 23, 33, 89, 106, 111, 114, 218, 236, 309*n*6
Churchill, Caryl 171
Churchland, Patricia Smith 293*n*55, 294*n*21
cingulate gyrus 36–37, 94, 107, 132, 282, 284, 292*n*34, 303*n*25, 307*n*15, 315*n*69
Ciompi, Luc 293*n*6
City Lights (Chaplin) 284
City of Angels 197, 199–201
Clark, Andy 305*n*43, 319*n*11
Clark, Robert L.A. 302*n*7, 303*n*27
Clash of the Titans 306*n*11
Clay, Jenny Strauss 303*n*26
Clements, A.M. 291*n*28, 294*n*19
Cless, Downing 304*n*10

Clottes, Jean 34, 36, 38–40, 42, 46, 297*n*56, 297*n*58–59, 297*n*61–62, 297*n*64, 298*n*68
Coatlicue 170
cognitive anchor 55, 56, 60, 62
cognitive blend 121, 126, 292*n*47
cognitive fluidity 45–46, 49, 55, 60, 61, 65, 122
cognitive philosophy 11–12, 305*n*43
cognitive science 4, 12, 136, 292*n*33, 295*n*28, 298*n*15, 300*n*49
cognitivism 4–5, 9–12, 26, 45–46, 49, 55, 60–62, 65, 103, 121, 135, 148, 284–86, 290*n*12, 290*n*15, 290*n*20, 292*n*33, 292*n*47, 292*n*49, 293*n*2, 293*n*10, 296*n*31, 296*n*32, 296*n*35, 298*n*15, 302–3*n*22, 305*n*43, 306*n*8, 321*n*39
Cole, Susan Guettel 63
collaboration 2, 4, 12–13, 80, 179–80, 213, 234, 268; *see also* cooperation
collective unconscious (Jung) 297*n*51
Collier, Jo Leslie 311*n*20
Collins, Francis S. 68–69, 299*n*38
commercialism 191, 232, 242, 244, 264, 269, 273, 275, 277, 318*n*106; *see also* consumerism
communal (or cultural or social) unconscious 82, 273, 277
communion 14, 19, 20, 28–30, 45, 48–49, 51, 62, 79–81, 86, 88, 116, 169, 176, 208, 301*n*72; *see also* mysticism
compassion 25, 66, 70, 103, 114, 116, 121, 123, 147, 154, 163, 165–66, 189, 204, 221, 224, 225, 231, 232, 241, 253, 255–57, 276, 281–82, 284, 286, 288, 295*n*28, 299*n*35, 306*n*12, 308*n*38, 311*n*27, 319*n*13, 319*n*17–18, 319*n*21; *see also* contagion; empathy; sympathy
complexity 17–18, 54, 193, 204, 253, 277, 281, 287, 294*n*21, 295*n*27, 314*n*57, 320*n*28
Condry, John 310*n*11
Congdon, Constance 305*n*50
consciousness 6, 9, 11–13, 18–19, 24–25, 32, 46, 51, 75, 77, 91, 93 105, 107, 122–24, 126, 134–36, 146–47, 164, 180, 238–39, 281–82, 289*n*7 290*n*22, 295*n*25, 295*n*30, 296*n*31, 296*n*43, 304*n*13, 320*n*28; theater of consciousness (Baars and Harth) 12, 306*n*10; *see also* altered state of consciousness; communal unconscious; higher-order consciousness; primary consciousness; self (and Other) consciousness; unconscious brain activity
Constantine 203, 206–7, 209–10, 228, 307*n*23
consumerism 3, 70, 95, 103, 171, 189–90, 227, 243, 279, 308*n*34
contagion (emotional) 73, 78, 116, 231, 239, 281–86, 295*n*26, 300*n*49, 303*n*31, 311*n*27, 315*n*69, 319*n*15; *see also* compassion; empathy; sympathy
cooperation 13, 25, 64–66, 70, 83, 92, 100, 105, 116, 147, 149, 179, 279–80, 290*n*17, 307*n*16, 321*n*38; *see also* collaboration

Copjec, Joan 311*n*26
copycat (or mimetic) violence 228–29, 239–40, 310*n*11, 315*n*71
Corballis, Michael C. 94, 291*n*26, 302*n*17
Cornford, F.M. 60
Corrigall, Jenny 290*n*15
Coveney, Peter 294*n*21
Cox, John D. 89, 302*n*10, 302*n*12, 303*n*27, 304*n*11
Cox, Ted 309*n*6, 310*n*9
Cozolino, Louis 7, 21, 72, 77, 291*n*23, 291*n*25, 291*n*30, 292*n*40, 298*n*19, 299*n*26–27, 300*n*61
Crabb, Peter B. 309*n*4
Creed, Barbara 308*n*25
cruelty 25, 66, 80, 85, 90, 92, 113, 115, 167, 190, 212, 218, 220–25, 240–42, 253–54, 256, 268–72, 274, 277, 280, 284–85, 287, 309*n*5, 316*n*89; *see also* Artaud, Antonin
cultural evolution 3–5, 18, 28, 36, 50, 84, 89, 99, 101, 119, 124, 152, 165, 189, 279, 286, 290*n*18, 299*n*37, 321*n*38
cultural selection 5, 177–81, 196, 222, 225, 295*n*29; *see also* natural selection
Curb Your Enthusiasm 284
Cymbeline 140–42, 147, 184, 304*n*28

daemons 75–77, 81, 84; *see also* demons
Dalcroze, Eugene 167
Damasio, Antonio 75, 147, 315*n*69, 318*n*5, 320*n*26
Damasio, Hanna 147
Dante's Inferno 307*n*19
d'Aquili, Eugene G. 25–33, 47, 51, 62, 79, 289*n*4, 291*n*25, 293*n*1, 296*n*36–37, 296*n*39–40, 296*n*43, 298*n*71
The Dark Knight 210–12
Date with an Angel 307*n*17
Davidson, Richard 149, 304*n*25, 317*n*93, 319*n*6, 319*n*17, 320*n*27
Davies, Paul 295*n*30
Davis, Michael 314*n*67
Dawkins, Richard 5, 24, 68, 279, 290*n*16, 295*n*25–26, 299*n*37, 306*n*4, 319*n*5
Day, Sean 120
Dayan, Daniel 309*n*2
Deacon, Terence W. 19, 292*n*43, 301*n*66, 318*n*1
Deal of a Lifetime 308*n*31
Dean, Carolyn 305*n*39
death drive 140, 236, 238–39, 244, 252, 262, 264, 266–67, 313*n*44–45, 313*n*48, 314*n*65, 318*n*119; *see also* drives
death penalty 253–56
deception 1, 35, 51–52, 55, 65, 71–72, 95, 152; *see also* Machiavellian Intelligence
Decety, Jean 7, 106–8, 291*n*31, 292*n*49, 303*n*24
Dehaene, Stanislas 290*n*18
Delasara, Jan 312*n*32
demonization 2, 85, 89, 105, 146, 231, 242, 279, 286

demons 2, 9, 15, 17, 23, 36, 91, 126, 136, 151, 152, 203–05, 207–12, 216–17, 221–22, 228, 231, 239, 246, 268, 283, 294*n*22, 301*n*1, 307*n*23, 308*n*28
Dennett, Daniel 11–12, 279, 293*n*50, 295*n*26, 319*n*5
DesAutels, Peggy 305*n*43
Detienne, Marcel 301*n*68, 301*n*71
The Devil and Max Devlin 212–13
The Devil's Advocate 215–17, 308*n*34
de Waal, Frans 56, 64, 100, 116, 280, 295*n*26, 299*n*22, 299*n*36, 302*n*21, 306*n*9, 319*n*7, 321*n*37
Diamond, Elin 318*n*117
Diamond, Jared 4, 68, 93
Diana 140–42, 144, 153, 304*n*20
Dionysian 21–23, 62, 65–66, 68, 76, 80–81, 86–87, 90, 92, 164, 169, 182–83, 195, 294*n*16, 298*n*71, 298*n*7, 298*n*12, 299*n*25, 305*n*46
Dionysus 21, 59–63, 65, 70, 75–76, 79, 81–82, 89, 97, 99, 168–71, 181, 185, 189, 191, 212, 224, 287, 298*n*7–9, 298*n*11, 298*n*13–14, 298*n*16, 298*n*29, 300*n*56, 300*n*58, 301*n*68, 301*n*73
Dissanayake, Ellen 294*n*17
divination 18–19, 293*n*4
divine judge 2, 13, 27–28, 63, 71, 78, 87–89, 96, 99–101, 104, 112, 114–15, 117–21, 123, 136, 142, 167–68, 181, 184–85, 195, 197, 221, 224–25, 232, 249–50, 276, 282, 286
divine messenger 115, 117, 127, 146, 193, 195, 197, 225, 253, 260, 263–64, 287, 301*n*1
divine warrior 49, 195, 203, 210, 221, 225, 253, 281, 286–87; *see also* rebellion; revolution; tricksters
Doane, Mary Ann 311*n*22
Doctor Faustus see Marlowe, Christopher
Dodds, E.R. 300*n*53
Dogma 203–7, 225, 253
Doidge, Norman 14, 290*n*15
Dollimore, Jonathan 313*n*48
Donald, Merlin 9–11, 31, 36, 49, 122, 292*n*35, 292*n*37, 292*n*41, 292*n*43, 292*n*46, 297*n*54, 318*n*1
Dooley, John A. 297*n*67
Douglas, John 312*n*34, 315*n*77, 315*n*80, 316*n*81–82, 316*n*84, 316*n*89
Dox, Donnalee 88, 301*n*3, 302*n*8
dream visions 50, 104, 171, 181; *see also* mystical visions; psychic visions; shamanic visions
dreaming 21, 23, 46, 58, 66–68, 70, 76, 98, 104, 112, 135, 139–44, 153, 155–57, 166, 170–71, 178, 191, 199, 201, 204, 207, 274, 280–81, 285–86, 289*n*7, 291*n*29, 294*n*22, 304*n*22, 316*n*82
drives 3, 27, 29, 38, 61, 71, 75, 84, 87, 89, 95, 108, 110, 121, 130, 136, 140, 144, 146–49, 158–61, 171, 188–90, 207, 212–13, 217, 236, 238–40, 244, 246, 248, 252, 262, 264, 266–67, 279, 282, 293*n*60, 299*n*28, 306*n*6,

Index

320n24, 321n34; *see also* animal drives; death drive; Freudian drives; good and evil drives; infanticidal drive; limbic and brainstem drives; sacrificial drive; theatrical drive
Dunand, Anne 314n63
Dunand, Françoise 58
Dunbar, Robin 292n39
Dyer, Richard 312n34, 312n38
Dyson, Freeman J. 296n45

Earl, James W. 313n55
Edelman, Gerald 5, 13, 24, 293n53, 293n3, 296n33, 299n33
Edwards, Emily D. 307n14
Edwards-Lee, Terri A. 291n25, 292n38
effects of mass media *see* audience effects
ego 1, 5–8, 10, 23, 30, 32, 61, 79, 81, 88, 90, 92–93, 95–97, 108, 124, 133–35, 144–45, 147, 149, 154, 175–76, 188, 192, 222, 234–35, 237, 240–41, 290n19, 295n27, 313n43, 314n66, 317n105
Egyptian coronation drama 55–59, 80–81
Ehret, Günter 291n24
Eisenberg, Nancy 319n9
Eliot, T.S. 166, 168
Elliott, John R., Jr. 91
Elliott, Rebecca 8
Emery, Ted 305n31
emotional contagion *see* contagion
emotional mimicry 164
emotional value operator (limbic system) 26, 79
empathy 94, 100, 103–05, 147–49, 151, 180, 232–33, 242, 295n28, 299n41–42, 305n42, 317n104, 321n37; *see also* compassion; contagion; sympathy
Enders, Jody 86, 301n5
environmental dependency syndrome 107
environmental influences (or input) 94–95, 293n54
environmental selection *see* natural selection
environmental theatre 2, 35, 48, 81, 168, 284, 289n5
epilepsy 20, 47, 301n72
Erinyes *see* Furies
Eshleman, Clayton 297n50
Eshu 169
ethics 53, 66, 85, 98, 153, 278, 280, 286, 288, 293n59, 303n5, 306n1, 309n5, 319n5–6, 321n39; *see also* morality
Euripides 60–64, 75, 79, 80, 119, 135, 168–69, 256, 300n47; *see also The Bacchae and Medea*
Eurydice 305n50
Evan Almighty 109, 194–95
Everett, Daniel L. 293n57
Everyman 112, 114–15, 117–24, 126–32, 135, 137, 142, 181, 186, 193, 195, 201, 253, 303n29, 303n34–36, 303n4, 304n6
evolutionary psychology 4, 9–10, 14, 17, 19, 46, 94, 97, 177, 290n12, 293n59, 298n73, 321n40

exhibitionism 243, 283; *see also* perversity
existential operator (limbic system) 79
exorcism 55, 182, 207–8, 261, 297n67, 307n24
The Exorcism of Emily Rose 208
The Exorcist 182, 207, 261, 307n24
Exterminating Angels 308n33
The Eyes of Laura Mars 312n34

facial expression and recognition 7, 24, 64–65, 107, 121, 164, 178–80, 196–97, 208, 282–83, 212, 219, 221, 244–45, 247, 254, 273, 272, 274, 290n19, 294n16, 296n38, 302n13, 305n42, 306n8, 319n21
facial expression neurons 178
facial identity neurons 178
fading (Lacanian aphanisis) 7
Fates (or fate) 14, 16, 18, 66–68, 71, 73–75, 78, 81, 96, 97, 112–13, 131, 139–41, 145, 155–57, 163, 165, 188–89, 194, 220–21, 224, 232, 236–38. 241–42, 252–53, 289n6, 300n47, 308n30, 313n43, 315n72–73
Father God 93, 101, 103, 111, 113–14, 131, 142, 162, 165, 197, 204, 217, 221–25, 281, 299n32, 307n16
Fauconnier, Gilles 4, 121, 292n47
Faustian bargain (or wager) 127–36, 185–87, 191, 212, 308n31
fear system 15, 28, 31, 36, 54–55, 85, 95, 125, 135, 184, 208, 222, 286
Feinberg, Todd E. 290n20
Felman, Shoshana 313n43
Ferré, John P. 309n2
fetishism 59, 126, 177, 220, 229–30, 242–44, 247–48, 269, 276, 283, 307n17, 312n40, 314n66; *see also* perversity
Fiddes, Paul S. 296n45, 321n42
filters (in the brain) 7, 13, 209, 221, 225, 231, 240–41, 250, 284, 288, 294n22; *see also* moral filter; stimulus barrier (in the brain)
Fink, Bruce 240–41, 250, 257, 283, 305n36, 309n5, 312n36, 312n40–41, 313n43, 313n45, 314n58–59, 314n61, 314n63–64, 314n66, 316n83, 317n97, 317n103, 317n105, 318n108, 318n111–12, 318n120, 320n23–25, 321n36
fitness indicator 177, 181
The Five People You Meet in Heaven 307n19
Fowkes, Katherine A. 307n14
Fowles, Jib 309n4, 310n14
free will 14–15, 24, 89, 129, 133, 136, 158, 182, 189, 192–93, 200, 217, 225, 295n27, 296n45, 299n27, 304n6, 304n13
Freeman, Anthony 296n31
Freeman, Morgan 190–94
Freeman, Walter J. 290n13
Freud, Sigmund 5–8, 72–73, 121, 132, 134–35, 158, 233, 236, 238–39, 290n15, 290n20, 292n34, 293n58, 298n13, 304n14, 312n40, 313n43, 313n45, 313n48, 313n55, 314n56, 316n92, 318n119–20

Index

Freudian drives 140, 236, 244, 264, 266, 298n15, 304n8, 313n44–45, 313n48, 314n65, 315n67, 318n119; *see also* death drive; ego; id; projection; repression; super-ego
Frey, Charles 149, 304n29
frontal lobes 6, 8, 20, 28–29, 33, 47, 57, 62, 69, 132, 221, 292n34, 293n7, 293–94n11, 303n25, 303n37, 319n12; *see also* prefrontal cortex
Furies (or Erinyes or Eumenides), 59, 67–68, 71–82, 91–92, 99, 109, 128, 134–35, 166, 181–82, 185, 224, 287, 300n46–47, 300n53, 300n62, 300n73, 301n64
Fuster, Joaquín M. 290n16

Gabriel (archangel) 108–10, 206–07, 307n22
Gaiman, Neil 318n1
Gallagher, Lowell 303n5
Gallese, Vittorio 282, 319n22
Gaster, Theodor H. 57–59, 298n2–3
gaze 58, 61, 145, 159, 168, 178–79, 201, 231, 235, 244, 289n11, 305n38, 307n17, 311n26, 311n29, 313n46
gaze direction 164, 178–79
Gazzaniga, Michael S. 293n51, 320n27
Geb 56, 58
gender 22–23, 233, 273, 291n28, 294n19–20, 297n55
Genet, Jean 167, 305n39
genocide 4, 85, 98, 126, 280, 301n2, 319n16; *see also* anti-Semitism; holocaust
Gerbner, George 309n6, 310n14
Geschwind, Daniel H. 52, 291n25, 291n27, 292n38
Gibbons, Reginald 298n12
Gibbs, Raymond W., Jr. 290n12
Girard, René 99, 300n49, 303n41, 304n28
Glenberg, Arthur M. 293n54
God Concept (Persinger) 22
God experiences 4, 19–22, 25, 30–33, 46, 49, 52, 176, 221, 294n22
God Is Brazilian 306n12
God of War 306n11
Godfrey, Bob 303n36
Godspell 217
Goethals, Gregor T. 309n2
Goldfarb, Alvin 311n19
Goldman, Alvin 282, 292n49
Goleman, Daniel 291n25
good and evil drives 155, 163
The Good, the Bad, and the Ugly 284
Goodale, Melvyn A. 319n19
Gordon, Robert M. 305n42
The Gospel According to Matthew 217
Gough, William C. 294n18
Gould, Thomas 311n19
Gozzi, Carlo 154, 305n31
Great Chain of Being 117, 120, 126, 129, 179, 303n1; *see also* hierarchies
The Greatest Story Ever Told 217

Green, Viviane 290n15
Greenblatt, Stephen 138
Greene, Joshua 148, 302n22, 303n30
Greenspan, Stanley I. 292n44
Gregory, Richard 122
Grigg, Russell 317n97
Groopman, Jerome 293n9
Grotowski, Jerzy 168
Guthrie, R. Dale 297n50
Guthrie, Stewart Elliott 293n10

Haidt, Jonathan 100, 286, 302n22
Halifax, Joan 297n50, 298n69
Halpern, Richard 304n8, 304n11, 304n17, 304n23
Hamilton, James T. 310n14
Hamlet 25, 118, 137–40
Hammill, Graham 304n12
Harper, Elizabeth 303n38
Harrington, Anne 319n6, 319n17
Harris, Sam 279, 307n15, 319n5
Harris, Thomas 309n7, 312n34, 316n92
Hart, F. Elizabeth 4, 304n20
Harth, Erich 12
Hasson, Uri 284–85, 293n59, 320n32
Hatfield, Elaine 319n14
Haught, John F. 293n8
heaven 16, 26, 84, 88, 90, 92–93, 96, 111, 114–15, 117, 120, 126–28, 130–33, 136–43, 145–47, 153–56, 160–63, 172–73, 190, 193, 200–01, 203, 207, 213, 221, 224, 227, 234, 253, 276, 281, 300n54, 302n12–13, 304n6, 304n16, 307n19
Heaven Can Wait 307n19
Hecht, Jennifer Michael 299n29, 299n39, 301n70
Heinrichs, Albert 298n9
hell 16, 26, 85, 90–92, 95–96, 111, 114, 119, 126–27, 129–35, 137–42, 146–47, 152–53, 155, 159–60, 164–65, 167, 200, 207, 209–13, 216, 224, 253, 281, 300n47, 304n6, 304n16, 312n34
Hellboy and *Hellboy II* 210–12, 308n28
Helm, James J. 300n53
Hera 70, 74; *see also* Juno
Herakles 70, 80
hierarchies 8, 13, 17, 51, 64–65, 79, 94–95, 98–100, 105, 115–17, 119, 123, 137, 157, 177–81, 279, 290n20, 293n57, 296n35, 306n7
higher-order consciousness (Edelman) 1, 10, 13, 16, 18, 24, 32–33, 51, 61, 69, 76, 79–80, 90, 93, 102, 128, 131, 135, 221, 279, 286
Highway, Tomson 170–71
Hill, John 309n2
Hinde, Robert A. 290n20
hippocampus 30, 296n40, 320n26
Hobson, J. Allan 289n7
Hogan, Patrick Com 290n12, 296n35, 304n27
holistic operator (right hemisphere) 26–28,

32, 33, 36, 49, 52, 72, 79–80, 85, 89, 100, 125, 132, 144, 146, 155, 302n17
Holland, Norman 290n15
holocaust 106, 167, 318n4; *see also* anti-Semitism; genocide
"holy war" 2, 23, 25, 85, 280, 313n50
hominids 3, 5, 9–13. 18, 26, 31–32, 45, 51, 66, 70, 82, 87, 92, 94, 106, 122, 291n27, 292n35, 292n39, 292n43–44, 292n46, 293n55, 294n22
Hood, Bruce M. 321n41
Horgan, John 20, 294n14, 295n28
Horsley, Jake 312n30
Horus 56–61, 67, 70, 81, 87, 181, 185, 189, 306n11
Houdé, Oliver 148
Hough, Kirstin J. 310n11
Hrdy, Sarah Blaffer 103, 144, 159, 280, 305n34
hubris 4, 61, 90, 96–97, 128, 130, 212, 227–28, 236, 303n27, 314n67
Huerta, Jorge 170, 305n49
Huesmann, L. Rowell 310n11, 310n13, 320n33
Huff, Richard 315n76
Hughes, Jan N. 309n4, 310n11
humanism 1–2, 15, 84, 127, 152–55, 171, 175, 208, 227, 278
The Hungry Woman 305n50
Hunk 308n31
Hunt, Maurice 304n22
Hurley, Susan 290n19, 319n12, 320n34
hybrids 34, 39–40, 47, 49–50, 54–57, 59, 71, 81, 123, 176, 178–79, 181, 183, 197, 210, 212, 304n24
Hymen 140
hysteric (or hysteria) 283, 301n73; *see also* neurosis

Iacoboni, Marco 52, 148, 282, 285, 290n19, 291n25, 291n27, 292n38, 319n22, 320n33
Ibsen, Henrik 165
id 8, 72, 147, 241, 281, 313n43
identification 1, 37, 50, 55, 106, 152, 164, 218–22, 224–25, 230, 235, 237–40, 246, 250–52, 267, 274–76, 282, 284–86, 288, 304n12, 310n11, 318n18
The Imaginarium of Dr. Parnassus 308n29
Imaginary order (Lacan) 5–14, 21, 26, 28, 31, 33, 36–37, 61, 66–67, 70, 77, 88–89, 93, 95–96, 101, 105, 107–8, 110, 122, 128, 135, 140, 144–48, 152, 155, 158, 171, 173, 208, 234, 236, 240–41, 249, 252–53, 277, 282, 289–90n11, 290n19, 291n29, 292n45, 295n25, 302n9, 303n39, 304n12, 307n13, 309n40, 309n5, 311n17, 313n44, 317n97, 320n23
Immortal 306n11
immortality 16, 18–19, 24–25, 60, 71, 82, 112, 124, 126–27, 130, 134, 135, 138, 154, 171, 178, 189, 193, 200, 225, 260, 275, 276, 279, 287
Immortals 306n11

imperialism 3–4, 13, 27, 85, 88, 105, 150, 181, 189–90, 227
infanticidal drive 105–6, 110, 216
infanticide 103, 119, 128, 216, 279–81, 303n23, 305n34, 306n7, 319n7
insula 132, 148, 303n25, 303n30, 303n37, 307n15, 319n21
Intelligent Design 23–24, 117
interagency attributions 135–36
Internet *see* the Web
intersubjectivity 10, 12, 21, 52, 54, 59, 70, 73, 77, 93, 107–08, 292n45
intuition neurons 132–33, 208, 303n25
Iris 146
Isis 56, 171
Islam 2, 27, 33, 85, 111, 227, 229, 276, 296n36, 301n3, 302n16, 306n11
It's a Wonderful Life 264, 307n14

Jablonka, Eva 289n9
Jacob, Pierre 7, 12
Jacob's Ladder 208–9
Jameson, Michael 298n7
Jarry, Alfred 166–67
Jaynes, Julian 298n19
Jeannerod, Marc 7, 12
Jesus 28–30, 84–87, 89, 93, 108–14, 116–19, 123–24, 157, 160, 169–70, 204, 217–25, 301n1, 303n28, 303n33, 308n38
Jesus Christ Superstar 217
Jesus of Nazareth 217
Jews 85–89, 97, 105–6, 110–16, 123, 125, 128, 187, 218–19, 221–24, 301n1, 301n3, 301n5–6, 302n7, 303n27–28, 308n30
Johnson, Mark 4, 121, 281, 296n34, 299n32, 305n43, 307n16
Joseph, Rhawn 22–25, 294n20–22
jouissance (Lacanian) 32, 110, 220, 230, 233–35, 240, 248–50, 264, 266–68, 276, 283, 308n33, 312n36, 312–13n40–41, 313n445, 314n58, 314n63, 314n67, 316n83, 316n88, 317n105, 321n36
Jourdan, Robert 293n59
Judeo-Christian tradition 1–2, 19, 49, 83–84, 99, 108, 116, 123, 125, 175–76, 181, 303n23; *see also* Christianity; Christians
judgment *see* divine judge; Last Judgment
Juno 146; *see also* Hera
Jupiter (or Jove) 139, 141–42, 147, 153–54, 184, 304n22; *see also* Zeus

Kagan, Jerome 289n3
Kahneman, Daniel 148
Kalamas, Alicia D. 310n14
Kalke, Christine M. 298n17
Kandel, Eric R. 37, 290n15
Kaplan, E. Ann 313n49, 315n70
Kaplan-Solms, Karen 8, 147, 290n15, 291n25
Kapogiannis, Dimitrios 196
Katz, Alyssa 311n29–30
Katz, Elihu 309n2

Katz, Leonard D. 302*n*21
Kellner, Douglas 309*n*2
Kennett, Jeanette 315*n*69, 320*n*26
Kernodle, George R. 298*n*4
kidnapping 203, 229, 233–34, 249, 251–53, 258, 271, 277, 281, 315*n*77, 316*n*87, 316*n*90, 317*n*96
Kiehl, Kent A. 315*n*69
Kihlstrom, John F. 298*n*15
Kimball, A. Samuel 303*n*23, 319*n*7
kin selection 20, 68–69, 74, 102, 104, 137, 149, 152, 159, 161–62, 210, 216, 227, 280–81, 299*n*35–37
King, Barbara J. 103
Kirby, E.T. 55, 297*n*67, 298*n*69
Kister, Daniel A. 297*n*67
Kleist, Heinrich von 155–56, 305*n*32–33
Kobialka, Michal 301*n*4, 302*n*8, 303*n*33
Koechlin, Etienne 8
Kordela, A. Kiarina 303*n*2
Kosslyn, Stephen M. 321*n*42
Krcmar, Marina 310*n*12
Kristeva, Julia 61, 70, 91, 292*n*42, 300*n*60, 308*n*25, 313*n*53, 314*n*65
Kubiak, Anthony 298*n*72
Kunkle, Sheila 309*n*5, 314*n*62
Kushner, Tony 172–73

Lacan, Jacques 5–6, 9, 61, 135, 138, 222, 228, 236, 257, 290*n*19, 292*n*41, 292*n*45, 303*n*2, 305*n*36, 305*n*39, 309*n*5, 311*n*26, 313*n*43, 313*n*45, 313*n*48, 314*n*63, 315*n*67, 317*n*92, 318*n*108, 319*n*22, 321*n*39
Lacoue-Labarthe, Philippe 314*n*56
Ladd, Roger A. 303*n*38
Lakoff, George 4, 121, 281, 296*n*34, 299*n*32, 307*n*16
Lang, Robert 308*n*26
Laplanche, J. 313*n*55
Last Judgment 89–90, 114, 119
The Last Temptation of Christ 217, 309*n*39
Laughlin, Charles D. 289*n*4, 293*n*1, 296*n*, 296*n*39, 296*n*43
Leader, Darian 314*n*61
leaders *see* alpha leaders
learned behaviors 6, 8, 12, 67, 84, 144, 286, 289*n*7, 307*n*18; *see also* teaching
Leary, Mark R. 289*n*4
LeDoux, Joseph 9, 291*n*23, 293*n*58, 301*n*66, 308*n*27
Lee, Meewon 297*n*67
Lefkowitz, Mary 66, 300*n*54
left brain (or hemisphere) 6–7, 9–10, 13, 18, 21–22, 25–27, 31, 33, 36, 52–54, 57–59, 62, 70, 72–74, 76, 78–81, 85, 88–89, 91, 93–94, 95–97, 99, 100, 102–3, 105, 107–10, 112, 125, 127–28, 130, 132, 140, 144, 146, 152, 155, 165, 171, 173, 187–89, 208, 234, 291*n*27–29, 293*n*7, 294*n*16, 296*n*41, 299*n*41, 300*n*60, 300*n*63, 301*n*67, 301*n*69, 320*n*27; *see also* neocortex; right brain

Legion 307*n*22
Lehrer, Jonah 285, 320*n*28
Lemoine, Jacques 297*n*67
Lesko, Barbara S. 56
Levesque, John 309*n*6
Levitin, Daniel J. 293*n*59
Levy, Neil 293*n*59, 305*n*37, 306*n*1
Lewis-Williams, David 34–36, 45–50, 176, 296*n*33, 297*n*49–50, 297*n*63–65, 298*n*68
Liebes, Tamar 309*n*2
The Life of Brian 217
The Life of Jesus: The Revolutionary 217
Lima, Robert 85, 91
limbic and brainstem drives 33, 54, 57–58, 64–65, 68, 80–81, 88, 128, 187–88; *see also* care system; fear system; lust system; panic system; rage system; seeking system
limbic brain system 6–10, 13, 15, 18–21, 23, 25–29, 31–33, 36–37, 50, 53–55, 57–58, 63–64, 70, 72–82, 84–85, 88–89, 91, 93, 95–96, 100, 102–3, 108, 125, 127–29, 132, 135, 140, 144, 146–49, 152, 155, 157, 161, 171, 173, 184, 187–88, 208–9, 221, 231, 234, 240–41, 250, 281, 284–86, 290*n*22, 291*n*23, 291*n*29, 294*n*12, 294–95*n*20–22, 296*n*38, 296*n*40, 299*n*41, 301*n*72, 304*n*26–27, 308*n*27
Lipton, Bruce H. 285, 289*n*9
Logan, Robert A. 304*n*23
Logan, Robert K. 292*n*41, 306*n*1
Lonsdale, Steven H. 314*n*60
Lorblanchet, Michel 40
Lovelock, James 25, 295*n*30
Lucifer 82, 84, 89–90, 92–93, 96–97, 99, 123, 126–30, 132–43, 136, 138, 143, 185, 195, 197, 203–4, 206–7, 217, 287, 302*n*12–13, 303*n*27; *see also* Belzebub; Mephistophilis; Satan
lust system 64, 208; *see also* seeking system
Lynch, Aaron 295*n*26
Lyon, Eleanor 310*n*14

Machiavellian Intelligence 51–52, 55, 71, 94–95, 97, 104, 108, 112, 207, 298*n*72; *see also* baboon metaphysics
MacIntyre, Alasdair 289*n*2
MacLean, Paul 291*n*23
MacLeash, Archibald 167–68
Malti, Tina 317*n*104
mammalian brain 6, 37, 65, 69, 74, 77, 84, 91, 103–4, 110, 121, 156–57, 179, 207–8, 280, 291*n*23, 302*n*22, 305*n*34
Manhunter 312*n*34
Margulis, Lynn 290*n*17, 295*n*30
Marks, Jonathan 298*n*20, 303*n*1
Marlowe, Christopher 15, 126–37, 143, 152, 193, 204, 279, 303*n*3–5, 304*n*6–12, 304*n*14, 304*n*16–17, 304*n*23
Marsalek, Karen Sawyer 304*n*21
Martin, Laura 305*n*38
The Matrix (film series) 306*n*12

Matthews, Caitlin 297*n*67
Maurer, Daphne 120
McConachie, Bruce 4, 289*n*11, 292*n*33, 293*n*60
McCullough, Malcolm 100
McGeer, Victoria 286, 319*n*18
McGrath, Alister 117, 295*n*26
McNamara, Patrick 293*n*11, 296*n*41, 298*n*70
Medea 23, 75, 119, 135, 256, 300*n*56
meditation 24, 28–29, 32–33, 46–47, 219; *see also* mysticism
Meeks, Dimitri 56, 298*n*5
Meet Joe Black 307*n*20
Mellars, Paul 297*n*47
Meltzoff, Andrew N. 306*n*8
memes 5, 20, 24–25, 27, 33, 52, 54, 58, 65, 68–70, 74, 81, 146, 183, 186, 188, 207, 290*n*16, 292*n*35, 295*n*25–26, 295*n*29, 299*n*37, 306*n*4, 308*n*33
Mephistophilis 127–31, 133, 136, 138, 187, 190, 212–13, 308*n*31
Merleau-Ponty, Maurice 319*n*22
metamind (or Metamind) 94–96, 102, 104, 106, 114, 123, 131, 141, 176, 181, 183, 187–89, 191–93, 196–97, 220, 278, 288; *see also* Theory of Mind
Metamorphoses (Zimmerman) 305*n*50
Metz, Christian 230
Meyer, Cheryl L. 280, 317*n*102
Meyer, Marvin W. 298*n*11, 299*n*29
Michael 202–03
Michael (archangel) 202–03, 307*n*22
Miller, Arthur 167
Miller, Geoffrey 51–52, 177, 299*n*25, 306*n*4, 306*n*7
Mills, David 303*n*29
mimicry 1, 12, 38, 107, 122, 132, 151, 164, 180, 249, 282–83, 285, 292*n*49, 306*n*6, 319*n*12
Minsky, Marvin 293*n*52
mirror neurons 12, 66, 94–95, 99, 107–8, 122, 132, 148–49, 177–78, 231, 239, 282–86, 291*n*29, 292*n*35, 292*n*49, 293*n*55, 302*n*17, 311*n*27, 319*n*12, 319*n*20–22, 320*n*34; *see also* intuition neurons
mirroring 6, 12, 95, 108, 149, 179, 284, 290*n*19
Mithen, Steven 11, 45, 49, 55, 292*n*46–47, 294*n*22, 296*n*41
mob violence 259–60, 262, 281
Moise, Jessica F. 310*n*11
Molière 153
Moll, Jorge 302*n*22, 320*n*31
Montelle, Yann-Pierre 35–37, 48, 50–51, 297*n*46, 297*n*54–55
Moody, Raymond A., Jr. 294*n*21
moral filter 177, 181, 183, 185, 220, 225, 240–41, 250; *see also* filters (in the brain)
moral intuition 148–50, 152, 284
moral order 185, 286, 321*n*38
morality 13–14, 16, 27, 30, 69, 72, 81, 83–84, 97, 100–06, 114–16, 118, 120–24, 127–28,
147–54, 157–59, 161–65, 177–78, 181–86, 193–94, 196, 203, 212–13, 220, 225, 229, 240–41, 250–51, 253, 277, 281–82, 284, 286–88, 289*n*3, 293*n*59, 293*n*8, 294*n*22, 295*n*28, 299*n*36, 299*n*41, 300*n*53, 301*n*1, 302*n*21–22, 303*n*29, 303*n*35, 303*n*43, 307*n*13, 307*n*16, 311*n*27, 315*n*69, 316*n*92, 319*n*5, 319*n*13, 319*n*16, 319*n*18, 320*n*26–27, 321*n*38; *see also* ethics
Morse, William R. 311*n*19
Mossman, J.M. 299*n*31
mother goddess 22–32, 109, 140, 142, 144, 146, 153, 171, 301*n*71, 304*n*20
Mother Nature 84, 101, 104, 106, 131, 144–45, 155, 165, 197, 216, 224, 236, 248–49, 317*n*105
Murphy, Nancey 8, 292*n*36, 293*n*50, 293*n*59
Murray, John P. 310*n*9, 310*n*11
mystical visions 2, 33–36, 38, 50–51, 125, 129, 133–35, 142, 153, 170, 172–73, 176, 180, 182–83, 185, 189, 231, 287, 294*n*22, 300*n*47, 300*n*56, 304*n*22, 313*n*42; *see also* dream visions; psychic visions; shamanic visions
mysticism 28–33, 47, 50–52, 62, 69, 79–81, 125, 166, 176, 183–84, 228, 236–37, 261, 264, 266, 297*n*67, 298*n*70, 301*n*72, 302*n*18

Nadel, Jacqueline 290*n*19, 292*n*35
Nanabush 171
The Nativity Story 217
natural (or environmental) selection 5, 12, 23, 51, 66, 92, 156, 177–82, 295*n*23; *see also* cultural selection; kin selection; neural group selection; sexual selection
Nelson, Katherine 296*n*44
neoclassicism 17, 136, 139, 141–42, 146, 152–55, 175, 306*n*11
neocortex 6–8, 15, 19, 21, 52, 64–66, 70, 75–78, 89, 125, 296*n*44; *see also* left brain; right brain
Nesse, Randolph M. 298*n*73
Nettle, Daniel 293*n*60
Neumann, Erich 109
neural evolution 5, 18; *see also* cultural evolution; plasticity (of the brain)
neural group selection (Neural Darwinism) 5, 8, 13
neuro-psychoanalysis 5, 53, 88, 144, 146–47, 228, 300*n*60
neurosis 171, 233, 238–39, 246, 250, 282–83, 305*n*36, 312*n*36, 312*n*40 313*n*45, 314*n*58, 314*n*66, 315*n*68, 316*n*83; *see also* hysteric; obsessional
neurotheatre 14, 57, 78–79, 81–82, 85, 158, 160, 185, 187, 189–90, 197
Newberg, Andrew 25–33, 47, 51, 62, 79, 291*n*25, 296*n*36–37, 296*n*40, 298*n*71, 301*n*67, 301*n*75, 304*n*26
Newitz, Annalee 312*n*34
Nickisch, Curt Wendell 305*n*35

Niedenthal, Paula M. 164, 292n38
Nietzsche, Friedrich 21, 69, 76, 79, 83, 169, 184, 217, 278, 301n68, 305n46, 314n56
9/11 227, 229, 276, 309n6
Nisse, Ruth 87
Nordmann, Alfred 305n32
Nussbaum, Martha C. 299n42, 314n67

Obatala 169
Obbink, Dirk 300n58
obsessional (neurotic) 171, 283; *see also* neurosis
occipital lobes 6, 291n26
Oedipus 99, 112, 212, 220, 228, 231, 233–39, 242, 248–53, 260, 265, 268, 299n41, 312n36, 312n40, 313n43, 313n45, 313n49, 314n61, 316n92, 317n105
Ogun 169–70
Oh, God! 181–83, 189, 194, 197
Oh, God! Book II 183–85, 189, 193, 194, 197
Oh God, You Devil 185–89, 191, 194, 197
Omen 217
O'Neill, Eugene 167
oracle 73–74, 112, 140–41, 236–37, 239, 245, 277, 279, 314n64
orbitofrontal cortex 8, 94, 121, 149, 315n69
The Oresteia 23, 71, 73–80, 82, 91, 109, 146, 166, 300n45, 300n47; *see also* Aeschylus; Apollo; Athena; Furies
orientation association area (in the brain) 7, 28–30, 47, 296n37
orishas 169–70, 307n21
Ornstein, Robert 7
Orphée 307n20
Osiris 56–60, 67, 70, 73, 81, 87, 181, 185, 189, 298n10–11
the Other 2, 5–8, 10–12, 18, 21–30, 32, 45, 47, 49, 51–53, 56, 60–65, 77, 79, 82, 84, 87, 95, 97, 102–3, 105–8, 110, 113, 125, 135, 141, 145, 153, 173, 176, 180–81, 197, 208, 216, 220–21, 224–25, 239–40, 248–51, 257, 260, 262, 264, 266–68, 277–79, 283, 285, 287, 308n34, 313n43, 313n45, 314n63, 316n83, 316n92, 317n105, 318n107, 318n111, 320n22, 321n36
Owen, Bradford 319n9

Padel, Ruth 75–77, 298n7, 300n52, 301n72–73
Pagels, Elaine 301n1
paleo-mammalian brain *see* mammalian brain
Palombo, Stanley R. 314n57
panic (or separation-distress) system 28, 31, 36–37, 50, 54, 96, 125, 208, 222
Panksepp, Jaak 28, 36, 98, 290n15, 291n23, 291n28, 292n34, 293n6, 294n19, 299n23
Papousek, Ilona 319n15
paranoia 135, 161, 175, 207, 221, 227–29, 232, 239–40, 242–43, 246, 250, 268, 272, 290n19, 312n33
parasympathetic nervous system *see* autonomic nervous system

parietal lobes 6–7, 9, 22, 25–26, 28, 30, 47, 62, 72, 79, 89, 107, 122, 128, 132, 148, 184, 291n26, 291n28, 293n7, 295n24, 296n40, 301n67, 304n26
Parker, John 303n34
Parker, Kim Ian 317n105
The Passion of the Christ 218–25
Paul, William 311n20, 311n23
Paulson, Julie 303n35
Penchansky, David 302n14
Percy Jackson and the Olympians 306n11
Pericles 140–42, 144, 153
Perry, Marvin 301n2, 301n5, 301n6
Persinger, Michael A. 19–22, 31, 52, 294n14, 294n22
perversity 65, 89, 95–96, 113, 133, 151, 154, 158, 160–61, 169, 173, 178, 201, 203, 208, 218, 222, 227, 229, 232, 234, 239–40, 243, 246, 248–52, 257, 275–76, 282–83, 286, 312n36, 313n41, 314n66, 315n68, 316n90, 320n25; *see also* exhibitionism; fetishism; sadomasochism; voyeurism
Pessoa, Luiz 293n59
Pfeiffer, John E. 35, 297n53
Philo, Greg 310n14
Pigafetta, Antonio 149, 304n29
Pineda, Jaime A. 319n20
Pirandello, Luigi 166
Piskorowski, Anna 317n105
plasticity (of the brain) 14, 18, 65, 124, 140, 205, 280, 317n93; *see also* pruning
The Play of the Sacrament 85–89
pleasure 31, 49, 64, 86, 131, 145, 147, 153, 159, 163, 178, 184, 186, 189, 192, 194, 203, 213, 217, 220, 225, 230, 233–34, 236, 238, 252, 260, 264, 281, 313n45, 315n70, 317n105
pleasure/reward (dopamine) system 69, 281, 285, 321n41
Poole, Kristen 304n6, 304n14
prefrontal cortex 8, 65, 70, 74, 95, 107, 137, 147–49, 196, 209, 231, 240–41, 250, 284–86, 288, 292n36, 292n43, 292n49, 299n27, 303n22, 306n10, 307n15, 308n27, 317n93, 319n28, 320n31, 321n34; *see also* orbitofrontal cortex; ventromedial prefrontal cortex
pretense (or pretending) 10, 94, 179, 191, 245, 292n49, 316n86
Price, Simon 289n8
primary consciousness (Edelman) 18, 293n53, 293n3, 296n33, 299n33
primates 5–6, 9, 12–13, 15, 51–52, 56, 64–65, 67, 71, 94, 100–03, 116–17, 119, 128, 143, 149, 156, 177–80, 185, 188, 195, 197, 203, 227, 279, 280, 282, 286, 290n14, 290n19, 292n38, 292n43, 293n55, 298n72, 299n36, 306n2, 306n8–9; *see also* apes
Prime Mover 25, 190, 288, 296n45
Prinz, Jesse J. 302n21, 320n26
Profiler 315n76
profiling 228, 232–33, 243, 245, 247, 254,

264–65, 268, 270, 273, 277, 309n7, 312n34, 312n38, 313n47, 315n76–77, 316n80, 316n82, 317n101
projection 1–2, 13, 23, 26, 47, 52–53, 59, 62, 73, 79, 83–84, 95, 99, 122, 125–27, 135–36, 145, 160–61, 176–78, 180–81, 183, 189–90, 196, 207, 228, 231, 234, 242, 259, 262, 286, 298n68, 298n73, 304n14
Prometheus 60, 65–71, 79–81, 92–93, 99, 181, 185, 189, 287, 299n31, 303n26
Prometheus Bound 65–71, 77, 79, 82, 299n31; *see also* Aeschylus
The Prophecy and *The Prophecy II* 307n22
pruning (of neurons) 14, 21–22, 49–50, 71–74, 77, 82, 100, 177, 217, 234, 247, 262, 276, 317n105; *see also* plasticity
psychic visions 74, 129, 158, 160, 178, 229–77, 283, 309n7, 312n34, 314n61; *see also* dream visions; mystical visions; shamanic visions
psychosis 35, 158, 160, 232–34, 239–40, 242–43, 245–46, 252–53. 268, 271–72, 276–77, 282–83, 305n36, 312n36, 312n40, 314n61–62, 314n66, 315n68, 316n83, 317n97, 320n23–24; *see also* paranoia; schizophrenia

Quartz, Steven R. 315n69
Quetzalcóatl 170

race 105, 163, 212, 289n8, 301n2, 302n7, 307n23, 308n30
Racine, Jean 153
rage system 15, 23, 28, 54, 85, 95, 125, 208, 222
Ragland-Sullivan, Ellie 7
Ramachandran, V.S. 18, 20, 23, 31, 52, 61–62, 72–74, 91, 121–23, 132–34, 136, 291n25, 293n50, 293n53, 293n59, 293n7, 294n22, 295n23–24, 297n57, 298n73, 299n24, 300n63, 301n72, 303n40
rats 97–99, 299n23
Read, Leslie Du S. 298n3
Real order (Lacan) 5–14, 21, 31, 33, 36–37, 61, 66, 70, 78, 88–89, 93, 96, 101, 105, 110, 112, 122, 140, 144–45, 147, 152, 155, 158, 170–71, 173, 234, 236, 237–42, 249, 252–53, 255, 261, 263, 275, 277, 282, 292n45, 295n25, 309n40, 309n5, 314n64
rebellion 13, 18, 24, 54–55, 57–58, 60, 66, 69–70, 79–81, 89–93, 96–97, 99, 101–02, 104, 123, 129, 139–40, 161, 163, 167, 169–73, 180, 185, 187–88, 207–8, 217, 225, 236, 238, 286–87, 298n7, 301n1, 302n12; *see also* revolution
reciprocal altruism *see* altruism
reciprocal violence (Girard) 13, 73, 99, 102, 300n49
recursion 94, 302n17, 306n10
redirected aggression 98–102
reductive operator (left hemisphere) 79, 85
regression 15, 18, 208, 240, 290n15

Rehm, Rush 62, 75, 299n30, 300n43, 300n56–57, 300n59, 301n74
Reid, Robert G.B. 290n17
Renfrew, Colin 292n48
Renfrew, Jane M. 294n22, 296n33
repetition 5, 17, 20, 23, 30, 32–33, 57–58, 78, 81, 86, 106, 113–14, 128, 133, 140, 142–43, 145, 159–60, 163, 171, 179, 183, 224–25, 229, 234, 236–37, 241–42, 245, 250, 253, 266, 268, 276–77, 281, 283, 285, 299n27, 307n23, 313n41, 313n45, 314n61, 314n67, 315n69, 315n71
repression 6, 13–14, 21, 61, 63, 72–74, 76, 78–79, 126, 145, 153, 170–71, 173, 194, 238–39, 252, 282–83, 287, 293n58, 298n13, 298n16, 299n41, 304n14, 312n40, 313n45, 316n92
"reptilian" brain (in humans) 6, 55, 77, 291n23
revenge 25, 62, 74, 78, 80, 92, 98, 100–01, 104, 106, 143, 169, 194, 203, 236, 241–42, 246, 249, 255–57, 276, 286, 300n48, 305n43, 318n3, 319n8; *see also* reciprocal violence; vengeance
reversal learning 149
revolution 38, 63, 69, 86, 117, 124, 153, 170, 172–73, 177, 204, 292n42
Revonsuo, Antti 98, 207
Reynolds, Bryan 298n72
rhesus monkeys (or macaques) 116–17, 119, 203, 280
Richerson, Peter J. 321n40
Ridge, Robert D. 318n3
Ridley, Matt 299n25–26
Rifkin, Jeremy 299n42
right brain (or hemisphere) 6–10, 13, 18–21, 23, 26–27, 31, 33, 36, 52–54, 57, 63, 70–73, 77–82, 85, 89, 91, 93–97, 100, 102–3, 105, 107–8, 110, 112, 15, 127, 132, 140, 144–47, 152, 155, 158, 165, 171, 173, 185, 187, 189, 234, 291n27, 291n29, 291n31, 292n38, 293n7, 294n16, 294n19, 296n39, 299n41, 300n61, 303n39, 310n9
righteousness 23, 25–27, 85, 114–15, 123, 125, 138, 150, 152, 212, 224–25, 228, 241, 280, 282, 286, 309n6, 319n16
Rilke, Rainer Maria 173, 305n52
Rivera, José 171–73
Rizzolatti, Giacomo 148283, 292n35, 292n38, 293n55, 302n17, 306n2, 319n12, 319n20–22
Roberts, Jeanne Addison 304n24
Rochat, Philippe 290n19, 299n24, 306n8
romanticism 15, 124, 126, 153–58, 160–61, 165–66, 175, 217, 278, 281, 287, 306n11, 311n20
Rosemary's Baby 207, 217
Roskies, Adina L. 315n69, 320n26
Roth, Melissa 303n39
Rothenberg, Jerome 298n2
Rothenberg, Molly Anne 318n113

Rotman, Brian 318*n*1
Russell, Gordon W. 309*n*4
Ryan, Robert E. 297*n*51
Ryle, Gilbert 11, 293*n*50

Sabom, Michael B. 294*n*21
Sacks, Oliver W. 290*n*15, 293*n*59
sacrifice 1–2, 13–18, 20, 25, 34, 36, 49, 53, 55, 58–59, 61–69, 71, 73–74, 76, 85–88, 92–100, 102–6, 108–17, 119–21, 123, 125, 132, 134–36, 140, 148, 150–52, 157, 161, 163, 165–66, 168–71, 173, 175, 181, 188, 191, 200, 203, 207, 213, 218–25, 227–29, 233–35, 238–42, 244, 248–57, 262, 269, 272–73, 276–77, 279–80, 282–83, 286–87, 293*n*3, 298*n*8, 299*n*28, 299*n*36, 301*n*6, 302*n*15, 304*n*28, 305*n*38, 312*n*36, 312*n*39–40, 313*n*45, 314*n*58, 314*n*63, 316*n*90, 316*n*92, 317*n*103, 317*n*105, 318*n*107
sacrificial drive 68–69
sadomasochism 98, 178, 217–20, 224–25, 241–43, 261, 272, 274–75, 283, 306*n*14, 312*n*40, 314*n*56, 314*n*66; *see also* perversity
Salkever, Stephen G. 314*n*67, 315*n*75
Sartre, Jean-Paul 167, 269
Satan 3–4, 82, 84, 89, 92–93, 96–97, 99, 112–15, 119, 125, 151, 167, 185, 195, 203, 207, 216–17, 221–23, 225, 242, 251, 253, 260, 269, 287, 301*n*1, 302*n*13, 302*n*16, 308*n*34, 316*n*85, 318*n*119; *see also* Lucifer
Satinover, Jeffrey 295*n*27
Saver, Jeffrey L. 294*n*12
Saxton, Alexander 305*n*41
scapegoating 17, 25, 73, 76, 80, 85, 87, 93, 97–100, 102, 105–06, 109–12, 114, 123, 125, 132, 144–45, 150, 158, 161, 167, 194, 213, 219–20, 225, 231, 238, 240, 259, 261, 282, 300*n*49, 302*n*7, 303*n*41, 305*n*38, 316*n*92, 319*n*16
Schacter, Daniel L. 293*n*56
Schafer, Elizabeth 305*n*38
Schechner, Richard 168, 289*n*5, 290*n*13, 296*n*39, 297*n*52
Scheff, Thomas J. 314*n*67
Scherb, Victor I. 86
schizophrenia 47, 158, 161, 175, 232–33, 301*n*72, 320*n*23
Schneiderman, Stuart 318*n*108
Schneiderman, William 311*n*28
Schore, Allan N. 290*n*19, 296*n*38, 299*n*41, 300*n*60
Schroeder, Gerald L. 24, 295*n*30
Sconce, Jeffrey 313*n*46
Scott, Robert A. 126
Scott, William C. 298*n*14
seeking system 28, 33, 37, 55; *see also* lust system
Segal, Charles 298*n*12–13, 298*n*16
self (and Other) consciousness 10, 58, 60–61, 79, 292*n*43, 302*n*17
Seltzer, Mark 309*n*3, 313*n*53

Sepinwall, Alan 311*n*29
serial bomber 269
serial killer 97, 203, 227–29, 231–33, 236–37, 241–50, 262, 268–70, 274–77, 281, 312*n*34, 312*n*38, 315*n*76–77, 316*n*81, 316*n*89, 316*n*92, 318*n*118
Set 56–60, 67, 70, 73, 81, 87, 92, 97, 181, 185, 212, 287
Setebos 142, 150, 304*n*29
Se7en 312*n*38
The Seventh Seal 307*n*20
sexual selection 5, 51, 131, 177–82, 193–94, 213, 221–22, 290*n*17; *see also* cultural selection; natural selection
Seyfarth, Robert M. 178–80, 290*n*14, 306*n*7, 306*n*9–10
Shaffer, Peter 168,
Shakespeare, William 3, 15, 85, 118, 137–47, 149–50, 152–53, 155, 169, 184, 236, 298*n*72, 304*n*20–23; *see also As You Like It*; *Cymbeline*; *Hamlet*; *Pericles*; *The Tempest*; *The Winter's Tale*
shamanic visions 11, 15, 32–36, 40–42, 45, 47–48, 50–52, 54, 59, 65, 77–78, 182, 291*n*29, 297*n*50–51, 297*n*56, 297*n*65; *see also* dream visions; mystical visions; psychic visions
shamanism 11, 14–15, 32–35, 38–40, 45–48, 50–51, 54–55, 59–60, 62, 65, 67, 77–78, 80–81, 99, 178, 180–82, 201, 291*n*29, 296*n*43, 297*n*47, 297*n*66–67, 298*n*69, 298*n*10; *see also* visions
Shapiro, Alan 300*n*51
Shariff, Azim F. 304*n*13
Sharratt, Bernard 311*n*17, 311*n*25
Shaw, George Bernard 164–65
Sheridan, Richard Brinsley 154
Shermer, Michael 293*n*10
Shim, Jung-Soon 297*n*67
Shlain, Leonard 299*n*26
Shrum, L.J. 229, 310*n*12, 310*n*14
Siegel, Daniel J. 290*n*15, 291*n*25, 294*n*16
Silence of the Lambs 312*n*34, 316*n*92
Simpson, Philip L. 312*n*34
simulation theories 98, 282–83, 286, 289*n*7, 292*n*49, 309*n*5
Sin Noticias de Dios 213, 215–16
Singer, Ben 308*n*26
Sissa, Giulia 301*n*71
Skura, Meredith Anne 304*n*30
Smails, John 293*n*59
Smith, Joseph H. 259–60
Sober, Elliott 299*n*35
Social Darwinism 4, 146,
social drama (Turner) 73, 300*n*50
social metaphysics 180–81, 185, 187, 192, 203, 209, 213, 219, 223, 225, 227, 242, 259, 277, 281–82, 285–88
Solms, Mark 8, 14, 29, 37, 147, 290*n*15, 291*n*25–26, 291*n*28, 293*n*56, 296*n*40, 298*n*73, 299*n*27

Sopcak, Paul 319*n*13
Sophocles 80, 235–37, 252, 313*n*43
Sor Juana Inés de la Cruz 15, 150–52
Soule, Lesley Wade 90–91
Soyinka, Wole 169–70
sparagmos 60, 87, 132–33, 298*n*12, 298*n*18
split subjectivity 61, 118–19, 124, 128, 137–38, 155, 236,
Spong, John Shelby 321*n*42
Stark, Rodney 303*n*32
Staub, August W. 302*n*18
Staub, Ervin 319*n*16
Steambath 305*n*46
stereotyping 18, 49, 85, 99, 113, 123, 125, 148–49, 164, 207, 210, 216, 224–25, 229–30, 240, 268, 276–77, 284, 301*n*5, 303*n*27, 305*n*43, 307*n*23, 308*n*34, 319*n*12, 320*n*23, 320*n*31
stimulus barrier (in the brain) 8, 147; *see also* filters (in the brain)
Strindberg, August 166
The String of Pearls 302*n*19
Subrahmanyam, Kaveri 321*n*35
Sullins, Ellen S. 319*n*14
superego 8, 10, 72, 81, 128, 114, 147, 209, 241, 249, 260, 286, 313*n*45
Surowiecki, James 318*n*2
Sweeney Todd 97, 302*n*19
Symbolic order (Lacan) 5–14, 21, 26–27, 28, 31, 33, 36–37, 61, 66–67, 70, 77–78, 88–89, 93, 95–96, 101, 105, 107, 110, 122, 128, 140, 144–48, 152, 155, 158, 171, 173, 208, 233–34, 236, 240–42, 245, 249, 250, 252–53, 277, 282–83, 289–90*n*11, 291*n*29, 292*n*42, 292*n*45, 295*n*25, 303*n*39, 304*n*12, 309*n*40, 309*n*5, 313*n*44, 314*n*62, 317*n*97, 320*n*23
sympathetic nervous system *see* autonomic nervous system
sympathy 66, 99, 103–05, 107–10, 112–13, 123, 130, 132, 145, 149, 152, 161, 163–65, 168, 210, 212, 218, 221, 224, 230, 234, 237–38, 240–41, 245–47, 250, 264, 266, 268–69, 275, 286, 291*n*32, 303*n*24, 308*n*29, 313*n*51, 315*n*72, 320*n*28, 321*n*38; *see also* compassion; contagion; empathy
Symptom (Lacan) 10, 36, 78, 236, 239, 240, 248, 249, 250, 290*n*17, 292*n*45, 301*n*65, 309*n*40
synesthesia 120–24, 127
synkinesia 122

Taçon, Paul S.C. 292*n*44
targeted help 100–01, 103, 105–06, 111, 114, 181–82, 188, 194–95, 287
targeted revenge 194
Taylor, Shelley E. 299*n*28
teaching 39, 86, 118–19, 179–80, 213, 240, 248; *see also* learned behaviors
Teilhard de Chardin, Pierre 19, 296*n*45, 321*n*42

The Tempest 142–47, 149–50, 169
temporal lobes 6, 19–23, 31–32, 39, 45, 47, 50, 52, 62, 69, 79–80, 89, 122, 128, 176, 178, 291*n*26, 291*n*29, 292*n*34, 294*n*13, 294*n*20, 296*n*40, 301*n*67, 301*n*72, 315*n*69
territoriality 1–2, 4, 49–50, 55, 57, 64–65, 74, 78, 87, 93, 104–5, 108, 111, 121–22, 126, 130, 139, 143, 152, , 152, 155–56, 168, 208, 279–80, 282, 287, 297*n*47, 305*n*34, 321*n*37
terrorism 2, 68, 83, 85, 87, 111, 212, 220, 227, 229, 232, 242–43, 269, 276, 279, 281, 289*n*6, 313*n*50
Tezcatlipoca 170
theatrical drives 1, 177
Theory of Mind (ToM) 65, 94, 102, 177, 179–80, 185, 196–97, 208, 231, 306*n*8; *see also* metamind
Tomasello, Michael 179, 290*n*18, 306*n*8, 306*n*10
Tonatiuh 170
top-down brain networks 8–9, 132, 320*n*28
Towneley Cycle plays *see* Wakefield Cycle plays
tragic flaw 2, 44, 53, 65, 102, 106, 112, 124, 128–29, 134–35, 195, 212, 221, 224, 235–36, 253, 262, 268, 310*n*10, 313*n*45
Traunecker, Claude 56
Treisman, Anne 120
Trevarthen, Colwyn 107, 291*n*24
tricksters 89, 99, 101–02, 104, 110, 114, 169, 171, 181–82, 184–86, 188–90, 193–95, 203–04, 211–13, 221, 224–25, 273, 281, 286–87, 308*n*29
Trivers, Robert 51, 299*n*36
The Truman Show 190
Turim, Maureen 317*n*100
Turnbull, Oliver 298*n*73
Turner, Mark 4, 121, 292*n*47
Turner, Victor 290*n*13, 296*n*39, 300*n*50

Uncle Tom's Cabin 161–65, 305*n*40–41
unconscious (or subconscious) brain activity 6, 8, 12–13, 25, 46–47, 51, 61, 73, 75, 77, 82, 122, 127, 132–36, 168, 175, 229, 238–39, 244, 249, 253, 282–83, 287, 290*n*15, 290*n*22, 293*n*55, 295*n*25, 295*n*27, 295*n*30, 304*n*13, 313*n*45, 314*n*66, 315*n*78, 320*n*28, 320*n*31; God as unconscious (Lacan) 300*n*61; patriarchal unconscious 315*n*70; subject as unconscious (Lacan) 135, 317*n*103

Valdez, Luis 170
Van Biema, David 69
van Itallie, Jean-Claude 168
Vardac, A. Nicholas 311*n*20
Varela, Francisco J. 295*n*28, 321*n*39
Vaughan, Susan C. 290*n*15
vengeance 23, 71, 73–74, 80, 96–97, 101, 104–6, 110, 112–14, 118, 123, 138–43, 145, 147, 149–50, 153, 160 , 194, 224–25, 229–30, 240–41, 245, 249, 256, 260, 262,

280, 309n6; see also reciprocal violence; revenge
ventromedial prefrontal cortex 8, 147, 209, 221, 284, 292n34, 293n7, 307n15, 315n69, 320n26, 320n27
videogames 2, 26, 49, 180, 201, 204–05, 239, 284, 286, 306n11, 321n35
virtual reality 2–3, 11, 18, 32, 49, 177, 201, 273, 276, 286, 321n35; see also the Web
visions see dream visions; mystical visions; psychic visions; shamanic visions
Volkan, Vamik D. 25
von Feilitzen, Cecilia 309n4
voyeurism 168, 178, 200, 230, 232, 243, 274–76, 283, 312n40; see also perversity

Wade, Nicholas 13
Wagner, Heinrich 154
Wakefield Cycle (or Mystery) plays 14, 90–114, 125, 128–29, 136–37, 142, 152, 181, 188, 190, 193–94, 200–1, 203, 207, 218, 221, 223–25, 260, 302n11, 304n16
Wall-Randell, Sarah 304n9
Wallace, Alfred Russel 23–24
war 25, 62, 65, 71, 80, 83, 92–93, 150–51, 166–68, 181, 199, 204, 242, 271, 276, 280, 306n11, 308n37; see also divine warrior; "holy war"
"war on terror" 4, 16, 230
the Web (or websites) 2, 49, 180, 230, 272–76, 286, 318n2, 321n35; see also virtual reality
Wegner, Daniel M. 134–36, 295n24, 304n15
Weimann, Robert 303n4
Weingarten, S.M. 294n13
What Dreams May Come 307n19
White, Stephen 299n40
Whiten, Andrew 51
Wilder, Thornton 167, 294n21, 307n14
Wiles, David 62, 75, 300n57

Wiles, Timothy J. 313n55, 314n67
Wilkinson, Richard H. 56–57, 59–60, 298n5
Williams, Linda 308n26
Williams, Raymond 309n2, 311n25
Wills, Christopher 295n23, 303n34
Wilson, Barbara J. 310n13, 311n16
Wilson, Edward O. 295n26–27
Wilson, Edwin 311n19
Winchell, Mark Royden 308n36
Wings of Desire 197–99
Wink, Walter 301n1
Winkeilman, Piotr 290n22
Winkelman, Michael 291n29, 297n66
Winkler, John J. 61, 82, 301n74
The Winter's Tale 140, 304n21
Wise, Jennifer 293n5
The Witches of Eastwick 214–15, 217
Wolf, Maryanne 7, 291n27
Womack, Peter 304n7
Woman on Top 307n21
Worldwide Web see the Web
Wrangham, Richard 64, 67, 103, 279, 298n20, 299n21, 321n37
Wright, Robert 321n38

X-Files 231, 309n6, 312n32, 312n37, 315n76, 318n121

Yeats, William Butler 166, 244–45, 315n79
Yemanja 307n21
York Cycle (or Mystery) plays 302n11–12

Zeki, Semir 293n59
Zeus 60–61, 65–71, 74–75, 78–79, 99, 181, 185, 298n8, 299n31, 300n54, 300n63, 301n70, 306n11; see also Jupiter
Zhang, Yehong 289n8
Zillmann, Dolf 310n10
Zizek, Slavoj 228, 292n45, 301n65, 309n5
Zupancic, Alenka 309n5

www.ingramcontent.com/pod-product-compliance
Lightning Source LLC
Chambersburg PA
CBHW051205300426
44116CB00006B/448